AF713287

Palgrave Perspectives on Process Philosophy

Series Editors
Wahida Khandker
Dept of History, Politics and Philosophy
Manchester Metropolitan University
Manchester, UK

Tim Flanagan
Philosophy
Murdoch University
Murdoch, Australia

The aim of this series is to provide monographs, edited collections and Palgrave Pivots from both established and early career scholars in Process Philosophy, with particular reference (but by no means exclusively) to the writings of Alfred North Whitehead, Henri Bergson, William James, and Charles Hartshorne. The series aims to promote new writing in this area that is innovative, rigorous, and yet accessible to readers both within philosophy and beyond disciplinary boundaries.

The expected focus of the books falls roughly into three areas, though it is likely that these areas will overlap:

- New readings of the major early twentieth-century process philosophers, Whitehead, Bergson, James, Hartshorne, et al, including studies of their influence on, or relevance to, contemporary philosophy
- The roots of process philosophy in ancient and early modern philosophy, as well as hitherto unexplored affinities with thinkers in these eras
- Future orientations: including cross-disciplinary explorations of the implications of process thought. We are particularly interested in studies of intersections with the arts, architecture, environmental humanities, and the sciences, but projects proposing other disciplinary intersections will be welcomed.

Dionysis Christias

Towards a Hyperstitional Process Nominalism

Outline of a Fractured Immanent Ontology

palgrave
macmillan

Dionysis Christias
Research Center for Greek Philosophy
Academy of Athens
Athens, Greece

ISSN 2524-4728 ISSN 2524-4736 (electronic)
Palgrave Perspectives on Process Philosophy
ISBN 978-3-032-06349-6 ISBN 978-3-032-06350-2 (eBook)
https://doi.org/10.1007/978-3-032-06350-2

© The Editor(s) (if applicable) and The Author(s), under exclusive license to Springer Nature Switzerland AG 2025

This work is subject to copyright. All rights are solely and exclusively licensed by the Publisher, whether the whole or part of the material is concerned, specifically the rights of translation, reprinting, reuse of illustrations, recitation, broadcasting, reproduction on microfilms or in any other physical way, and transmission or information storage and retrieval, electronic adaptation, computer software, or by similar or dissimilar methodology now known or hereafter developed.
The use of general descriptive names, registered names, trademarks, service marks, etc. in this publication does not imply, even in the absence of a specific statement, that such names are exempt from the relevant protective laws and regulations and therefore free for general use.
The publisher, the authors and the editors are safe to assume that the advice and information in this book are believed to be true and accurate at the date of publication. Neither the publisher nor the authors or the editors give a warranty, expressed or implied, with respect to the material contained herein or for any errors or omissions that may have been made. The publisher remains neutral with regard to jurisdictional claims in published maps and institutional affiliations.

Cover illustration: © Irena Sowinska / Getty Images

This Palgrave Macmillan imprint is published by the registered company Springer Nature Switzerland AG.
The registered company address is: Gewerbestrasse 11, 6330 Cham, Switzerland

If disposing of this product, please recycle the paper.

Acknowledgments

Thanks to Costas Markou, John Mollindris, Vasso Kindi, Stelios Virvidakis, Costas Filippakis, Georgios Evangelopoulos, Dimitris Michiotis, Costas Pagondiotis, Doukas Kapantais, Athanassios Gatzaras, Thodoris Dimitrakos, John Efstathiou, Costas Morphis, Carl Sachs, Daniel Sacilotto, Ray Brassier, Paul Giladi, Luz Seiberth, Jim O'Shea, Bill deVries, Stephanie Dach, David Landy, Cathy Legg, Ryan Simonelli, Kevin Temple, Curtis Allen, Kyrill Potapov, J-P Caron, Antonio Nunziante, Pietro Salis, Luca Corti, Matthias Egg, László Kocsis and Krisztián Pete. Discussions with all these friends and colleagues in private, in reading groups, social media, seminar rooms, zoom meetings and conferences were pivotal for formulating the basic outlook and ideas that gave birth to this book. Thanks also to three readers for *Palgrave* for very helpful comments and suggestions, and to Amy Invernizzi and Arunaa Devi for their invaluable help and support in the editing process.

Most of all thanks to Demi Raftopoulou for being a constant source of love and support.

Most of the material in this book is previously unpublished, though there is some overlap between parts of Chaps. 4, 7 and 10 and my former book *Normativity, Lifeworld and Science in Sellars' Synoptic Vision*. I am grateful to *Palgrave* for permission to re-use some of that material here.

Competing Interests The author has no competing interests to declare that are relevant to the content of this manuscript.

Contents

CHAPTER 1

Introduction

The main aim of this book is to articulate a novel metaphysical framework, which I characterize as a 'fractured immanent' one. This framework emerges as a response to a transcendental investigation into the possibility of thinking about the world while being situated within a world that we did not create. In other words, the fundamental problematic out of which our metaphysical framework emerges as a response is the issue of how thinking can be *in* the world while simultaneously being *about* the world. Insofar as thinking is *about* the world, in the sense of being directed toward the world in a way that makes it rationally responsive and answerable to the world, it cannot be *in* the world as one of its parts but must stand over against it as a standard of measurement as it were (of measuring and be measured by the world). Yet, at the same time, thinking cannot be about the world unless it is also somehow *within* it. If this were not so, it would be hard to see how it is so much as possible for thinking to be about *this* world which we did not create, and which materially supports (and undermines) our continuing bodily and mental existence. Thus, ultimately the task is to understand how thinking can be both inside and outside the world.

As was mentioned above, this issue is approached from a decidedly *metaphysical* point of view which yet, hopefully, avoids being naively 'pre-critical' in that it is actually the result of the above kind of an essentially *transcendental* investigation of the conditions that enable thinking to be

© The Author(s), under exclusive license to Springer Nature Switzerland AG 2025

D. Christias, *Towards a Hyperstitional Process Nominalism*, Palgrave Perspectives on Process Philosophy,
https://doi.org/10.1007/978-3-032-06350-2_1

simultaneously a material process within the world and something intentionally directed toward and answerable to the world. In particular, I attempt to understand how the natural world and our relation to it can appear in two conflicting ways to ourselves: on the one hand, it seems rather uncontroversial (at least from modernity onward) to say that human minds are *ontologically immanent* to nature. Our minds and their distinctive characteristics (normativity, freedom, conceptually articulated thought) are not a 'kingdom within a kingdom', separate ontological regions within the world. They are part of nature, they are ontologically within (dependent on) it, and whatever their distinctive or emergent character consists in, it is not such as to exert any kind of transcendent causality to nature (including our bodies). On the other hand, this very natural world, precisely in its immanent expression *as* mind, appears as *always already correlated* with the categories of mind (our categorial perspective) or even as projected in the form of mutually incompatible categorial perspectives of ourselves-in-the world (normativist, causal-naturalistic[1]). Thus, it seems that the immanence of mind to the world is such that it can change its own 'topological orientation' or 'direction of fit' and turn the world to revolve immanently around the mind (where the conditions of possibility of the experience of objects are conditions of possibility of the objects of experience). How must the (unconditionally existing) natural world be related to the (conditionally existing) mind so as to immanently appear as always already correlated to the latter?[2] And how can this mind-world correlation take the form of mutually incompatible perspectives (causal vs normative) of one and the same underlying world which both these perspectives somehow express?

I shall argue that this very general metaphysical question as to the nature of this asymmetrically immanent relation between mind and world which appears in essentially perspectival form can be illuminated by

[1] According to the normativist self-image, we are essentially normatively individuated persons, whose perception, thinking and action are intentionally directed and rationally responsive to external reality, while according to the causal-naturalistic self-image, inspired by growing scientific achievements, we are essentially non-normatively individuated complex physical systems which are non-rationally—that is, only causally or lawfully—connected to other complex physical systems.

[2] By 'unconditional existence' I mean a manner of existence that is self-sufficient, that does not itself depend on something other than itself for its existence, while 'conditional existence' is a way of existence which is not self-sufficient and depends on something other than itself for its existence.

introducing and developing the concept—and the corresponding metaphysics and epistemology—of *fractured immanence*. On the metaphysical side, this viewpoint leads to a resolute acceptance of the univocity of being (inspired by Deleuze) and to a process ontology (inspired by Sellars, Peirce and Deleuze), while on the epistemological side, it commits us to a resolute rejection of the myth of the Given (inspired by Sellars). Specifically, it will be argued that a univocal process ontology which eschews both normative and causal categories provides a platform for ontologically *integrating* these two equally indispensable viewpoints of the relation between thinking and being, namely, thinking and being as *normatively* correlated (commonsensical understanding of mind and world) and as *non-normatively* correlated (causal-naturalistic understanding of mind and world). At the same time, the process framework proves more fundamental in explanatory scope in that it can also give a process-evolutionary *account* for the nature, utility and limits of the respective commonsensical and natural-scientific mind-world correlations involved. In this way, both the indispensability and the incommensurability between the reasons and causes framework are explained in practical and ultimately evolutionary terms, as indispensable tools for navigating the 'phenomenal world' of the ordinary lifeworld. By being accounted for in this manner, the indispensability and the incommensurability between reasons and causes cease to appear as unexplained brute facts. Indeed, it will turn out that these viewpoints are *complementary*, and, in the context of a fractured immanent metaphysics, they can even be integrated into a single (though essentially fractured) immanent process vision of ourselves-in-the-world.

From this metaphysical standpoint, mind is immanent to nature, but this immanence is essentially fractured: what is most proximal in the order of being—our own experience of ourselves-in-the-world—is most distant in the order of understanding. What is immanent is not the identity nor the difference as such between mind and world, but the ontological identity-cum-logical difference between them. The book defends and explores this peculiar kind of immanent relation between mind and world and argues that it results once we resolutely reject the myth of the Given (Sellars), namely, the view that the relation between mind and world is experientially, intuitively, phenomenologically or categorially transparent to the mind. It is precisely because it comes with a resolute rejection of Givenness in all its forms that the mind-world immanence is essentially fractured. Yet, this does not mean that thought fails to reach into the world as such, as if it is destined to be about something less than it, for example, the 'phenomenal world', or the world as always already correlated to our categories, experience, individual, historical, socio-political perspective, etc. Rather, it means that thought's *conception* of the world as it is in itself is not *itself* Given and is always subject to self-correction. In this connection, I attempt to draw the implications of a resolute rejection of Givenness not only in epistemology but also in semantics and ontology. The most important of them are that a rejection of the Given entails a) an anti-representationalism in semantics and ontology and b) a post-Kantian, critical and transcendentally informed process metaphysics. It is precisely from this radically anti-givenist non-traditional metaphysical standpoint—which proves to be at once non-representational, transcendental and critical—that the proposed process ontological framework can successfully account for both the normative and the causal-naturalistic dimension of our understanding of the relation between mind and world without reproducing the reifications and dogmatism of old-school pre-Kantian metaphysics and without presupposing an impossible epistemic 'God's eye view' from which the totality of possible worlds or conceptual frameworks can be antecedently surveyed.

Lastly, as will be shown in the overview, fractured immanence is not just a metaphysical diagnosis of the mind-world relation, but a proposal for rethinking grounding, emergence, normativity and freedom in a processual register, bypassing both substance metaphysics and semantic representationalism.

1.1 Overview

The overall argument of the book is developed in nine chapters.

In Chap. 2, I provide a clarification of the concept of immanence. I distinguish between different kinds of immanence, I clarify the precise sense in which I use this term and I provide a rationale for considering immanence a central concept for the development of a resolutely non-dualist and non-epiphenomenalist ontology, which accounts for the creative evolution of being without introducing transcendent causes or reasons for understanding it. Immanence is the view that there is no transcendent ground, cause or viewpoint that explains the existence of the world (or the mind, or the mind-world correlation). There are different conceptions of immanence depending on a) what kind of transcendence is to be rejected (ontological, epistemological, explanatory) and b) on what is taken as an immanently primary—that is, unconditional—unexplained explainer. Group (a) refers to *formal* conceptions of immanence while group (b) to *substantial* conceptions of immanence. Regarding the latter, think, for example, of Hegel's 'thought-being' immanence or of Husserl's 'phenomenological immanence' revealed through the transcendental bracketing of the natural attitude. I clarify that my substantial conception of immanence takes being to be unconditional: being's existence is not conditional upon a transcendent metaphysical ground (e.g. God, nothingness, spacetime, laws of nature) nor upon a transcendental consciousness (and the special kind of openness to the world it makes possible through the categorial or existential structures of the transcendental subject). I also highlight how I understand immanence in relation to thinking: thinking about the world is immanent in that it necessarily unfolds and self-corrects '*in medias res*', in the midst of our practices—thus rejecting 'the view from nowhere' conception of thought.

Chapter 3 further elaborates this latter point. What does it mean for thinking to reject the 'view from nowhere' and to embrace its in medias res and self-correcting character? I argue that the key point here is the resolute rejection of the myth of the Given (Sellars, 1956, 1981). Indeed, it is suggested that immanence can provide an illuminating account of the relation between mind and world only if it goes hand in hand with a complete rejection of the Given (Brassier, 2016). From this anti-Givenist point of view, what is immanent is the *identity-necessarily-presented-as-difference* between thought and world. Thought is immanent to being, but this immanence is essentially *fractured*. I also explore the consequences of resolutely rejecting the Given for ontology and the nature of our access to the world through thought. I argue that the rejection of Givenness implies that, at the level of semantics, thinking attempts to grasp the world *as such*: there are no fixed limits to thinking—for example, transcendental conditions of possibility of experience—that prevent it from being semantically *about* the world as it is in itself. Yet it also implies that the form or content of the world as such are never *themselves* Given to thinking in its semantic dimension and are always subject to self-correction. In this connection, I also focus on the anti-representational implications of the rejection of Givenness in our conception of the mind-world relation. It is argued that a resolute rejection of Givenness entails anti-representationalism in semantics and ontology and a post-Kantian, critical and transcendentally informed metaphysics. In this spirit, it is suggested that a Sellars-inspired non-representational semantics coupled with a Deleuze-inspired non-representationalist ontology can provide a viewpoint from which we can launch a positive critique of traditional representationalist (substance-property) metaphysics, by constructing an evolutionary process metaphysics with anti-representationalist yet realist implications. From this standpoint, the realist view that there is no (semantic, epistemic, transcendental) barrier in thinking being about the world as it is in itself is rehabilitated but is given an important twist: there is indeed no such barrier provided that this direct access of thinking to being in the *semantic* sense of the term (in which conceptual thinking is intentionally directed to the world *tout court*) is itself understood as an expression of a non-semantic, non-representational direct relation between mind and world. Ultimately, the semantic aboutness of thought to a common, public, objective world through reference relations is revealed as an indispensable yet non-representationally functioning tool for collectively coordinating thought, considered as natural process, to the world, considered as natural process, where the relation between them at this level is not referential, logical or causal. Rather, it is a non-semantic expressive relation of 'resonance' (Deleuze) or 'picturing' (Sellars) between dynamically co-evolving environmental, bodily and brain processes, whose degree of adequacy is measured by the degree of empowerment expressed in the collective conative processes involved.[3]

In Chap. 4, based on Christias (2023), and inspired by Sellars and Deleuze, I suggest that a process-oriented categorial framework can provide the most promising successor of the 'substance-property' categorial framework at work in our everyday lifeworld. The former is what results from an immanent critique of the latter. Moreover, I attempt to show that if a process ontology is to supersede the substance-property ontological framework in a consistent manner it must endorse a nominalistic metaphysics which eschews all kinds of abstract universals, such as substances, properties, relations, kinds, facts, and construct a nominalistic process language without recourse to Fregean contents or predicates as referring expressions.

Chapter 5 further investigates important ontological implications of process ontology and attempts to find a place for emergence within a

[3] Interestingly, certain problematically 'Givenist' positions about the relation between mind and world, such as direct realism (according to which in favorable circumstances, conceptually articulated perceptual capacities give us direct access to the world as it is in itself) and transcendental phenomenology (according to which we only have access to the correlation between mind and world), can be understood as expressions of the failure to draw this distinction between two radically different ways in which thinking can be said to have direct access to the world as it is in itself.

univocal process nominalist framework. Emergence in this context is understood as a self-fracturing of being. The account on offer, whether it involves the emergence of qualia of sensory and affective experience or the emergence of symbolic thinking, must be such as to avoid the nearly inevitable epiphenomenalist implications of non-reductive naturalism or an appeal to non-explanatory relations of 'supervenience' between different ontological levels (Heil, 2021; Kim, 1993). Having shown (Chaps. 3 and 4) that experiential qualities can be emergent yet non-epiphenomenal only if we switch from a substance-property metaphysics into a monistic process framework, I here examine whether and in what sense process nominalism can account for levels-emergence. On the face of it, strict nominalism seems to allow only the emergence of non-universal structures and properties based on combinations of a finite stock of actual basic particulars. How can these limited resources account for emergence and novelty? Here I use Deleuze's notion of 'intensity' (Deleuze, 1968), stressing its topological properties, as a way of developing a non-traditional 'intensive processual nominalism' which can account for emergent levels of being in a nominalistically kosher way. In this connection, I highlight an important implication of this view about physicalism and panpsychism. Both physicalism and panpsychism are reductive positions that rightly reject overly simplistic (e.g. non-reductive physicalist, supervenience-based) accounts of emergence while wrongly concluding that this amounts to a rejection of emergence *tout court*. My suggestion here is that this is because both physicalism and panpsychism are blind to the possibility of emerging discontinuity from underlying continuity (e.g. the emergence of discontinuous actual levels from topologically continuous deformations or bifurcations of intensive processes).

The philosophical cash value of the novel ontological-modal process framework developed in the book is also tested by utilizing its basic principles to solve core problems in contemporary analytic metaphysics of powers. In Chap. 6, based on a (Sellars, Peirce and Deleuze-inspired) evolutionary rendering of the principle of sufficient reason, I provide a novel account of grounding, which questions background assumptions about the form of explanation that ontological grounds must offer to count as such, shared by contemporary neo-Humeans, neo-Aristotelian realists and analytic neo-Hegelian absolute idealists.

In Chap. 7, I turn my attention to the issue of the seeming incommensurability between two equally indispensable ways of understanding ourselves-in-the-world: our normativist self-image (space-of-reasons intelligibility) and our causal-naturalistic self-image (space-of-causes intelligibility). My suggestion is that both these frameworks are best understood as fractured projections of processual reality. To this end, I first stress the irreducibility, isolation yet also 'parallelism' of the space of reasons (norms) and the space of causes (facts), and suggest that these relations as a package deal are an essential part of a metaphysical view (itself a sophistication of our common-sense view of the world after modernity) that Brandom calls 'bimodal hylomorphic conceptual realism'[4] (Brandom, 2019). I then propose that Brandom's conceptual realism and modal realism (that articulate our ordinary descriptive metaphysics about reasons, causes and their relations) can be understood as indispensable 'ladder languages' that can yet be eventually transcended by a naturalistic process metaphysics—where both causes and reasons are understood as different ways of being processes.

Chapter 8 takes up the challenge of articulating a successor concept for 'normativity' that construes the causal efficacy of the latter in a resolutely immanent naturalistic manner (yet without reducing it to mere physicalism). I mobilize Land's (2011) notion of 'hyperstition' as such a successor concept and draw a picture of normativity as an emergent intensive regime which, by way of we-attitudes as its manner of causally affecting the world, institutes a novel kind of hyperstitional dynamics which can turn 'lies into truths'.[5] This is a way of illustrating and developing my basic immanent materialist view of normativity according to which its causal power lies exclusively in the *enactment* of normative statuses in the corresponding (we)attitudes. In this connection, I note an unexpected convergence of

[4] According to this view (which Brandom attributes to Hegel), the world is literally conceptually articulated and is bifurcated into two different yet deeply isomorphic modal orders: the alethic order, namely, the causal realm of worldly objects, their properties and nomological relations, and the deontic normative order, the realm of reasoned thought as articulated by our norm-governed practices.

[5] Hyperstitions are cybernetically functioning sign-making practices, typically expressing ideas, speculations or narratives able to bring about their own reality through the workings of positive feedback loops, generating new socio-linguistic-economic-psychological attractors. Stock-market speculation in capitalism is a prime example of hyperstitional feedback circles. The gradual entrenchment of Judeo-Christianity through trust in the truth of gospels is another. It is important here to stress that, as I see it, hyperstitional phenomena are not limited to the cultural sphere, but also find application to the workings of non-conceptual systems of organisms (Chap. 8) and even to the inorganic realm as a result of an evolutionarily entrenched variety of iterative (repetitive, recursive, replicating) processes (Chaps. 5 and 6).

viewpoints with a conception of normativity articulated in Sellars (1949) according to which norms are embodied generalizations which tend to make themselves true, by inhibiting the occurrence of the events that would falsify them (Sellars, 1949, pp. 123–124). Building on Kukla's (2000) suggestion that at the core of the function of ideology lies the misrecognition of a norm for correctly perceiving, thinking and acting as always already binding (thereby concealing the fact that it would not be valid independently of its enactment in corresponding normative attitudes), it is further argued that this way of construing normativity can throw light on the phenomenon of ideology. From this point of view, ideology in its most general form consists in construing our fundamental lifeworld categories as representational, that is, as cutting nature and society at its categorial joints. Moreover, I suggest that, in line with my general 'fractured immanent' model of ourselves-in-the-world, hyperstitional dynamics is best seen as bifurcated in a non-conceptual and conceptual dimension: non-conceptual-level hyperstitional dynamics can be understood in terms of predictive processing[6] (Clark, 2016, 2019), whereas conceptual-level hyperstitional dynamics can be construed in terms of Brandom's neo-Hegelian model of content determination and conceptual change (Brandom, 2019; Christias, 2025). Lastly, it is argued that the notion of hyperstition can also be an essential component in providing an illuminating empirical explanation of currently much-discussed phenomena as the proliferation of sealed-off echo chamber communities and the viral status and increasing causal impact of fake news.

In Chaps. 9 and 10, I develop some important ethico-political implications of a fractured immanent account of ourselves-in-the-world.

Chapter 9 argues that freedom is also expressed in a fractured form: (1) as normative freedom, that is, freedom as self-determination (with Kant, Sellars and Brandom as its main representatives) and (2) as non-normative freedom, that is, as an increase of the power to act, as opposed to the 'power to dominate' (the main representatives here being Spinoza and Deleuze). It is suggested that freedom as self-determination has as its regulative ideal the emancipation from whatever hinders the exercise of our individual and collective epistemic and practical capacities, the ensuing

[6] According to predictive processing, the brain is continually engaged in formulating predictions, or hypotheses (called 'generative models'), about the causes of its sensory inputs, and in testing these predictions against incoming streams of sensory signals, thereby shaping perceptual content, guiding action and driving learning.

expansion of our capacity for perception, thought and action, and the creation of novel desires, interests and purposes that go hand in hand with this expansion. Freedom as an increase of the power to act points toward a maximization of our collective *Conatus*. Normative freedom and non-normative freedom are intimately related in that the former is epistemically and practically necessary for achieving the latter, while the latter is the ultimate *raison d'être* of the former. In this connection, I suggest that the well-known link, emphasized by German Idealism, between freedom and reason (epitomized in Brandom's conception of freedom as 'constraint by norms' (Brandom, 1979)) expresses a profound truth albeit only in connection with a non-metaphysical, normative and critical rendering of the universal-particular dialectic in our reason-giving practices. Specifically, building on Srnicek and Williams (2015) and Christias (2023), I propose that the surplus 'ideal' dimension of universals, stripped of its metaphysical connotations and understood in normative and practical terms, can be thought of as an indispensable subversive tool of immanent critique of the existing semantic *status quo*. From this point of view, universals are understood as projections of emancipatory ideals regarding our collective epistemic and general welfare that function as an axis for coordinating collective projects in the direction of fulfilling these ideals.

Building on the above discussion, Chap. 10 starts by noting that normative freedom and non-normative freedom prescribe divergent regulative ideals that cannot be algorithmically synthesized at the level of collective ethico-political theory and practice. The regulative ideal at the heart of our conception of normative freedom is expressed by the universalist impulse for collective (semantic, epistemic, moral) self-mastery while the regulative ideal of our non-normative conception of freedom is expressed by the 'particularist' impulse for unhinged monadic material empowerment. It is suggested that in order to achieve an optimal balance between these regulative ideals we must opt for what I call 'democratic prometheanism'. I examine the implications of this view on the issue of climate change: the democratic prometheanist view, among other things, implies that humans are essentially normatively defined as persons and not as a distinct and hierarchically superior metaphysical category called 'humanity'. From this point of view, humans as complex natural processes are ultimately conceived non-hierarchically, as being in a mutually symbiotic relationship with the social and environmental *niches* to which they are intertwined. This change of focus comes with a higher sensitivity to the strong interdependence relations that exist between our form of life, other living beings and the environment, and this necessitates reconceiving promethean freedom in terms of controlled interdependence.

1.2 Some Terminological Clarifications: Immanence, Conceptuality, Process, Nominalism, Appearances and the in-Itself

To avoid confusion, some terminological clarifications are in order. I begin with one of the most central concepts of this book, namely, that of 'immanence'. It is true that immanence is not the clearest concept one could use in the context of metaphysics, as it (along with its contrary, namely, 'transcendence') is ambiguous and subject to many different—and sometimes incommensurable—interpretations. This is correct so far as it goes; but then, at this level of abstraction, the same goes for concepts such as 'naturalism' and 'normativity': they are accordion words the contraction and expansion of which can often lead to philosophical confusion. The main reason why I make extensive use of the concept of immanence in the book, despite its being an accordion word with all the dangers of confusion and equivocation which that entails, is that it is a succinct way of expressing a fundamental guiding principle of this book. This metaphysical principle is intimately related to a version of *the principle of sufficient reason* (PSR) according to which to fully account for a phenomenon we have to understand *why*—not just *that*—it is the way it is. Based on PSR we can argue for the *unconditionality* of being (nature) in the order of explanation: the ultimate unexplained explainer is *being itself*. All beings (*natura naturata*) are conditional upon the existence of being (*natura naturans*). This does not make the relation between being and beings transcendent because beings are themselves immanent ways in which being morphogenetically expresses itself. It just emphasizes the fact that beings are perishable in a way that being is not. Being's existence is not conditional upon something else; thus, the question 'why is there something rather than nothing?' does not make sense. Of course, this is not to say that the unconditionality of being is here understood as just a brute fact (it is not asserted just *that* it unconditionally exists), as this would be at odds with our contention that this view is an outcome of PSR. Rather, it amounts to the view that only if we construe being as unconditional and immanent to beings can we understand the self-differentiation of worldly becoming into conditionally existing beings.

Further, our focus on being's immanence is a way of stressing the fact that the existence of worldly modes of being with all their diverse and distinctive characteristics (inorganic beings, life, consciousness, thought, rationality), including the existence of being as a whole, should not be understood (explained) by appeal to *transcendent* structures such as Platonic Ideas, Aristotelian Forms, Augustinian *adaequatio rei et intellectus*, absolute spacetime or their transcendental correlates, such as transcendental subject (Kant), transcendental consciousness (Husserl) or the pre-reflective body-world correlation (Merleau-Ponty), which in one way or another view as unconditional the *relation* between being and transcendental consciousness. But, to the extent to which transcendental consciousness seems to be a privileged distinctive feature of a specific mode of being, namely, the human way of being-in-the-world, it seems that this kind of transcendental immanence raises something conditional (the existence of the mind-world transcendental correlation) to the status of the unconditional. And to the extent to which it allows for the existence of being outside the transcendental correlation, it ends up reinstating a spurious dualism, and hence, a transcendent relation, between humanity and nature. By contrast, from our point of view, immanence entails that humanity (and its way of relating to the world) is not a kingdom within a kingdom (Spinoza) endowed with special causal-ontological powers (e.g. free will, rationality, pre-understanding of being) which qualitatively transcend the powers of all other beings and can exert transcendent top-down causality upon the world. All qualitatively emergent structures of beings must be accounted for in such a way that the emergent powers in question are not conceived in an ontologically dualistic or epiphenomenalistic manner. This also includes special 'transcendental' powers that disclose in distinctive ways the world and ourselves. To the extent that they are emergent powers of 'world-making', they must be understood as *immanent* to *non*-correlational being if they are to avoid making the transcendental mind-world immanence itself unconditional or something that stands in a *transcendent* relation to unconditional being.

Another important terminological clarification concerns the—notorious and frequently misunderstood—Kantian distinction between the phenomenal world and the world as it is in itself. I shall make use of this distinction, but not in the Kantian sense in which the world as it is in itself is constitutively unknowable to us. I shall develop a position that is similar in spirit to Sellars' view to the effect that conceptual articulation both in our discourse about reasons and in discourse about worldly facts is a feature of the phenomenal world, but not of the world as it is in-itself, which lacks logical form, and is best understood in processual terms. What I want to stress here is that on my understanding of the terms, the phenomenal world—or the world of what we might call 'conceptual appearances'—is no less real than the world as it is in itself. They are immanently connected and thus equally real, but they are explanatorily asymmetrical in that the world as it is in itself is the explanatory ground of the phenomenal world, and the existence of the latter depends on the existence of the former but not vice versa. If there is a sense in which the phenomenal world is a world of 'appearances' it only involves our knowledge or *understanding* of it and not its very being as such: it turns out that the categorial character of 'outer sense' (the external world standing over against the subject) and 'inner sense' (the subject's bodily, sensory and inner psychological states) as well as that of space, time, color and extension are more intimately related to our sentient nature and evolutionary history that we thought to be, but this does not abolish the reality of the latter. It just reconceives their categorial place in the scheme of things. Whenever I use this—notorious, potentially confusing but to my mind still philosophically useful—Kantian distinction between the phenomenal world and the world as it is in itself, it is always meant as above.

An equally important clarification in this connection is that, to my mind, the reality that grounds the phenomenal world of appearances does not do so in the sense of being its (independent) *cause*. This would make the phenomenal world in an important sense illusory and epiphenomenal, thus not deserving to be called really existing world at all. The phenomenal world is a world of appearances in the sense that its true categorial character is not the one we thought it to be. That is, its true categorial character does not match our 'manifest' understanding of it in the context of our everyday lifeworld. Thus understood, the sense in which the ordinary lifeworld is 'phenomenal' or a world of appearances is much the same as the sense in which the Freudian real meaning of the manifest content of a dream is revealed by its latent content (Freud, 1899). Obviously, no one would think that the psychoanalytically revealed latent content of the dream *caused* its manifest content. The dream, as far as its existence is concerned, is one and the same. What changes is its categorial understanding, its categorial place in the overall scheme of things. I contend that the same goes for the relation between the ordinary lifeworld and the process world: the latter reveals the latent categorial content of the former. It changes its categorial place in the world, but it is not its hidden cause, nor does it make it illusory or epiphenomenal.

A note also is in order about the frequently used terms 'conceptual' and 'non-conceptual': throughout this work, the term 'non-conceptual' is used as an essentially negative *epistemological* characterization, not as a negative ontological one. Moreover, *contra* McDowell, Brandom and other adherents of *philosophia perennis*, to my mind, the term 'non-conceptual' does not refer to a hierarchically lower, more primitive level of being or understanding on the basis of which our higher rational capacities are built as a superstructure. The view propounded in this book to the effect that the world as it is in itself has non-conceptual (non-logical/categorial) form is essentially the rejection of the (e.g. Hegelian, Aristotelian, McDowellian) view that the world has logical or categorial structure. At most, the term 'non-conceptual' points toward *another* non-logical/non-categorial structure but does not positively characterize it. By this I do not mean that the 'non-conceptual' denotes worldly processes or events that cannot, as a matter of principle, be captured by concepts. I take it that such a positive ontological characterization *can* be provided by a process metaphysics. And, from this point of view, conceptual form can be said to be enveloped in non-conceptually yet positively characterized processes.

I call the metaphysical framework presented in this work a 'process nominalist' one, and I want to clarify how *not* to understand these terms. In developing a process metaphysics, as opposed to a metaphysics in which objects are treated as substances, I use the term 'process' as a shorthand for a whole *system* or cluster of categories that aim to supersede a corresponding system of categories of the substance framework. Thus, the categorial difference between the process and the substance framework does not merely come down to the fact that in the latter processes and events are derived from the fundamental category of objects (substances) undergoing change, whereas in the former objects are abstractions of more fundamental relatively stable processes. The substance framework is a holistically articulated system of categories and goes hand with hand with a specific conception of modal categories (in which, e.g. possibility is understood roughly along the lines of Aristotelian potentiality), spatio-temporal categories (which view space and time as antecedently existing frames for individuating objects), universals (kinds, properties, qualities, relations) necessary for understanding the 'nature' or 'essence' of particular objects and fundamental logico-metaphysical concepts (identity, difference, negation), which distinguish between compatible and incompatible combinations of the universals of the framework. In my understanding, the process framework is the expression of a rejection of the above categorial framework not in a piecemeal fashion but as a *whole* and represents the conscious construction of successor post-objectual categories in all domains of interest (modal, spatiotemporal, logical, metaphysical categories). If we realize that the substance-property framework is a tightly interconnected whole, we also see that a piecemeal rejection of it will involve tensions or inferential loose ends that will diminish the coherence of the resulting hybrid view. Thus, in my view, this framework can be best challenged by the articulation of another equally tightly interconnected whole built around different categories.

As for the nominalist part of the process view of the world endorsed in this book, it must be emphasized that it is a non-traditional form of nominalism that, unlike classic empiricist nominalism, does not eliminate universals altogether and thus does not hold that only particulars exist. Rather, it *reconceives* universals and their relation to concrete particulars in a way that (a) universals cease to be abstract or general entities inhering or exemplified in particulars, (b) universals do not stand in a relation of logical mediation to concrete particulars while particulars are not conceived as actual instances logically subsumed under universal essences and (c) the fact that the semantic content of universals goes beyond how universals are actually used to describe concrete particulars in the context of our everyday practices is fully acknowledged, yet stripped of its metaphysical connotations: this surplus, 'ideal' dimension of universals is not understood as a metaphysically existing unrealized potential or essence, but as a way of representing, calibrating and critically adjusting the structure of the inferential space of the universals in question in accordance with projections about the epistemic and collective welfare conferred in our actual social practices of implementing particular instances of these universals. Thus, since this view does not attribute causal powers to abstract entities (universals, reconceived in the above manner, are not abstract entities representing metaphysically necessary facts) and strictly abides by the dictum that 'the abstract does not explain, but must itself be explained' (Whitehead), it is fully compatible with nominalism. Yet, it is obviously, a non-traditional form of nominalism since it does not eschew universals as such, but only the view that they exert causal influence on the actual world due to their being general essences (forms, kinds, ends), that is, potentialities that contain in abstract form the essential properties of a concrete individual, leaving to the actual world only the role of determining the individual's accidental properties. Importantly, our nominalist view makes room for *two* (intertwined yet distinct) non-traditional forms of (hyperstitionally functioning) universals: (a) evolutionarily developed intensive multiplicities which structure becoming and ground the field of real possibilities (Chap. 6) and (b) normatively instituted universals that navigate and reshape this field through norm-governed projection in the context of our symbolic practices (Chaps. 8 and 9). These kinds of universals are not ontologically distinct (normative universals are distinct *ways* of being intensive multiplicities) but differ in that non-normative intensive multiplicities do not ontologically depend on normatively existing ones, while the latter do ontologically depend on the former.

Lastly, I want to say something about the unabashedly metaphysical orientation of the present work. I make a conscious effort to avoid metaphysical dogmatism (a risk every metaphysics sooner or later faces) by attempting to draw a post-Kantian, transcendentally informed, metaphysical framework, which is itself put forward as fallible and as having an essentially self-correcting dimension. Indeed, especially in the case of a *process* metaphysics according to which the relation between mind and world is one of *fractured* immanence, one must be keenly aware that in order to be self-referentially consistent, this metaphysics must not view itself as finalized doctrine but as a projected research *program*; not as a finished product but as an ongoing self-correcting project in the course of which novel answers to our metaphysical questions open up yet further questions that cannot be antecedently anticipated and often delegitimize old questions[7] (Rescher, 2000). Be that as it may, I am aware that this work has little chance of succeeding in avoiding the dangers of dogmatism, of reifying abstractions and of making metaphysical claims that seem vacuous, vague or both. The reason for taking these risks is that I think that however

[7] This has the important implication that a question cannot arise before its time has come. Certain questions cannot even be posed until others have been already resolved, because the resolution of the latter is presupposed in the very articulation of the former (Rescher, 2000, p. 61). I submit that this holds for all questions, including metaphysical ones. To believe otherwise is to think that the meanings of certain questions are static, certain and transparent to the mind, independently of the vicissitudes of inquiry, a view that is committed to mythical Givenness. Moreover, the rejection of Givenness in this regard also implies that a philosophical system cannot be refuted on the basis of formal logic or direct observation, but only on broadly pragmatic criteria of explanatory coherence, scope, fertility and relevance. From this point of view a philosophical system is 'refuted' when 'the questions it seeks to answer are no longer asked' (Sellars, 1948, p. 601), in other words, when it becomes a 'degenerate research program' (Lakatos, 1978).

transcendentally informed, no philosophical viewpoint can consistently avoid any reference to a view of the world and our place in it. This view, through transcendental reflection, can avoid dogmatism and the reification of our parochial means of representing the world into features of the world itself, but it cannot do this without minimally sketching a way of understanding the world and our place in it—which is to say, a metaphysical view. I thereby consider metaphysics as non-optional and thus as the lesser evil compared to (supposedly) resolutely anti-metaphysical viewpoints. In particular, I take it that even those whose proclaim to avoid metaphysics altogether or leave room only for a 'negative' metaphysics, without any positive (even if very abstract) commitment as to how the world is at the level of being and its relation to beings or to the mind, either succumb to one or another version of the myth of the Given or surreptitiously invoke metaphysical views that do not directly state. And if they really and consistently do abstain from every such underlying metaphysical framework in articulating their philosophical views, I do not see how they can avoid ending up being skeptics, whose 'fear of error turns out to be a fear of truth' (Hegel, 1977, p. 47). In any case, in view of the above, I think it is preferable to bite the bullet and attempt to directly engage with (an epistemologically sophisticated) metaphysics, trying to do one's best to delineate the framework that guides one's thinking and expose oneself to criticism, while hoping that even the shortcomings and downright falsehoods of one's metaphysical views will not be uninteresting, and will motivate others to do better, hopefully leading metaphysics to more insightful, fruitful and creative directions in understanding ourselves-in-the-world.

References

Brandom, R. (1979). Freedom and constraint by norms. *American Philosophical Quarterly, 16*(3), 187–196.

Brandom, R. (2019). *A spirit of trust: A reading of Hegel's phenomenology*. Harvard University Press.

Brassier, R. (2016). *Transcendental logic and true representings*. Glass bead. http://www.glass-bead.org/article/transcendental-logic-and-true-representings/?lang=enview

Christias, D. (2023). *Normativity, lifeworld and science in sellars' synoptic vision*. Palgrave Macmillan.

Christias, D. (2025). The place of spirit in nominalistic metaphysics. In C. Sachs (Ed.), *Interpreting sellars*. Cambridge University Press.
Clark, A. (2016). *Surfing uncertainty: Prediction, action and the embodied mind*. Oxford University Press.
Clark, A. (2019). Replies to critics. In M. Colombo, E. Irvine, & M. Stapleton (Eds.), *Search of the embodied, extended, enactive predictive (EEE-P) mind* (pp. 266–302). Oxford University Press.
Deleuze, G. (1968). *Difference and repetition*. Columbia University Press.
Freud, S. (1899). *The interpretation of dreams*. Basic Books.
Hegel, G. W. F. (1977). *The phenomenology of spirit*. Cambridge University Press.
Heil, J. (2021). *Appearance in reality*. Oxford University Press.
Kim, J. (1993). *Supervenience and mind*. Cambridge University Press.
Kukla, R. (2000). Myth, memory and misrecognition in sellars' 'empiricism and the philosophy of mind'. *Philosophical Studies, 101*, 161–211.
Lakatos, I. (1978). *The methodology of scientific research programmes*. Cambridge University Press.
Land, N. (2011). *Fanged noumena: Collective writings 1987-2007*. Urbanomic.
Rescher, N. (2000). *Process philosophy*. University of Pittsburgh Press.
Sellars, W. (1948). Realism and the new way of words. *Philosophy and Phenomenological Research, 8*(4), 601–634.
Sellars, W. (1949). Language, rules and behavior. In S. Hook (Ed.), *John Dewey: Philosopher of science and freedom* (pp. 289–315). Dial Press.
Sellars, W. (1956). *Empiricism and the philosophy of mind*. Harvard University Press.
Sellars, W. (1981). Foundations for a metaphysics of pure process. *The Monist, 64*(1), 3–90.
Srnicek, N., & Williams, A. (2015). *Inventing the future: Postcapitalism and a world without work*. Verso.

PART I

Immanence, Metaphysics and Givenness

CHAPTER 2

Different Conceptions of Immanence

As was mentioned in the introduction, one of the main aims of this work is to understand how thinking can be *in* the world while simultaneously being *about* the world. Insofar as thinking is *about* the world, in the sense of being directed toward the world in a way that makes it rationally responsive and answerable to the world, it cannot be *in* the world as one of its parts but must stand over against it as a standard of measurement as it were. Yet, at the same time, thinking cannot be about the world unless it is also somehow *within* it. If this were not so, it would be hard to see how it is so much as possible for thinking to be about *this* world which we did not create, and which materially supports (and undermines) our continuing bodily and mental existence. Thus, ultimately the task is to understand how thinking can be both inside and outside the world.

To see how and in what sense being can exist unconditionally yet internally connected with conditionally existing actual beings and to understand how thinking is in the world yet not in the sense of being one of its 'parts', we need a clarification of the concept of immanence, one of the most crucial concepts used in this work. And to see how thinking can be both in and about the world we need a concept of fractured immanence, which will be developed in Chap. 3.

© The Author(s), under exclusive license to Springer Nature Switzerland AG 2025

D. Christias, *Towards a Hyperstitional Process Nominalism*, Palgrave Perspectives on Process Philosophy,
https://doi.org/10.1007/978-3-032-06350-2_2

2.1 Formal Conceptions of Immanence: Ontological, Epistemological and Explanatory Immanence

What is immanence? There is no single way of defining immanence, but perhaps its main feature in all its forms is its opposition to *transcendence*. But, as transcendence itself comes in many forms, I think that at the most general level we can say that at first approximation immanence is the view that there is no transcendent *ground*, *cause* or *viewpoint* that explains the existence of the world, of worldly modes of being (be they inorganic or living beings), or of the mind-world relation. Based on this preliminary definition, we can distinguish between three different types of immanence that express a rejection of three respectively different conceptions of transcendence:

1. *Ontological* immanence (*contra* ontological transcendence)
2. *Epistemological* immanence (*contra* epistemological transcendence)
3. *Explanatory* immanence, or immanence as self-grounding (*contra* explanatory transcendence, according to which we cannot understand or explain being, mind and their relation by recourse to space-of-reasons rational self-reflexive procedures).[1]

2.1.1 *Ontological Immanence*

Ontological immanence resolutely rejects ontological transcendence, namely, the view that the existence of the world as a whole or of actual entities or types of entities is ontologically grounded (i.e. is conditional for its existence and structure) upon essences whose being is independent of the actual world. Well-known examples of transcendent essences from the perspective of classical metaphysics are Platonic Ideas, Aristotelian forms

[1] Of course, this classification between different conceptions of immanence is not meant to be exhaustive. It attempts to throw light on problematic aspects of the mind-world relation as the latter are formulated in the context of this book. A properly scholarly contribution to the topic of immanence in philosophy (especially regarding its neo-Platonic and religious origins) would demand its own book. For useful analysis of the concept of immanence in its ontological dimension see Deleuze (1968) and Fourtounis (2023). The concept of immanence as self-grounding is thoroughly examined in Kerslake (2009). See also Baltas (2012) for an interesting account of immanence in relation to thinking (epistemological immanence).

and the Thomistic Divine Intellect. All these are examples of independently existing and self-sufficient essences which are necessary for explaining the existence and being of actual entities. One could object here that Aristotelian substantial forms are an example of ontological immanent universals since, by contrast to Platonic Ideas, they appear only *in re*, in the experience of particular actual beings. While this is undoubtedly the case, Aristotelian forms remain ontologically transcendent grounding essences in the sense that their being is *independent* of the being of actual entities and not essentially affected by it. Aristotelian forms are instantiated or exemplified in actual particular beings; yet their being is not situated in the realm of actuality, but in the realm of potentiality, and in this sense, it is transcendent to the being of actual entities. A grounding essence does not count as ontologically immanent to actual beings just by being actualized or instantiated in them, especially if it is defined in such a way as to be completely independent in its essential nature of the becoming of actual entities and processes.

Interestingly, given this criterion of ontological immanence, even some contemporary scientifically or naturalistically informed metaphysical grounds of reality, such as Newtonian absolute space and time or fundamental laws of physics, would count as strictly speaking transcendent grounding principles, to the extent to which they exist (or 'subsist') independently of the becoming of actual entities and processes. From this point of view, it is instructive to see that while early modern (and many strands of contemporary) philosophy, in the wake of the scientific revolution resolutely rejected the Aristotelian conception of the universe as populated by objects with qualities and dispositional properties that account for what these objects are and do, and replaced it with a picture according to which powerless objects are what they are and do what they do not because of their natures but because of their conformity to the fundamental laws of nature, the essentially *transcendent* status of the metaphysical grounding principles which account for the being and behavior of actual worldly entities was not itself put into question.

Moreover, ontological immanence has important implications for the relations and ontological interconnections between categorially different kinds of beings as well as between mind and world. Humanity, and its distinctive way of relating to the world through rational means which pervade perception and action, is not a 'kingdom within a kingdom'. Free will and rationality are not special kinds of causal powers which qualitatively transcend the causal powers of all other beings. All qualitatively emergent structures of beings (including the non-human ones that characterize living beings as opposed to inorganic structures) must be accounted for in a way that the emergent powers in question do not involve some kind of transcendent causality.

One might wonder, of course, why is that a (philosophical) problem? As we shall see in more detail in Chap. 5, a non-univocal conception of being[2]—based on irreducible categorial distinctions in being—for example, between mind-body, sensibility-understanding, reason-nature, animate-inanimate—draws the distinctions in question in such a way that the ontological relation between the categorially distinct levels of being becomes unintelligible. These relations are construed in terms of brute irreducible emergence (emergent causality) or brute supervenience relations that are either non-explanatory (epiphenomenal) or reintroduce transcendent forms of causation between ontological levels. At bottom, all these irreducibly distinct ontological-categorial regions reflect a mode of understanding the world in it which our practices or ways of experiencing the world are destined to be irreducibly incommensurable domains of *intelligibility* which in turn reflect categorially incommensurable domains of *being*. Again, the problem here is one of intelligibility: the incommensurability of the above domains is understood in such a way that the relations between these domains become unintelligible. Moreover, this equivocal incommensurabilist understanding of being also contradicts *explanatory* immanence, one expression of which is the philosophical quest for an intelligibility of how our multifarious practices and ways of experiencing the world *hang together*—as opposed to just being incommensurable modes of intelligibility or ways of being-in-the-world. Thus, we see here an important connection between ontological and explanatory immanence.

[2] From the standpoint of a non-univocal, 'equivocal' understanding of being, the term 'being' is said of beings in several senses, and these senses have no common measure: 'God is' does not have the same sense as 'man is', for instance, because God does not have the same type of being as humanity. Construing being as characterized by different and mutually incommensurable ontological categories (*à la* Aristotle) expresses an equivocal understanding of being. By contrast, to say that being is univocal, as Duns Scotus affirmed, means that being has only one sense, and is said in one and the same sense of everything of which it is said, whether it be God or man, animal or plant (Smith, 2001). A univocal understanding of being eschews talk of incommensurable ontological categories. This is not to say that it is expressed by only one category, but that all its fundamental determinations are situated on the same plane of immanence and affect—and are affected by—all actual beings.

This does not mean that one cannot endorse ontological immanence while rejecting explanatory immanence. For example, arguably, Heidegger's philosophy does just that. The ensuing position boils down to some kind of de-absolutized correlationism[3] in which the transcendental correlation between thought and being is considered as primary (as unexplained explainer) yet radically contingent (it does not necessarily exist; it is an inexplicable 'gift' from self-relating being). Yet, as we shall see, what is problematic with these transcendentally informed positions is that the very *bringing forth* of the subject-world correlation becomes a complete mystery unless it presupposes that the conditions of access to the world are themselves conditioned by a pre-correlational world. In other words, the transcendental subject in all its guises (including the pre-subjective yet person-like Heideggerian *Dasein* or the Merleau-Pontian body-schema) and the manifest world correlated with it cannot be intelligible as 'world-making' transcendental correlational structures, unless they are themselves immersed *within* a world which is not categorially correlated with the transcendental subject. Thus, a transcendentally informed thought-world correlation that endorses ontological immanence but rejects explanatory immanence, while coherent at first glance, is shown to be an ultimately unstable position.

Lastly, let me briefly examine the implications of the rejection of transcendent causality from the standpoint of ontological immanence. Whatever else ontological immanence is, its trademark is the rejection of transcendent causality in all its forms. But what replaces transcendent causality from the standpoint of ontological immanence? How does ontological immanence account for the relation between being and beings? The short answer is that it replaces transcendent causation with *immanent* causation. But what does this mean? We shall discuss causation in detail in Chap. 7, but we can point out right away that immanent causation between being and beings is essentially *expressive*, as opposed to ordinary transitive causation. This means that in the context of an immanent expressive model of the relation between being and beings, the expressed (unfolding of actual beings) cannot be related to what expresses it (being) on the model

[3] 'Correlationism' is a term for a whole family of views according to which we only ever have access to the correlation between thinking and being, and never to either term considered apart from the other (Meillassoux, 2008, p. 5). According to correlationism, we can never grasp an object as it is in itself in complete independence from its relation to the subject, but at the same time, we cannot grasp a subject which would not be always already related to an object.

of transitive causation, namely, *as a prior and separately identifiable cause is to its subsequent effect.* Moreover, relatedly, as the expressed is not something already fixed and self-enclosed in its form prior to its expression, it cannot be correlated with that expression in the manner in which antecedently fixed laws of nature are exemplified in the behavior of actual beings. The expressed is not an independently existing realm of laws or possibilities incarnated or realized in actuality while remaining unaffected by it. The expressed is ontologically inseparable from its expression and is progressively determined (though not exhausted) in it. Further, according to this model of expressive causation, the expressing cause is *'in'* its expressed effect, *as in something else* while still being and remaining *in* it. For example, being is expressed in an actual being—for example, a tree, in the mode of a tree while still retaining its identity as being. In this way, being and beings, the expressor and the expressed, a thing and its effects, are *ontologically inseparable* yet *non-resembling* with one another (Bowden, 2017, pp. 220–223). Using neo-Platonic and Deleuzian terminology we can say that expressions progressively *explicate* and constitute what is expressed, but equally, what is expressed complicates or remains *implicated* in its expressions. What the expression expresses can be called the 'sense' of its immanent cause, provided that 'sense' here is sharply distinguished from its conceptualistic rendering as semantic, propositional or judgmental content.[4] Yet, the expression not only expresses the sense of its immanent cause, but also *dynamically constitutes* the latter through the particular way in which what the expression expresses is progressively articulated in the actual world: for example, the unfolding of actual beings, by expressing the unfolding of the actual history of the world, progressively forecloses certain possibilities of expression and enables others. This dynamic process of expressive causation between being and beings, and between beings-'causes' and beings-'effects' are operative at all levels of beings (including inorganic beings), but we can perhaps better understand it by means of an example from our own case. My embodied behavior 'says something' about (non-semantically expresses) the desire that animates it, but the sense of what my behavior 'says' about my desire can be understood in a variety of ways—as when someone 'reads' in my actions a desire that I was not consciously aware of or could not admit to myself. This, in

[4] From this point of view, both reasons and causes can be understood as reifications, situated at the level of conceptual representation of the world, of this immanent notion of causation which expresses the 'sense' of beings. See Chap. 7.

turn, can influence to a very great degree our future behavior (Bowden, 2017, pp. 223–225). Eventually, what my behavior says about my desire is a matter of ongoing explication which is itself implicated in everything I do and say about what I do, and where the complete unpacking of these latter facts about myself would require a complete account of my interactions with others and the surrounding world that have left their causal traces in my behavior and my understanding of it. Analogous points can be made about non-human animals and even about inorganic beings: the behavior of a rock's body non-semantically expresses its range of powers (its *conatus*) in its interaction with other things (air, water, fire), but this 'rock-sense' can be explicated in a variety of ways by other bodies as the interaction of the former with the latter's powers unlocks different dimensions of the rock's powers. Eventually, what the rock's actual behavior in its interaction with other bodies expresses about its powers is a matter of ongoing explication which is itself implicated in this particular rock's actual causal history (occurrently expressed in the causal traces left to this particular rock by its interactions with its environment and its conditions of genesis).[5]

[5] Of course, a big difference of the immanently expressed senses in the case of inanimate beings such as rocks and humans is that in the latter case, but not in the former, the immanent expression in question involves an *understanding* of the immanently evolving situation. Yet, note that understanding (especially in Deleuze's version of immanence) counts as immanently expressed only if it is *non*-semantically articulated: understanding in this sense is not properly expressed in propositions or thoughts occurring in the subject's mind that represent a world standing over against the subject (Deleuze, 1968). It is more akin to an unconscious grasp on our part, rather than a conscious, explicitly understood one. One could insist here that even on this non-semantic model of understanding there still are sharp categorial differences between the immanent senses expressed by animate and by inanimate beings since the latter do not seem to be capable of grasping anything, even unconsciously. However, the above objection seems to presuppose a semantical conception of understanding since it conceives the unconscious dimension of understanding in semantic terms, as being directed to semantically significant intentional objects (e.g. the father in Freudian Oedipus complex). Given a resolutely non-semantical conception of sense, we do not need to construe the distinction between human, animal and inanimate immanently expressed senses as one of ontological category, but as one of difference in degrees of freedom and of disengagement of their respective expressive powers from immediate survival needs and contexts of interaction. From this point of view, it can be argued that what the rock 'unconsciously grasps' consists is the information contained in a compressed form in the causal traces of the physical interactions of the rock with its environment. This information indirectly shapes the features of the environment to which the rock is explicitly sensitive but is not itself an element differentially responded to by the rock itself. An important more general issue here is that the notion of conscious or unconscious grasping seems to presuppose the existence of an interior perspec-

Let me here reiterate perhaps the most central feature of expressive causality. *Contra* emanative Plotinian immanence in which the cause (being) remains in itself, while the effect it produces (beings) is not and does not remain in it, expressive immanent causality implies that the expressing cause (being) can remain in itself (be a self-relation of being) *only* through its morphogenetic expression in actual beings. It also implies the *non-resemblance* of expression (actual beings) and what is expressed (being). Non-immanent ontologies positing transcendent causes as grounds of the being of beings—including neo-Platonic emanative ones—leave mysterious and ultimately unintelligible the connection between being and beings and presuppose a *resemblance of form* between the structure of being and beings which is ultimately circular and uninformative (i.e. tautological) as an explanation of how being and beings can *so much as be* the kinds of things that are capable of being related to one another. Relatedly, these non-immanent ontologies also posit a similarly circular and thus explanatorily uninformative 'resemblance of form' between essence and existence or between *possibility* and *actuality*—where the former (e.g. the essence of horseness or a possible horse) has the exact same form with the latter except that it lacks actual existence (we shall discuss this issue in more detail in Chap. 4). These problems can be dissolved by thinking of the relations between being and beings, essence and existence or possibility and actuality, as relations of expressive causality, where the expressed (beings) explicates what expresses it (being), is implicated in it, does not resemble it in its form, and whose dynamic unfolding is not the actualization of an always already subsisting field of possibilities or essences, but rather, amounts to the transformation of this field itself based on the actual causal history of actual beings expressed in this unfolding.[6]

tive which in turn seems to presuppose the existence of sense organs, namely, something that cannot be literally attributed to inanimate beings. And since inanimate beings do not have sense organs, the only way to salvage a non-panpsychist sense in which inanimate beings 'informationally grasp' the world is by holding that they possess a dynamically evolving *non*-sensory perspective of the world. In Chaps. 4 and 5, I shall argue that we can make sense of the notion of monadic perspective that does not depend on the literal existence of sense organs, and only presupposes differential responsiveness to (a 'contracted openness' of) the environment and a relational conception of spacetime in which objects are processual and adverbial modifications of spacetime itself.

[6] As we shall see (Chap. 6), this expressive relation between being and beings does not relinquish the metaphysical distinction between ground (being) and grounded (beings), but dynamicizes it.

2.1.2 *Explanatory Immanence*

I turn now to the second type of immanence, namely, explanatory immanence. As was briefly mentioned earlier, this kind of immanence is not necessarily implied by ontological immanence, but together they sketch a more coherent and stable picture of the world and our place in it in relation to de-absolutized transcendental correlationism that rejects explanatory immanence and opts only for a transcendentally informed ontological immanence. Explanatory immanence is opposed to what we might call explanatory transcendence, according to which we cannot understand or explain the relation of being to beings or the mind by recourse to space-of-reasons rational self-reflexive procedures. Explanatory immanence functions as a condition of intelligibility and has an intrinsically *self-reflexive* element. Thinking is explanatory immanent when it aspires not only to understand the world and ourselves in a comprehensive manner (i.e. in a manner that makes intelligible how 'things in the broadest possible sense of the term, hang together in the broadest possible sense of the term' (Sellars, 1963, p. 1)), but to *self-grounding*, namely, to account for the status of the philosophical (metaphysical, epistemic) grounding principles *themselves*. This makes explanatory immanence—or immanence as self-grounding—distinctively post-Kantian, placing it in the tradition of *immanent critique* or *metacritique* (Kerslake, 2009). Explanatory immanence, for example, attempts to account for the validity of the Kantian grounding (transcendental) principles through which reason criticizes the dogmatic metaphysical image of thought and concludes that these principles can be accounted for only by drawing a comprehensive and self-referentially consistent metaphysical picture of ourselves in the world. The key to such a move lies in the realization that it is not possible for reason to reflect on its own nature or function without already engaging in the attempt to know how the world is *tout court*. Otherwise, one ends up to a self-refuting skepticism which puts into doubt the very capacity of our representings to *refer* to something external to the mind (their representeds), thus putting into doubt their very objective purport (Brandom, 2019; Hegel, 1979).[7]

[7] Immanent critique is arguably rooted in Kant's idea of a tribunal of reason, in which reason must criticize itself in order to avoid dogmatic metaphysical pitfalls. Yet, it was soon observed, mainly by Schelling and Hegel, that the Kantian system did not possess the means to account for the kind and status of knowledge of the very principles that are necessary for all possible experience. At certain crucial junctures, Kant seems to appeal just to the 'fact of reason' as a justification of its validity, but this is plainly inadequate for a philosophy that aims

This lack of an absolute foundation of reason or knowledge (it is not possible for reason to justify its own validity without presupposing substantial worldly knowledge) coupled with the latter's semantic openness to the world *tout court* and its self-referential character (one cannot examine the means of knowledge without at the same time expressing knowledge about the world *tout court* and vice versa) makes possible—and indeed necessitates—the provision of a self-critical *self-grounding* of reason's basic principles from an explanatory standpoint that immanently connects it with the world. In this way, self-grounding works a) by providing a provisional yet comprehensive system of understanding the world and ourselves, and b) by attempting to ground the basic principles of that system in an immanent way, that is, by the use of the system's own conceptual resources. Note here, importantly, that metatheoretical, structural closure does *not* entail closure in the system for which it accounts; metatheoretical closure is fully consistent with essential openness in the *resulting* philosophical system[8] (Kerslake, 2009, pp. 21–22).

Moreover, this demand for immanent self-grounding amounts to a resolute implementation and adherence to the *principle of sufficient reason* (PSR) (for every thing, fact or occurrence there must be a ground that explains *why* it is thus and so as opposed to merely registering *that* it is thus and so). Explanatory immanence goes beyond PSR in that it has an essentially reflexive and self-referential element that does not seem to be essentially included in PSR. Yet, though distinguishable from PSR, it is inseparable from it. We shall see in what way PSR can be used for the construction of a comprehensive self-grounding metaphysics in Chap. 6. For the time being, it is worth mentioning that an interesting consequence of explanatory immanence, to the extent that it uses PSR as a condition of

at immanent self-grounding. Schelling and Hegel concluded that this could be achieved only by utilizing a genetic approach (as opposed to the procedure of finding conditions) in the context of a comprehensive system that would not only offer us an adequate picture of the world and ourselves, but would also be capable to account for its own emergence and function.

[8] Deleuze comes to mind as a paradigm case here. For example, the fundamental metaphysical concepts of 'difference' and 'repetition' in his *Difference and Repetition* (1968) are such (what repetition repeats is difference, not sameness) that do not only enable the construction of an essentially open metaphysical system, but also necessitate the occasional revision of the basic principles of the metaphysical system itself in response to novel circumstances that trigger thinking. Indeed, arguably, the central concepts of Deleuze & Guattari's *A Thousand Plateaus* (1980) can be understood as 'repeating' the founding gesture of *Difference and Repetition* in a transposed key.

intelligibility, is that the commonsensical categorial understanding of the world in terms of individual things (objects) having properties (expressing capacities, abilities, dispositions) such as to make the objects that have them belong to certain kinds, characteristic of their nature, cannot adequately account for the *individuality* (*haecceity*) of existing particular things (e.g. Socrates as opposed to Plato, this rock over here as opposed to that rock over there) save only by way of inessential, *accidental* properties, such as different spatiotemporal location (i.e. different 'matter' which 'enmatters' the same form ('human', 'rock')). In other words, from this point of view, a conception of the individual as defined by general essential properties (what it has in common with other things of the same kind) plus specific accidental properties (its contingent differences from other things of the same kind) is problematic since it cannot *essentially* individuate particular actual beings. That is, it cannot provide a PSR for them, namely, a ground that explains *why* an individual is the individual it is as opposed to another individual of the same kind. Only a conception of *singular essence*,[9] which immanently expresses the 'highest' universal that stands above every genus (being) directly within absolutely concrete individuals (actual beings) that cannot be subsumed under any general kind, is in a position to account for the concrete singular 'thisness' of existents in a non-subsumptive and non-accidental manner.

2.1.3 Epistemological Immanence

Let us now turn to the *epistemological* dimension of immanence, which is defined in opposition to what might be called 'epistemological transcendence'. Epistemological immanence is immanence as applied to thinking in its dimension as an epistemic medium of accessing (knowing, being intentionally directed to) the world. To say that thinking is immanent, and thus to endorse epistemological immanence, is to say that thinking necessarily unfolds and self-corrects in medias res, in the midst of our practices, thus rejecting 'the view from nowhere' conception of thought. According to this conception, essential aspects of which can be found in Neurath (1921), Quine (1960), Sellars (1956) and Lewis (1986), our knowledge

[9]The notion of singular essence stands in opposition to the Aristotelian conception of essence in which a thing's essence is what it has in common with other things of the same kind (Deleuze, 1992, p. 300). For example, in Spinoza, the *Conatus* plays the role of singular essence.

of the world and ourselves, as expressed in thought, is not constructed from a given or a priori foundation of from first principles (provided by reason or sensibility) but always already presupposes that we live within a world in which we act and of which we have some provisional understanding and knowledge. This understanding and knowledge is provisional in that it is always subject to self-correction in response to pressing incompatible commitments in our overall system of understanding ourselves-in-the-world. Thus, to say that the conceptual system that expresses our current conception of ourselves-in-the-world is provisional is to say that it is also thereby reflective and critical, in that no element of our current conceptual framework is absolutely beyond question. It always remains open to new experiences and new ways of organizing experience. In Sellars' words: 'empirical knowledge, like its sophisticated extension, science, is rational, not because it has a foundation but because it is a self-correcting enterprise which can put any claim in jeopardy, though not all at once' (Sellars, 1956, §38). Sellars' point here holds for thinking in general. Thought, including its symbolic activities that seem less prone to change in the course of empirical inquiry, such as mathematics and logic, is rational not because it has some kind of a priori, axiomatic or transcendentally justified foundation, but because of its self-reflexive and self-critical character in response to incompatible commitments in our overall system of beliefs.[10] Importantly, this epistemologically immanent picture of thought has consequences for philosophy: philosophy must not presuppose that there is a static, once-and-for-all vision of ourselves-in-the-world, since we are constantly learning new things in all domains of experience (mathematics, natural-social sciences, history, ethics, politics, aesthetics) that necessitate revisions in our overall view, and also because there is no reason to think that the way we fit into the world is itself static (deVries,

[10] As Neurath famously puts this point: 'We are like sailors who on the open sea must reconstruct their ship but are never able to start afresh from the bottom. Where a beam is taken away a new one must at once be put there, and for this the rest of the ship is used as support. In this way, by using the old beams and driftwood the ship can be shaped entirely anew, but only by gradual reconstruction' (Neurath, 1921, p. 199). Similarly, Quine sketches an essentially epistemologically immanent picture of thought: 'We all must start in the middle. Our conceptual firsts are middle-sized, middle-distanced objects, and our introduction to them and to everything comes midway in the cultural evolution of the race. [...] We cannot strip away the conceptual trappings sentence by sentence and leave a description of the objective world; but we can investigate the world, and man as a part of it, and thus find out what cues he could have of what goes on around him. Subtracting his cues from his world view, we get man's net contribution as the difference' (Quine, 1960, p. 4).

2005, p. 8). Indeed, as we shall see throughout this work, there is ample reason to believe that the way we fit in the world is dynamic and can be more adequately approximated by equally dynamic categories of thinking that emphasize unfolding processes over ready-made objects, finished products and fixed, unchanging essences.

As mentioned above, epistemological immanence is opposed to an epistemologically transcendent conception of thinking. What is involved in this latter transcendent way of thinking about thinking? According to this view, thought can survey the totality of possible worlds or conceptual frameworks independently of its immersion within the world and the semantic and epistemic practices that are developed and unfold within it as a means for thinking about it. This view thereby operates with a picture of the mind as something Given, namely, as a capacity to directly and transparently grasp universals (serving as unchanging metaphysical—or transcendental—grounds for understanding the actual world of becoming) with the 'mind's eye', independently of and antecedently to our social space-of-reasons practices in the context of which (according to the immanentist) we construct, internalize, transmit and self-correct these universals. The Sellarsian myth of the Given and its multifarious consequences for thinking, being and immanence will be more thoroughly explored in Chap. 3. For the time being, it will suffice to point out that the standpoint of epistemological transcendence cannot be avoided just by rejecting patently transcendent conceptions of mind's access to the world such as the Thomistic notion of divine Intellect, the Platonic notion of mind as grasping the eternal realm of Ideas, the Spinozistic conception of the mind as capable of surveying the world *sub specie aeternitatis* or Leibniz's conception of the mind of God as capable of accessing the totality of possible worlds. The standpoint of epistemological transcendence—what Quine aptly called the standpoint of the 'cosmic exile'—remains intact even if the mind's access to the world is construed in a non-metaphysical manner, for example, transcendentally (Kant), 'dialectically' (Hegel) or phenomenologically (Husserl). In effect, all these views conceive thought from a cosmic exile's standpoint, although they all resolutely reject pre-critical dogmatic metaphysics.

Consider, for example, Kant's and Husserl's transcendental subject. They both have transparent access to contents independently of the mind's immersion in the actual world (the latter is needed only for 'activating' this content, as it were, not for constituting it as such). In Husserl's case, things are even clearer. His transcendental phenomenology is based on the suspension of the 'natural attitude' which assumes that objects have a consciousness-transcendent existence. This suspension of the independent existence of the external world is precisely what reveals the pure structures of constituting consciousness, namely, noetic structures and their noematic correlates (Husserl, 1931). Ironically, this transcendental correlational realm revealed by the phenomenological *epoche* is sometimes itself called (and it has every right in doing so) a realm of phenomenological *immanence*, namely, immanence to phenomenologically revealed transcendental consciousness. Yet, as is obvious, this plane of phenomenological immanence is itself the corollary of an epistemologically *transcendent* standpoint, since it is conceived entirely independently of and antecedently to thinking and intentionality as articulated by our practices of giving and asking for reasons. And this shows that, just as there is no direct entailment between ontological immanence and explanatory immanence, there is also no such entailment between a transcendental-phenomenologically informed immanent ontology *à la* Husserl and epistemological immanence.[11] Consider, moreover, Hegel's conception of the mind (or, better, the absolute) in terms of a self-relating negativity capable of unveiling the conditions under which alone can thinking and being have determinate content. While at first glance Hegel's view, in contrast to Kant's and Husserl's, seems to endorse an essentially world-involving, historically evolving and self-correcting model of thought as immanently expressed in our practices, it actually betrays this model by insisting that the metaphysical principles derived from the 'dialectical movement of the Concept' (in the *Science of Logic*) are themselves a priori and unrevisable (Hegel, 2010). It thus seems that while this dialectical movement of the Concept is indeed *expressed* in our semantic and epistemic practices, it is not essentially affected by the contingencies of the latter. A Hegelian might retort that the unfolding of the absolute is sensitive to the contingencies of our practices, and this sensitivity is expressed by the tensions and incompatibilities within and between our central metaphysical concepts that push us to self-correct them in the dialectical manner articulated in Hegel's *Logic* (being, nothing, becoming, essence, concept). But again, this supposed sensitivity

[11] Merleau-Ponty and Heidegger rejected Husserl's transcendental subject and attempted to place the latter within the world while retaining its transcendental (non-empirical) character. They thus made a move toward placing phenomenological immanence within an epistemologically immanent framework. The extent to which they succeeded is debatable (see Christias, 2023 for more on this issue).

to the contingencies of our practices and our experience is in a sense already included in the essence of the absolute's unfolding, and it cannot be taken by surprise, as it were, by unanticipated contingencies in the world. Actual things or experiences may well be contingent, unanticipated or incomprehensible, but the very *form* or this unanticipatedness or incomprehensibility is always already anticipated or comprehensible by Hegel's fundamental metaphysical concepts of his *Logic*. Using an existentialist idiom, we could say that Hegel is still in the grip of a metaphysical picture in which 'essence precedes existence'. To this extent he inadvertently retains epistemologically (and ontologically) *transcendent* elements in his metaphysics.[12]

2.1.4 Fractured Immanence

In this book I shall defend a view that combines all three types of immanence discussed above, namely, a view that adheres to ontological, epistemological and explanatory immanence. As we saw, this is not the only possible view in the vicinity, but it will be argued that it is the most promising and comprehensive one in accounting for the relation between being and being and between being and beings capable of representing the world. As we shall see in more detail in Chap. 3, the combination of ontological, explanatory and epistemological immanence results in a picture where immanence comes in an essentially fractured form, in which mind and world stand in a direct unmediated immanent ontological relation whose intelligibility nevertheless presupposes a transposition of mind and world as semantically and epistemically correlated poles of a diachronic self-correcting process that ultimately aims to align our conception of mind and world to mind and world as natural processes, without any

[12] While this might be true about the historical Hegel, it must be stressed that there are readings—or, better, creative reconstructions—of Hegel's philosophy that, by interpreting his logico-metaphysical categories as meta-conceptual tools for making explicit the structure of first-order empirical concepts, can do justice to their a priori character while at the same time allowing for their potential revisability. This is because by functioning as pragmatic meta-concepts whose *raison d'etre* is the regulation of first-order empirical concepts, metaphysical categories remain indirectly dependent on the contingencies of empirical inquiry and thus to potential changes in our empirical concepts (Brandom, 2019). From this point of view, which to my mind is very close to the spirit of epistemological immanence, the only criterion of adequacy of our higher-order philosophical meta-concepts is the illumination they provide to our various practices of using empirical concepts.

guarantees of achieving full transparency or full success. Another way to state this fractured immanent relation between mind and world is to say that it expresses the fact that thought and being are ontologically identical yet asymmetrically so (what we might call *asymmetric ontological immanence* between thought and being) *and* epistemologically different yet symmetrically so (what we might call *symmetric epistemological immanence* between thought and being). While being and thinking are ontologically identical, they are asymmetrically so (being is unconditional, thinking is conditional). And while being and thinking are epistemologically different (being is conceptually expressed by modal categories while thinking is expressed by normative categories), they are symmetrically so (the normative categories of thought 'mirror' the modal categories of the world).[13] It will be argued (Chap. 5) that this can be possible only if we conceive of being itself as essentially *self-fractured*, in a way that the immanence of mind to the world can, through a series of 'topological deformations' in being's self-expression (effected by what we might call 'productive-cum-subtractive imagination'), change its orientational structure as a whole and turn the world to revolve immanently around the mind (where the conditions of possibility of the experience of objects are conditions of possibility of the objects of experience).

2.2 Differentiating Fractured Immanence from Phenomenological, Object-Oriented, Deleuzian and Hegelian Immanence

As was mentioned in the introduction, besides the formal way of classifying different kinds of immanence, depending on what kind of transcendence is being rejected (ontological, explanatory, epistemological), there is another, *substantial*, way of classifying different kinds of mind-world immanence, this time depending on what functions as an unexplained explainer that immanently grounds the mind-world relation. Different conceptions about what functions as such an unconditional unexplained explainer result in different substantial conceptions of mind-world immanence. Some of these are phenomenological immanence, object-oriented immanence, Deleuzian immanence and Hegelian immanence. In this

[13] Importantly, this asymmetric ontological identity and symmetric epistemological difference between thinking and being must be conceived as a *productive metastable tension*, rather than a failure of synthesis between mind and world.

subsection, I will attempt to differentiate my position from these views, so as to make clear what fractured immanence is *not*.

2.2.1 Phenomenological Immanence

If we take as unconditional unexplained explainer *the mind-world correlation* as revealed from a transcendental-phenomenological standpoint, we thereby draw a plane of phenomenological immanence which always already discloses the mind as correlative to the (life)world and vice versa. We get very different versions of phenomenological immanence, for example, in Husserl, Merleau-Ponty and Heidegger, depending on whether the transcendental-phenomenological plane which discloses the mind-world correlation is revealed through the phenomenological *epoche* (which brackets the commonsensical belief in the independent existence of the natural world) or through a transcendental standpoint that eschews the phenomenological reduction. According to these latter viewpoints (Merleau-Ponty, Heidegger) we are always situated within an already existing surrounding world which is largely out of our control: our transcendental standpoint is not theoretical, disinterested and all-encompassing, but is essentially embodied, existentially involved, perspectival and finite: we are thrown within a natural and cultural world we never fully control or comprehend, and in which our being is an issue for us, something that we essentially care about. This caring is in turn expressed in a temporal manner, in the form of an anticipation of an open future in conditions partly determined by our cultural-biological past.

Yet, these admittedly significant differences in developing the transcendental-phenomenological plane of immanence share some important features that are objectionable to say the least from our point of view. While the Husserlian version of the immanent mind-world correlation is theoretically oriented and functions as a kind of absolute ground (justified by categorial intuition), Merleau-Ponty and Heidegger opt for a de-absolutized and unobjectified correlationism between a pre-objective subject (*Dasein* in Heidegger's case) and a pre-objective world of affordances and solicitations to act to which we are affectively and pre-reflectively attuned and immersed in the context of our everyday practices. But in all cases, the transcendental correlations involved (however unobjectified, contingently existing, non-transparent, finite or perspectival they might be), to the extent that they function as immanent grounds of *bringing forth* the mind-world correlation, construe this transcendentally-phenomenologically revealed mind-world correlation as unconditional, as a rock-bottom unexplained explainer. From this point of view, to say that the mind-world correlation needs to be *itself* immanently grounded to pre-correlational being (or to a pre-correlational self-relation of being) is sheer nonsense. The only way in which we can make sense of a pre-correlational mode of existence of the world is in the case of the *empirical* world, which yet can be considered as real, and thus as existing independently of its correlation with the mind, only in conjunction with its *transcendental* sense, and thus, only by *reinstating* a transcendental correlationist viewpoint from which the independent existence of the empirical world would be more perspicuously stated by saying 'the world *gives itself* as existing independently of all correlation between mind and world' (Meillassoux, 2008, pp. 14–15). In this way, the empirical unconditionality of being over thought is effectively subordinated to the transcendental unconditionality of the being-thought correlation over its terms. But this move makes the very possibility of the emergence of the mind-world correlation as a perspectival point of view (unconcealment) of being, a complete mystery, a constitutively inexplicable 'gift' from being. We must either take it that the mind-world correlation has God-like powers, bootstrapping itself into existence in the manner of a metaphysical *causal sui*, or we must believe that the mind-world correlation and the finite unconcealment it makes possible is a self-relation of being, which is yet constitutively beyond comprehension; its emergence and function transgress the boundaries of sense, since we can comprehend being only through the lens of the mind-world correlation. I find this position, where the mind-world correlation is itself radically contingent (it could well not exist), thereby pointing to something beyond itself that might exist (the world as it is in itself) which is constitutively incomprehensible for us (since we has had access to it through the lens of the mind-world correlation), profoundly unsatisfactory, itself incomprehensible, a conversation-stopper and an artificial block of the way of inquiry. According to this view, it seems that thought *does* have access to the absolute: the very fact that the mind-world correlation is conceived as radically contingent or 'factical' shows this. The possibility of it not existing is *not* relative to us, its very conceivability presupposes that we have access to the 'in itself', even if this access serves only to make us aware of our radical finitude. Yet at the same time, thought's access to this absolute is such as to render constitutively impossible any meaningful assertion about it.[14]

On the other hand, I also believe that transcendental-phenomenological immanence contains important insights that need to be incorporated into an account of immanence if the latter is to be a sound one. Heidegger's (and Merleau-Ponty's) emphasis on the priority of existence over essence and the phenomenological discovery of the mind-world correlational structure are only two of those important insights. Any sound account of immanence must abide by the dictum that 'existence precedes essence' as it is only in this way that genuinely novel structures can emerge in the actual world without being already presupposed as potentialities awaiting their realization. Moreover, any sound account of immanence must admit the irreducible reality and unobjectified character of the mind-world correlation, while at the same time (*contra* phenomenological immanence) explaining its possibility and function as a dimension of self-relating (self-fracturing) being. While transcendental phenomenology correctly emphasizes that the mind-world correlation cannot be explained by objectifying one of its terms (mind or world), this does not preclude the possibility of accounting for the emergence of the mind-world co-constitution as an expression of an *unobjectifiable* dimension (self-fracturing) of *being itself*.

2.2.2 Object-Oriented Immanence

Another philosophy of immanence lately on offer, which has been the object of much discussion and controversy in contemporary continental speculative realism, is 'object-oriented' immanence, associated with the works of Harman (2018, 2022), Bryant (2011) and Morton (2013).[15]

[14] This fact gives rise to the aura of fathomless profundity which has been invested to questions such as 'why is there something rather than nothing' by Heideggerians. I agree here with Meillassoux (2008, p. 63) and Brassier (2007, p. 73) that by disqualifying traditional metaphysical responses to this question (which invariably evoke the transcendence of a supreme being or a *causa sui*), the existential-phenomenological correlationist thereby leaves the door open for a more pernicious variety of transcendence: the unobjectifiable transcendence of the 'infinitely Other'. A fully consistent immanent viewpoint cannot accept such (explanatorily transcendent) views, and this must be reflected on the way it handles the question 'why there is something rather than nothing'. For example, to say that this question is meaningful yet insoluble amounts to being in the grip of explanatory transcendence (Meillassoux, 2008, pp. 72–73). I attempt to show how this issue might be dealt with from a resolutely immanent viewpoint in Chap. 6.

[15] For reasons of space, in what follows I shall focus on Harman's version of this view.

Object-oriented immanence accepts Husserl's phenomenological insight to the effect that the world is disclosed as an intentional correlate of noematic acts, but it does not give primacy to the human subject-world intentional correlation. Instead, it generalizes the noesis-noema correlational structure to all objects in the world, including inorganic or artificial ones. Objects intentionally interact by unlocking and decoding each other's sensual qualities. Yet, every such disclosure thereby hides from view the real core of the intended objects (this is Harman's metaphysical realist reappropriation of Heidegger's disclosure/withdrawing structure). Reality is replete with objects intending ('alluding' to) one another, but these objects only unlock each other's sensual qualities, never their real ones. Thus, object-oriented philosophy draws a picture of transcendence at the heart of immanence by way of a close inspection of the immanent 'sensuous' relations of worldly objects with other such objects. Yet, it seems to retain its immanent credentials in other respects: it does not privilege any intentional correlation between objects over other objects (or subjects), and it endorses a univocal view of the universe according to which dogs, trees, factories and governments enjoy the same ontological status as protons or the human mind. We might call this a *flat* immanence.

Evidently, this view cannot be acceptable from our viewpoint since it rejects both epistemological and explanatory immanence and only accepts an idiosyncratic version of ontological immanence. Yet, obviously, what is problematic in object-oriented philosophy is not its rejection of resolute immanence as I define it; it is rather that its extension of the subject-object intentional correlation to all objects and that its outright rejection of the view that the real properties of objects can be intelligible by thinking leads to insuperable difficulties.

In his attempt to escape anthropocentric correlationism in all its forms, Harman interprets the intentional relation between objects in terms of his concept of *allure*, which denotes a supposed non-epistemic aesthetic way in which objects grasp, intuit and are related with one another (even in the case in which they fall outside the causal-spatiotemporal nexus). Obviously, this concept is highly metaphorical and deliberately so. There is no problem in using metaphorical language in our attempt to construct a novel philosophical concept. But unless the metaphors are somehow cashed out in the development of this concept in the context of the philosophical system of which it is part through the confrontation with criticism, then metaphorical language becomes increasingly problematic, as the claims involving the concept in question are so vague that they are very hard to be pinned down and critically assessed. We might think of Brandom's 'reliable responsive differential dispositions' (RDRDs)[16] or of Sellars' 'picturing' and Adorno's 'mimesis' (a non-representational, non-conceptual relation of 'affinity' between objects) as concepts that function in a way that structurally resembles Harman's allure. Yet, *contra* Harman, both Sellars and Adorno are clear that while picturing and mimetic relations are non-epistemic, they cannot be grasped as such directly, through non-conceptual intuition (as this would immediately fall foul of the myth of the Given), but only through mediation from concepts. Harman's whole project is based on the explicit rejection of the view that epistemic or conceptual mediation is necessary for understanding the world and ourselves, but I do not think he has done enough to provide determinate meaningful content to his view that objects 'intuit'—without consciousness, non-anthropomorphically but in an aesthetically interesting sense—other objects simply by virtue of being objects.

A related problem here for Harman's account is that by ontologically distinguishing between the real (essential) and merely sensible properties of objects, where the first are forever beyond our—and all other objects'—grasp, he thereby makes our thoughts about objects completely indeterminate. The 'sensual chair' I am sitting on and the 'sensual tree' I am staring at are sensually distinct but might not at all be really distinct. Harman's radical dissociation of the individuation of sensual objects from the individuation of real objects leaves open the possibility that there may be just one object in the 'really real' world. And then, with such a degree of indeterminacy, why even call this an 'object' (in the categorial sense) at all? Moreover, Harman's allure, besides its vague and highly metaphorical character, inherits this more radical indeterminacy.

I think the above problems are symptomatic of two basic features of the object-oriented approach: a) Its attempt to overcome epistemic and sensory mediation by appeal to the perspective of the objects themselves, which, moreover, b) is conceived by analogy with categorial concepts drawn from the lifeworld in the context of which we lead our everyday lives. The very core of Harman's position ('objects intuit other objects

[16] All objects, including inanimate, merely physical ones, are disposed to respond differentially in the presence of certain stimuli in the environment, as in the case of a piece of iron, which reliably responds to the presence of oxygen by rusting. For Brandom, the specific ways in which objects (inanimate, animals, humans) are causally connected with the world is captured by their respective RDRDs (Brandom, 1994).

through allusive relations') testifies to this analogical lifeworld thinking, which, by abjuring certain necessary mediating lifeworld categories—especially modal notions, such as dispositions or capacities, that link the sensible to the real properties of objects—without providing successor categories for directly characterizing the intrinsic nature of objects, ends up with an extremely implausible view that has been aptly called 'modal mysterianism' (Wolfendale, 2019, p. 87).

These problems ultimately spring from the fact that object-oriented flat ontologies wrongly assume that conceptual mediation subordinates ontology to epistemology and always leads to anti-realist consequences. This is why these ontologies typically attempt to altogether bypass the epistemological conditions of access to the world, and account for objects themselves or their relations to other objects by postulating direct, conceptually unmediated, real or sensual properties and relations for understanding them. Yet, this can only lead to a spurious and ultimately dogmatic ontology. In my view, only a thorough epistemological investigation into the conditions of access to the world could lead to a sophisticated ontology which avoids reifying the categorial framework we use to make sense of the world in our everyday life. *Contra* object-oriented ontologies, we can maintain the primacy of being over knowing, and acknowledge that being is irreducible to knowing, while insisting that epistemology is a necessary condition of access to this ontological primacy and semantic irreducibility of being to knowing (see also Brassier, 2014). This is the standpoint of fractured immanence that will be the object of investigation throughout this book.

2.2.3 *Deleuzian Immanence*

What about Deleuzian immanence? Deleuze's immanence has also been taken to imply a 'flat' ontology, albeit one that differs importantly from Harman's version of it. Moreover, Deleuze's conception of immanence has been an explicit inspiration for the delineation of the concept of immanence in the present work, and thus merits closer examination. Deleuze clearly endorses ontological and explanatory immanence; he also endorses epistemological immanence, yet in a version which is not entirely unproblematic. We have already mentioned the basic features of Deleuzian immanence in our discussion of ontological immanence and immanent causation in Sect. 2.1. Roughly, the Deleuzian 'plane of immanence' consists of pre-individual intensive fields of energy flows in various stages of individuation. Importantly, this plane of immanence is not immanence to a subject or to consciousness, not to already individuated objects with extension and properties. Deleuze does not begin by taking individuals as ready-made, pre-constituted or always already unified entities (objects, subjects, consciousness, forms stamped on matter). Instead, Deleuze, following Simondon (2020), draws our attention to the *processes* of individuation through which unified individuals are constituted. These processes as progressively explicated in intensive fields[17] morphogenetically produce subjects with consciousness and objects with extension and qualities in the actual world.

Importantly, this conception of immanence is resolutely anti-representationalist. The manner in which being is expressed in actual beings is not that of a conceptually unfolding form or essence relating propositionally articulated beliefs with states of affairs comprising individuated objects with extension and qualities. From a Deleuzian standpoint, representation remains at the level of the actual; it deals with a ready-made world, with concepts of individuals that are tailored to already individuated subjects or objects. Yet the plane of immanence, being essentially pre-individual, is inaccessible to representation. But how can we gain epistemic access to the plane of immanence without conceptual mediation and linguistic representation? In Deleuze, the contemplative mirroring relation between mind and world through linguistic representation is replaced by a dynamic pragmatic *identity* relation between thinking, considered in its non-representational intensive and pragmatic dimension, and the world, again considered in its non-representational, intensive dimension. This conceptually unmediated relation amounts to a kind of 'intuition' of this pre-individual plane of immanence.

Does this mean that Deleuze disavows representation only to lay claim to a superior faculty of intuition? Yes and no: Deleuze does indeed think that we have a non-conceptual and non-representational access to the world, which does not make any essential use of the semantic framework of meaning, truth and reference, but he does not construe this alternative means of access to the plane of immanence in terms of a static,

[17] Deleuzian intensive fields are essentially continuous, non-metrical and cannot be divided without changing in kind. Intensive differences are at work in energetic systems characterized by differences in potential, which undergo a phase transition when reaching a certain threshold, losing symmetry and changing the dynamics of the system involved. For a more detailed discussion of the notion of intensity, which is crucial for the formulation of our own process ontology, see Chaps. 4 and 6.

unmediated, diaphanous or 'clear and distinct' intuitive grasp of the latter. According to Deleuze, semantically contentful propositions, along with their objective referents (states of affairs), emerge as 'solutions' through the working of an underlying 'problem' which activates and implicates all the faculties involved in the production of our total experience of the world, namely, sensibility, memory, imagination and thought (Deleuze, 1968). These faculties communicate in a discordant manner: there is no direct communication between them; there is no particular object or form that is transmitted from one faculty to the other. The only thing that is communicated is a difference or problem. For example, sensibility does not grasp, perceive or intuit some worldly fact; it rather grasps an 'intensive difference' that is not a particular 'something', and confronted with such intensive differences, each faculty, one after the other, is set in motion (is being forced to deal with it).[18] In this sense, the empirical and socially articulated determination of the sense of propositions, while taking place entirely within the space of reasons, takes place only because of the communication of a larger problem from one faculty to another. From this point of view, fully formulated propositions with a determinate sense, reference and truth conditions express a surface level of an attempted *resolution* of these intensive differences between the faculties, such that the experiencer ceases being 'shocked' by difference and successfully copes with his environment (Bowden, 2014, pp. 192–194).

In this way, Deleuze does not fall prey to any version of mythical Givenness, as his intuitive, non-representational method of accessing the plane of immanence is not presuppositionless or unmediated. He replaces the conceptualist model of understanding mediation between mind and world, where our space-of-reasons practices conceptually mediate our access to the world and ourselves, with a non-conceptual dynamic relation and mediation between them. This is best described in terms of increasing mind-world resonance effected by 'learning' through encounter with singularities—as opposed to knowing though subsuming experience to generalities. This learning process is in turn itself animated by affectively and cognitively concrete problematic (i.e. incomprehensible, disorienting,

[18] For example, in being provoked by sensibility to recollect, the faculty of memory will be forced to recall some 'immemorial' source of its recollecting. The 'solution' to this problem comes in the form of the faculty of imagination, which in turn will fantasize about the essentially unimaginable source of the fantasizing it is forced to undertake. This problem is solved by the faculty of thought, which in turn reaches its limit in attempting to think the essentially unthinkable source it forces it to think.

disturbing) encounters with the world, and is self-corrected by constant experimentation, namely, by the creative implementation and updating of the proposed 'solutions' to these concrete problems. Moreover, even if we want to call this dynamic, holistic, mediate relation between mind and world an 'intuitive' one, we must always bear in mind that this intuitive relation is radically different from Cartesian intellectual intuition which provides a transparent unmediated access to the ultimate categorial structure of the mind though the 'natural light' of reason, composed of 'clear and distinct' ideas. Deleuze resolutely rejects this foundationalist model of thinking about thinking and replaces it with a Leibnizian model according to which increasing clarity goes hand in hand with increasing confusedness (as is the case in 'white light' which 'perplicates' all distinct colors), while increasing distinctness is revealed only on the background of increased obscurity (the distinct sound of sea waves is revealed as such by obscuring the contribution of the minute unconscious perceptions of the sound of each drop of water that slides over and through another). This creative reappropriation of Givenness in non-Cartesian terms leads to a notion of Givenness purged of its connotations to transparency, and coupled with the fact that this non-conceptual relation of mind and the world is dynamic, holistic, mediated and self-correcting, makes a strong case for believing that Deleuze does not endorse a view of thinking (or of the mind-world relation) as mythically Given in the Sellarsian sense.

However, at this point another related problem rears its head. The representational level where worldly states of affairs are understood as the referents of semantically significant propositions is not just an optional extra or a superstructure added to a more fundamental non-representational base of mind-world 'resonance' relations. However ontologically fundamental the non-representational level might be, it seems to epistemically presuppose for its very *intelligibility* as such the existence, function and internalization through language learning of a conceptually articulated semantic level of representational relations between mind, language and world. To see this consider that a necessary condition for something to function as a language expressing determinate thoughts is the digitalization of the audiovisual linguistic continuum and its organization and stable placement into discreet meaningfully articulated units (propositions) that can function semantically, namely, as signs pointing toward (referring to) a reality external to them that is not always directly open to perception. This representational function of thought and language, namely, its ability through its fundamental categorial concepts to refer to an independent reality that may or may not be directly perceptible, is made possible by the fact that the linguistic audiovisual intensive continuum is cut up into stable, nested, hierarchically structured and easily reidentifiable and recombinable units, such as phonemes, morphemes, logico-syntactic-categorematic subject-predicate structure with recursion. To function as such in a stable and reliable manner, all these structures must be understood normatively, as involving correct or incorrect ways of identifying and using them. More importantly, these linguistic functions are involved in our ability to *abstract* from the context in which propositions are uttered, and to distinguish between the *contentual* and purely *intensive* and *associative* dimension of propositional utterances (as reflected in the Fregean distinction between the content and the force of a proposition). And without these normative constraints and abstractive abilities that make the representational dimension of language possible, it is difficult to see how we could establish a minimally intelligible (stable, shared and reidentifiable) distinction-cum-relation between *words* and *things*, propositions and worldly states of affairs, concepts and their objects. Correspondingly, we could make no sense of the fact that propositions sometimes *fail* to refer and are connected to the states of affairs to which they refer not just causally or through associative connections but through *inferentially* articulated *justificatory* relations, where these worldly states of affairs can provide for a suitably trained and placed subject good reasons to accept the proposition's content. And in the absence of the above semantical word-world relations and distinctions and the syntactical structure of language it is not at all clear how we can so much be able to self-consciously *grasp* the non-representational expressive element of language, thought and being in the sense of making it a sufficiently stable and internally coherent object of study and assessment. This is not to say that the representational level must be ontologically reified. In fact, I am inclined to endorse a version of Deleuze's view to the effect that the non-representational/intensive dimension of the mind-world relation is ontologically primary. Yet, this does not mean that the representational dimension of the mind-world relation is *epistemically* or *semantically* derivative. Throughout this book it will be argued that this semantic (but not ontological) primacy of the representational over the non-representational mind-world relation can be illuminated through an ontology and epistemology of fractured immanence. In the Chaps. 3, 7, and 8 to come, I shall attempt to situate the element of truth of (neo)Hegelian immanence, which to my mind best accounts for this semantic and epistemic primacy of the representational level of discourse in understanding the world and ourselves, within the larger framework of fractured immanence. As we shall see (Chap. 8), from this point of view, the mind-world correlation itself, in its conceptualist-representationalist dimension, is understood as the manner of appearance of the self-unfolding of a non-conceptual and non-representational mind-world correlation, where both terms of the correlation are conceived as processes standing in relations of resonance or projective picturing whose adequacy is estimated by their contribution in the allostatic balance[19] of the organism in its environment.

2.2.4 Hegelian Immanence

Lastly, let us briefly examine a representationalist variant of mind-world immanence, which, in a sense, is the polar opposite of Deleuzian immanence. This is the Hegelian immanence between mind and world, as is succinctly exemplified by the dictum 'to be is to be intelligible'. This view amounts to an immanence between thinking and being, which, in contrast to the Deleuzian one, builds the plane of immanence at the *conceptual* level where the expressive relation between mind and world is understood in *representational* terms. We saw earlier (Sect. 2.1.3) that there are reasons to doubt about whether Hegel—despite his best intentions—can avoid introducing transcendent a priori elements in his epistemology, and we shall discuss neo-Hegelian absolute idealist immanentist views in contemporary analytic philosophy at length in Chap. 6 (Rödl-Kimhi) and Chap. 7 (Brandom). For the time being it will suffice to briefly mention the main features of 'orthodox' Hegelian immanence and some of its problematic ontological-explanatory consequences from our point of view.

As is well known, Hegel was the first to understand the essential role of the community as a mediating factor in explaining the relation between the intelligible order and individual minds. For Hegel, subject (the individual mind) and object (the world standing over against the subject) are relata that exist only with and in their relations, so that their mediation can no longer be conceived in individualistic mentalistic terms. They exist only as expressions of objective Spirit, the collective spirit embodied in a

[19] Allostasis is a process that continuously reestablishes stability through change and differs from homeostasis in that it involves a feedforward mechanism (rather than the negative feedback mechanisms of homeostasis), with continuous re-evaluation of need and continuous readjustment of all parameters toward new set points.

community of persons, held together by relations of mutual recognition, which form the basis for understanding ourselves as social beings capable of forming intersubjectively binding judgments referring to the objective world (i.e. one and the same world for everybody, existing independently of us). In this way, objective spirit expresses the immanent principle through which individual minds are always already semantically and epistemically open to an external world. Mind and world have the same essential form, namely, *conceptual* form, articulated in two different ways: the mind is expressively articulated in deontic normative terms (normative commitments and entitlements) while the world in alethic modal terms (objective modal facts)[20] (Brandom, 2019). This expresses the essentially immanent representational relation between mind and world. Of course, for Hegel, this relation is not static but is itself articulated historically in a self-correcting manner. We can justify our own standards only if we can see them as removing conflicts and contradictions in former worldviews, by making explicit what were hitherto implicit presuppositions of worldviews that have become dysfunctional by their own standards (Habermas, 2005; Hegel, 1979; Pinkard, 1994).

However, Hegel conceives of the relation between the world, the conceptual order and individual minds by analogy to the capacities and abilities of person-like entities. The fact that Hegel explains this mediating role of the community as an expression of the socio-historical unfolding of the world Spirit avoids the pitfall of attributing a direct efficient causal action of some person-like entity on the individual. Yet, it ends up endowing this communal mediating factor with person-like characteristics that exhibit the rational unity of a mind, and are conceptually and explanatorily irreducible to more elementary (non-conceptual, a-rational, non-teleological) human activities. Thus understood, the plane of thought-world immanence construed as a representational relation between persons and the world through semantic relations of intentionality, sense and reference presupposes the rational unity of persons-in-the-world, and hence cannot account for the possibility of its own emergence as a semantic, representational and rational person-world structure. Any such explanation would necessarily be unilluminating, as it would *presuppose* the network of our basic categorial (person-like, normative, intentional) concepts in terms of

[20] Objective modal facts (possibilities, necessities and impossibilities) are expressed, for example, in the fact that an object that is pure copper cannot be pure aluminum, that the object in question melts at 1084 °C, etc. This theme will be developed in Chap. 7.

which we collectively make sense of ourselves-in-the-world, and hence could not account for the possibility of their *own* emergence and function as a social-collective medium of representing the world and ourselves. This is Hegel's blind spot, which eventually leads to his controversial and ultimately idealist (even if not in the subjective sense) spiritualization of nature, as the latter finds expression in the quasi-teleological causal powers that Hegel attributes to the socio-historical unfolding of world Spirit.[21] From our point of view, this absolutization of the representational mind-world relation, while avoiding many of the pitfalls of the anti-representationalist rendering of the mind-world 'flatly' immanent relations discussed above (in object-oriented and Deleuzian ontology), it does so only at the cost of elevating the representational-cum-rational mind-world immanent relation into an unconditional, ontologically self-sufficient principle, which seems capable of bootstrapping itself into existence only because of a surreptitious reading of the principle 'to be is to be intelligible' as identical with the principle 'to be is to be *conceptually* intelligible'. It might well be that conceptually articulated language, thought and self-consciousness are transcendentally necessary conditions of our rational access to a common objective world. Yet, the conditions of access to the world, instead of being problematized as to their very possibility of emergence as world-making structures and critically assessed as to their empirical and transcendental presuppositions, are transformed into something unconditional, the ultimate horizon of meaning, mind, truth and being. And this makes these conditions, and the self-consciousness that 'unifies' them, not only possessors of inexplicable God-like bootstrapping powers, but also conversation stoppers, a block to the road of inquiry. A related point in this connection is that in deducing the conditions of the intelligibility of things, Hegel presupposes their factual existence. Indeed, it is only by presupposing the 'facticity' of existence, the brute 'that-ness' of things, that these conditions are more than empty forms and can account for their 'what-ness' (their essence). But this shows that Hegel's 'logic as metaphysics' (Pippin, 2019) is not completely presuppositionless as it

[21] Of course, Hegel is more subtle than this: His 'Spirit' is only the highest degree of organization and development of all the powers of nature. But, conversely, according to Hegel, seemingly inert matter is only the lowest degree of organization and development of all the powers of the mind. Thus, in this Hegelian monism of reason and nature, of the subjective and the objective, both poles are conceived as different degrees of organization of a single *vital force*, whose spatiotemporal unfolding can only be understood in terms of principles exhibited in the rational unity of a mind (Beiser, 1993).

cannot account for the 'facticity' of existence (Bruno, 2025). Thus, whatever the merits of Hegelian 'conceptualist' mind-world immanence (and there are many), it cannot be accepted as a package deal. As we shall see in the Chaps. 3, 7, and 8 to come, from our fractured immanent point of view, the very representational coordination of the world to the categories of the mind—their *symmetric differentiation* in the context of our semantic practices—is itself the expression of a topological change ('folding') of their *asymmetric identity* at the 'factical' non-representational and non-semantic dimension of immanence.

References

Baltas, A. (2012). *Peeling potatoes or grinding lenses: Spinoza and young wittgenstein converse on immanence and its logic.* University of Pittsburgh Press.

Beiser, F. (1993). Hegel and the problem of metaphysics. In F. Beiser (Ed.), *The Cambridge companion to hegel* (pp. 1–24). Cambridge University Press.

Bowden, S. (2014). Antirepresentationalism and objectivity in rorty, brandom and deleuze. In S. Bowden, S. Bignall, & P. Patton (Eds.), (pp. 180–198). Routledge.

Bowden, S. (2017). The intensive expression of the virtual: Revisiting the relation of expression in difference and repetition. *Deleuze Studies, 11*(2), 216–239.

Brandom, R. (1994). *Making it explicit: Reasoning, representing and discursive commitment.* Harvard University Press.

Brandom, R. (2019). *A Spirit of trust: A reading of Hegel's phenomenology.* Harvard University Press.

Brassier, R. (2007). *Nihil unbound: Enlightenment and extinction.* Palgrave Macmillan.

Brassier, R. (2014). Deleveling: Against 'flat ontologies'. In C. van Dijk, E. van der Graaf, M. den Haan, R. de Jong, C. Roodenburg, D. Til, & D. Waal (Eds.), *Under influence - Philosophical festival drift* (pp. 64–80). Omnia.

Bruno, A. (2025). *Facticity and the fate of reason after Kant.* Oxford University Press.

Bryant, L. (2011). *The democracy of objects.* Open Humanities Press.

Christias, D. (2023). *Normativity, lifeworld and science in sellars' synoptic vision.* Palgrave Macmillan.

Deleuze, G. (1968). *Difference and repetition.* Columbia University Press.

Deleuze, G. (1992). *Expressionism in philosophy: Spinoza.* Zone Books.

deVries, W. (2005). *Wilfrid sellars.* Acumen.

Fourtounis, G. (2023). *Deconstruction 'in the last analysis': Althusser's spinozistic structuralism and early derrida.* (in Greek). Nisos.

Habermas, J. (2005). From Kant to Hegel and Back again: The move toward Detranscendentalization. In J. Habermas (Ed.), *Truth and justification* (pp. 175–211). MIT Press.

Harman, G. (2018). *Object-oriented ontology: A new theory of everything*. Penguin.
Harman, G. (2022). *Skirmishes: With friends*. Punctum Books.
Hegel, G. (1979). *Phenomenology of spirit*. Oxford University Press.
Hegel, G. (2010). *The science of logic*. Cambridge University Press.
Husserl, E. (1931). *Cartesian meditations: An introduction to phenomenology*. Kluwer Academic Publishing.
Kerslake, C. (2009). *Immanence and the vertigo of philosophy*. Edinburgh University Press.
Lewis, D. (1986). *On the plurality of worlds*. Wiley-Blackwell.
Meillassoux, Q. (2008). *After finitude: An essay on the necessity of contingency*. Continuum.
Morton, T. (2013). *Hyperobjects: Philosophy and ecology after the end of the world*. University of Minnesota Press.
Neurath, O. (1921). Anti-spengler: Dedicated to the young and the future they shape. In M. Neurath & R. Cohen (Eds.), *Empiricism and sociology* (pp. 158–213). D. Reidel Publishing Company.
Pinkard, T. (1994). *Hegel's phenomenology: The sociality of reason*. Cambridge University Press.
Pippin, R. (2019). *Hegel's realm of shadows: Logic as metaphysics in the science of logic*. University of Chicago Press.
Quine, W. V. O. (1960). *Word and object*. MIT Press.
Sellars, W. (1956). *Empiricism and the philosophy of mind*. Harvard University Press.
Sellars, W. (1963). Philosophy and the scientific image of man. In W. Sellars (Ed.), *Science, perception and reality* (pp. 1–40). Routledge.
Simondon, G. (2020). *Individuation in light of notions of form and information*. University of Minnesota Press.
Smith, D. (2001). The doctrine of univocity: Deleuze's ontology of immanence. In M. Bryden (Ed.), *Deleuze and Religion* (pp. 167–183). Psychology Press.
Wolfendale, P. (2019). *Object oriented philosophy: The noumenon's new clothes*. Urbanomic.

CHAPTER 3

Immanence without Givenness

In this chapter, I will introduce and develop the concept of fractured immanence, and I shall connect it with the epistemological conception of immanence, discussed in Chap. 2. It will be argued that the epistemological conception of immanence, namely, the view that thinking about the world necessarily unfolds and self-corrects in the midst of our practices (thus rejecting 'the view from nowhere' conception of thought), is best understood as an expression of a resolute rejection of the myth of the given (Sellars, 1956, 1981). It is suggested that immanence can provide an illuminating account of the fact that thought is both about the world and within it only if it goes hand in hand with a complete rejection of the Given (Brassier, 2016). From this anti-Givenist point of view, what is immanent is the identity-necessarily-presented-as-difference between thought and world. Thought is immanent to being, but this immanence is essentially fractured.

In this context, I will explore the consequences of resolutely rejecting the Given for the nature of our access to the world through thought. I argue that the rejection of Givenness implies that at the level of semantics thinking attempts to grasp the world *as such*: there are no fixed limits to thinking—for example, transcendental conditions of possibility of experience—that prevent it from being semantically *about* the world as it is in itself. Yet it also implies that the categorial structure or content of the

© The Author(s), under exclusive license to Springer Nature Switzerland AG 2025

D. Christias, *Towards a Hyperstitional Process Nominalism*, Palgrave Perspectives on Process Philosophy,
https://doi.org/10.1007/978-3-032-06350-2_3

world as such and of our relation to it are never *themselves* Given to thinking and are always subject to self-correction.

I shall also focus on the anti-representational implications of the rejection of Givenness in our conception of the mind-world relation. It is argued that a rejection of Givenness entails a resolutely anti-representationalism in semantics and ontology and a post-Kantian, critical and transcendentally informed metaphysics (which will be developed in detail in Chap. 4). In this spirit, it is suggested that a non-representational semantics coupled with a non-representationalist ontology can provide a viewpoint from which we can launch a positive critique of traditional representationalist substance metaphysics, by constructing an alternative evolutionary process metaphysics with anti-representationalist yet *realist* implications. From this standpoint, the realist view to the effect that there is no barrier in thinking being about the world as it is in itself is rehabilitated but is given an important twist: ultimately, the semantic aboutness of thought to a common, public, objective world through reference relations is revealed as an indispensable yet non-representationally functioning tool for collectively coordinating thought, considered as natural process, in the world, considered as natural process, where the relation between mind and world at this level is not referential, logical or causal. Rather, it is a non-semantic expressive relation of resonance between dynamically co-evolving environmental, bodily and brain processes, whose degree of adequacy is measured by the degree of empowerment expressed in the collective conative processes involved.

3.1 The Myth of the Given and Fractured Immanence

3.1.1 The Rejection of Givenness as the Denial of Transparent Relations and of Necessary Boundaries Between Mind and World

In this section, I shall attempt to show the intimate connection of epistemological immanence, and the fractured immanent relation between mind and world which I advocate, with a resolute rejection of the Sellarsian myth of the Given.

As is well known, the concept of the Given is supposed to give content to our conviction that our only guarantee of a non-circular, solid (infinite-regress-stopper) and objective (non-arbitrary) access and knowledge about reality is the existence of a point of *immediate contact* between mind and world, subject and object, that enables the former to have immediate (non-inferential) knowledge of the latter. This immediate epistemic connection between the knowing subject and the object of knowledge makes the cognitive contact between them immediate and transparent, independent of any other knowledge, and is precisely what constitutes Given knowledge, which, by serving as a foundation on which other bits of knowledge could be justified, guarantees the very possibility of knowledge. (If we substitute 'meaning' for 'knowledge' and 'semantic' for 'epistemic' in the last sentence above, we arrive at a notion of 'Given meanings', i.e. of the semantic Given.) More precisely, a certain (usually mental) state functions as a epistemic Given if 1) it is *epistemically independent*, that is, has an epistemic content independently of any (formal or material) inferential relations to which it may stand in with other such contents, and 2) it is *epistemically efficacious*, that is, capable of transmitting its epistemic status to other states or contents, thereby accounting for the possibility of objective knowledge (deVries & Triplett, 2000, pp. xxv–xxvi). (Again, if we substitute 'semantic' for 'epistemic'—and 'purport' for 'knowledge'—in the above definition, we arrive at a definition of the semantic Given.) It is important here to note that by characterizing the Given as a myth, we do not object to the notion of immediate, non-inferential knowledge—that is, knowledge that is not inferred by other knowledge. What one thereby objects to is the idea that the immediate, non-inferential knowledge in question is *presuppositionless*, that is, that it is 'out there' (in the world) or 'in here' (the mind) independently of its functional role within a historically evolved categorial framework transmitted through language from generation to generation.

In his later work (1981), Sellars suggests that the most general form of the myth of the Given is that of the categorial Given, namely, the view that 'if a person is directly aware of an item which [in fact] has categorial status C, then the person is aware of it *as* having categorial status C' (Sellars, 1981: I §44). The categorial Given is intimately related to the epistemic or semantic Given: it is what *explains why* and *how* certain mental states can be semantically or epistemically independent and efficacious. That is, it provides the ontological foundation for the semantically or epistemically 'Givenist' functions of those states. The Given can play its foundational role just because the categorial structure of reality is imposed on—or discloses itself to—the subject who 'receives' it as it really is in perception, thought and action. One falls prey to this version of the Myth if one takes it that there is a direct classificatory awareness of categories such as 'sorts', 'resemblances', 'facts', 'universals' or even 'particulars', independently of our acquisition of a categorial framework through our initiation to language (Sellars, 1956, §29–31). The categorial structure of the world or of the mind cannot be somehow directly 'read off' or intuited on the basis of the basic descriptive and explanatory concepts of the categorial framework we actually employ in our everyday language. Those concepts, though capable of yielding direct knowledge of the world and our place in it in the context of categorial system of which they are part, are nonetheless not unchangeable essences which put us in direct conceptual contact with the essential categorial structure of the world or the mind. As Sellars puts it: '*To reject the myth of the Given is to reject the idea that the categorial structure of the world -if it has a categorial structure- imposes itself on the mind as a seal imposes an image on a melted wax*' (emphasis original) (Sellars, 1981: I: §45). Another way to put this point is that immediacy must not be equated with transparency. An experiential or mental state is transparent if by just *undergoing* such an experience, its mode of being is thereby given in propria persona (as it really is), or, in other words, if by just *having* an experience one is thereby aware of its *categorial* status. I do not deny that an experiential or mental state can be directly present to non-conceptual consciousness in an unmediated manner. The sense of a transparent access to these states that is in dispute is the further view that this kind of direct, unmediated awareness amounts to a direct awareness of the categorial status of the state in question.

Now, as Sellars has convincingly shown, the Given is ultimately a myth. In the last analysis, the supposed secure and transparent immediacy of semantic, epistemic or categorial content revealed by sensuous, intellectual or categorial intuition cannot function in an epistemically foundational manner: by being deprived of any logical (presuppositional, inferential) connection to the categorial and empirical knowledge articulated in our ordinary and scientific epistemic practices, the Given ultimately undermines itself as it lacks the resources to distinguish between cases in which the supposedly transparent intuitional content *is* actually thus-and-so and cases in which it only *seems* to be thus-and-so (Sellars, 1956, 1975). And in such a case, as Wittgenstein puts it 'whatever is going to *seem* right to me *is* right. And that only means that here we cannot talk about right' (Wittgenstein, 1958, §258). The atomistic insulation of foundational epistemic content from all categorial and empirical presuppositions of our space-of-reasons practices ultimately deprives it of its normative power since this 'content' cannot be meaningfully constrained by the authority of something external to its own conditions of enunciation (e.g. worldly objects), and is thus deprived of the resources for recognizing the authority that objects have as the arbiters of the legitimacy of its claims.[1] Conversely, the inclusion of those categorial and empirical presuppositions in this foundational epistemic content deprives it of its privileged foundational character, as in such a case it becomes another space-of-reasons content or principle which however central it might be in our worldview is subject to self-correction and is ultimately 'one among many which compete for adoption in the market place of experience [empirical inquiry]' (Sellars, 1953, p. 138).

Thus, the bottom line of the critique of the semantic, epistemic and categorial Given is that for something to function in semantically or epistemically significant ways is for it to be a node (play a functional role) within what Sellars calls the 'logical space of reasons' (Sellars, 1956) or the language game of 'giving and asking for reasons' (Brandom, 1994). The space of reasons is a socially articulated logical space in which persons have their being. To be a denizen of the space of reasons is to be sensitive to certain collective pattern-governed behavioral habits regarding 'language-entry' transitions (perception), intra-linguistic transitions (inference) and 'language-exit' transitions (action), in a way that places them in a network of reciprocally recognized rights and duties ('cans' and 'oughts') pertaining to meaning, knowledge and action (Sellars, 1954). And the important point regarding the categorial status of claims placed within the space of reasons is that the rationality of the latter—and thus, derivatively, of the categorial status of these claims—does not depend on its having any synchronic foundations, but in its diachronic *self-correcting* character, namely, the fact that, within it, *any* claim can be put into jeopardy, though not *all* at once (Sellars, 1956, §38).

The connection between the rejection of Givenness and epistemological immanence thus becomes obvious. The latter is one of the most important implications of the rejection of the former. For to reject the Given is to realize that our knowledge of the world and ourselves is not constructed from a given or a priori foundation of from first principles provided by

[1] In other words, for a subject's utterance to have normative force, the subject 'must be able to meaningfully posit a gap between things and experiences, and he must understand enough about how epistemic authority works and when it becomes binding to know when he can claim that this gap is closed' (Kukla, 2000, p. 178).

reason or sensibility, but always already presupposes that we live within a world in which we act and of which we have some provisional understanding and knowledge. This understanding and knowledge is provisional in that it is always subject to self-correction in response to pressing incompatible commitments in our overall system of understanding ourselves-in-the-world. Thus, to say that the conceptual system that expresses our current conception of ourselves-in-the-world is provisional is to say that it is also thereby reflective and critical, in that no element of our current conceptual framework is absolutely beyond question. It always remains open to new experiences and new ways of organizing experience. There is no privileged 'God's eye view' from which we can antecedently survey the space of possibilities, the totality of possible worlds, and compare it with actuality. This implies that our understanding and knowledge of the categorial structure and empirical content of the world and our minds are never *themselves* Given to thinking and are always subject to self-correction.

Yet, crucially, this does not mean that thinking falls short of semantically referring to the world *as such* (the world as it is in itself). Although this might not be directly evident, it is one of the most important implications of the rejection of Givenness. If this is so, it follows that transcendentally informed positions according to which thinking is intentionally directed toward something else than the world as such, such as Kant's phenomenal world or Husserl's lifeworld, avoid metaphysical dogmatism only at the cost of surreptitiously reintroducing some form of controversial Givenness. The same goes if in place of Kant's phenomenal world or Husserl's lifeworld, we put the world as horizontally given and transcendentally bounded by the socio-historical embeddedness of our concepts, our social upbringing, our class interests, our gender, our biological constitution as a species, etc. At best, these are broadly causal-factual limitations of our knowledge of ourselves-in-the-world, not transcendentally necessary epistemic boundaries that preclude access to the world as it is in itself.

To see this let us briefly recall why transcendental philosophy feels the need to transcendentally delimit thinking about the world within the bounds of experience or intuition, and hence intentionally correlate it with something less, more, or, in any case, other than the world *tout court*. In Kant's case, thinking about the categorial structure of the external world (and of the mind) is said to be intentionally directed to the phenomenal world, namely, the world of actual and possible representings, as opposed to the world as it is in itself, because thought about the world is strictly delimited by the Kantian forms of intuition and the categories of understanding that synthesize the former. And since the spatiotemporal forms of intuition are contingently projected to the world by peculiarities of the spatiotemporal forms in which we humans process external stimuli, they do not disclose to us the world as it is in itself independently of those forms, but only the world as an object of actual and possible representings (the phenomenal world). Similarly, in Husserl's case, thinking about the categorial structure of the world and the mind is delimited by—and thus cannot transcend—lifeworld categories which are essentially 'intuitable'. An example here is the fact, revealed by categorial intuition, that color and extension are necessarily features of external objects. Non-intuitable categories of thinking about the world, such as scientific ones used in explanations in physics, biology or psychology, can be legitimized only by referring them back to the originary sensuous lifeworld intuitive structures that evoked them (Husserl, 1970, pp. 127–129). Another, more recent, view exemplifying a similar way of thinking about the relation between thinking, experience and the world, and which does not officially belong to the transcendental camp, is the Wittgensteinian one, according to which the world disclosed to thinking can be meaningfully conceived as independently existing and as independently possessing the features attributed to it by our empirical-scientific practices only on condition that we recognize in the concept of the independent world as revealed in our practices 'the trail of the human serpent' (James, 1906). This is because, according to this Wittgensteinian view, the content of the concept of the 'objective world' is a function of our historically evolved and diachronically evolving rule-following practices of using this concept, and these rule-following practices are not grounded by invariant metaphysical essences, but are ultimately response-dependent and species-specific: they exist only because we as a 'form of life' happen to have relatively similar responses on what counts as criteria of similarity and difference in our use of this concept (Cavell, 1969; Kripke, 1982; Wittgenstein, 1958).

What all these views—as well as many other post-Kantian correlationist ones discussed in Chap. 2—have in common is that in their effort to protect thought from its dogmatic speculative forays into the absolute, they attempt to put forward transcendentally secured cognitive-cum-experiential conditions that avoid metaphysical dogmatism by anchoring abstract thinking about the metaphysical structure of the world to more concrete cognitive and experiential patterns of ours, such as the categories of our everyday lifeworld (Husserl), the (species-specific) forms of our spatiotemporal intuition (Kant) or the de facto agreement in our patterns of response in using a concept in the context of our form of life (Wittgenstein). It is precisely because they conceive of these transcendentally revealed experiential patterns and conditions as semantically more transparent and epistemically more secure than the contents of thinking about the world that abstracts from these conditions that make them feel entitled to say that if thinking is to be about the world it must be exercised within a *strictly* and *antecedently* delimited domain (possible experience, categorial intuition, normative attitudes that constitute the content of our concepts in our form of life). Otherwise, they would not have any reason to believe that the view that thinking is intentionally directed toward the world as such independently of these antecedently delimited conditions necessarily commits one to a dogmatic, experientially unconstrained metaphysics. Yet, in this way they surreptitiously reintroduce a form of *endogenous Givenness* in their thinking by taking it for granted that we have a semantic and epistemically secure access to the form of our spatiotemporal intuition (Kant), to the categorial structure of the lifeworld by distilling its essences from the framework of perceptual experience through 'categorial intuition' (Husserl), or to the structure and function of our semantic and epistemic practices (Wittgenstein). In all these cases, it seems that the fact that these structures (spatiotemporal intuitional form, phenomenological intuition of essences, structure and function of our semantic and epistemic practices) are 'ours', an essential feature of our mindedness, entitles us to believe that we thereby have privileged access to their essential nature independently of empirical inquiry into their structure and content.

Here I want to make clear that whereas I take it that the above kinds of views are deeply problematic, I also think that they have a kernel of truth that must be retained as part of any sound philosophy. What is problematic with these views is not their demand for an experiential anchoring of thinking or their focus on the ways in which the evolutionary and socially induced contingencies of our subjective constitution (sensibility, imagination, understanding) shape thinking's objective access to the world and ourselves. The problem is the reification of the evolutionarily-cum-socially induced *'coloring'* of thought's access to the world by our evolutionarily-cum-socially shaped experience and similarity responses into an antecedently understood a priori *epistemological boundary* between thought and world. It is only if we construe this evolutionary experiential coloring as an epistemological boundary that we end up endorsing a form of endogenous Givenness and feel the need to say that if thinking is to be about the world, this world must be something *other* than the world as it is in itself. For while it is true that there is such an evolutionarily induced experiential influence and limitation of thought's access to the world (arguably, the spatiotemporal form and qualitative content of the proper and common sensibles have evolved through the millennia of evolution for survival purposes), it is also true that the categorial structure of this very experiential boundary between thought and world is not self-intimating and is not transparent to itself. It can only reach 'semantic self-consciousness' by schematizing itself through categories of thinking. Indeed, the historical development and self-correction of empirical concepts and categories in the course of empirical-scientific inquiry can itself be seen as an expression of a dynamic schematizing function of experience in its effort to restore incompatibilities in its total experience of itself-in-the-world. From this point of view, while it is true that experience provides epistemic friction to thinking about the world, it is also true that experience does not do so by being antecedently known to itself, but by being epistemically distinguished by thinking through criteria immanently constituted and self-corrected in the course of inquiry by our 'total system of beliefs' (Quine, 1960). The same goes if in place of our evolutionary experiential coloring of the world we put its socio-historical coloring as expressed in the 'historical *a priori*' categories that simultaneously enable and limit (in ways that are synchronically invisible to us) our access to the world in different historical epochs (Foucault, 1970; Heidegger, 1977; Spengler, 1926). These enabling and limiting categorial conditions of our access to the world that function as historically a priori conditions of intelligibility are not antecedently known to themselves but are determined as such (as enabling and limiting conditions) only by being retrospectively understood as *appearances* of a *reality* as the latter is conceptualized by the lights of each community's *current* categories. This, of course, does not mean that our present categories of making sense of the world and our place in it are metaphysically privileged; every categorial system is based on background assumptions that can be potentially challenged and retrospectively revealed as placing arbitrary and unjustified limits on intelligibility. But while these limits are *non-conceptually expressed* in the very application of any conceptual system to the world, they can be conceived *as* limits only by becoming determinate contents of thought within a semantic and epistemic space, the space of reasons, where there are no a priori transcendental barriers or boundaries between thinking and being. To determine a limit in thought, to place it within the space of reasons is already to transcend it (if even initially only in thought). Talk about limits or enabling conditions of intelligibility has a point only as part of a self-correcting dialectic between what counts as appearance and what counts as reality at each point of inquiry, and not as something that can be a priori conceived through non-empirical (transcendental, phenomenological) methods of inquiry (see also Chap. 7). Thus, the fact that thought is ontologically constrained by the evolutionarily sedimented structure of experience and epistemically constrained by conceptualized experience in the (historically mutable) context of our total system of beliefs does not imply that thought is epistemically *delimited* by experience, history or society in a way that *prevents* it from being about the world as such (i.e. independently of its being an object of actual or possible experience by us). To believe so is to reify the evolutionary and socio-historically induced shaping of thought's access to the world through experience into epistemological boundaries between mind and world and to mistake the natural and social mechanisms that shape our access to the world for transcendental necessities.

3.1.2 *Fractured Immanence and Objectivity*

Moreover, this resolute rejection of Givenness has important implications for the clarification of the sense in which the relation of thought to the world through experience can be said to be an *immanent* one. As has been already mentioned earlier, I take it that a resolute rejection of Givenness implies the mind-world immanence is essentially a fractured one: mind and world stand in a direct unmediated immanent ontological relation whose intelligibility yet presupposes a transposition of mind and world as semantically and epistemically correlated poles of a diachronic self-correcting process that ultimately aims to align our conceptual understanding of mind and world to mind and world as natural processes. This unmediated direct relation between world and mind is Given in a non-epistemic sense of the term (in the form of a pre-subjective and pre-objective 'monadic' perspectival openness to the world (see Fig. 3.1)), but this experiential Givenness is not epistemically Given to itself as to its categorial structure and function. It is a givenness which is neither conceptually nor non-conceptually transparent to itself.

This view is very close in spirit (though not identical) with what Brassier calls 'deleveled' immanence, according to which what is immanent is the (conceptual-semantic) *difference* between thoughts and worldly things, not their identity[2] (Brassier, 2014). Yet, our view does not completely coincide with Brassier's deleveled conception of immanence. The epistemological critique of Givenness points toward an immanent mind-world ontology where both mind and world are considered as naturalistically construed absolute processes that can make themselves explicit as such only by the self-fracturing of this immanence and its transposition into a representational relation between mind and world considered as distinct kinds of substances (see Chap. 1 and Sect. 3.2.2). In view of this fact, I suggest that what is immanent is the *(asymmetric) identity presented as (symmetric) difference* between thought and things. More specifically, it can be argued that what is immanent is the identity-presented-as-difference between the world, the mind and experience viewed (a) as *conceptually* represent*ed* contents and (b) as patterns of occurrence in *non*-conceptually articulated processual representi*ngs*.[3] Immanence is thus *fractured* because what we know about the mind, the world and their experiential tie as conceptually represented contents is immanently conditioned in ways that currently elude us by the manner in which the world, the mind and their experiential tie non-conceptually exist in our representing acts. But this fracture between the mind-world relation as conceptually represented and as non-conceptually registered in acts of representing, being precisely *immanent* and itself dynamically unfolding as such, far from secluding the fractured realms of conceptually represented contents and non-conceptually representing acts, actually *internally relates* them as distinguishable yet inseparable poles of cognitive inquiry that calibrate one another and mutually self-correct so as to maximize explanatory coherence in our total system of beliefs. Indeed, it can be argued that the very

[2] Note here that Brassier's Sellars-inspired view is consistent with holding that thought and being are *identical* in the 'order of being'. In other words, the logical difference between thoughts and things is not an *ontological* difference: thoughts are just things considered in terms of their logical or epistemic powers, rather than their material powers alone.

[3] In his *Science and Metaphysics* (1967), Sellars introduces the distinction between the world and the mind as conceptually represented and as non-conceptual states of representings and interprets this distinction in a 'transcendental naturalist' manner, as expressing a naturalist mind-world relation and general ontology which however is transcendentally justified from within the conceptual order. Obviously, the above fractured immanent account of the relation between mind and world is inspired by Sellars. However, Sellars never discusses the issue of immanence, and does not distinguish between different kinds of immanence. Thus, I am not sure that he would concur with the views about immanence developed in this book. Yet, since my purpose here is not interpretative but systematic, this issue can be put aside for now.

end of inquiry consists in incorporating ever more facts about the non-conceptual structure of representings that subjectively 'color' our access to the world and ourselves into conceptually represented facts. The manner in which we conceptually represent the world is always conditional upon the categorial structure of our language, our biology, ideological factors, etc., in ways that can potentially distort our conceptual representation of the world (as their subjective, biased ways in which they filter information remain invisible and pass for impartial objective knowledge). But since we can represent these conditions that subjectively mediate what we take to be an objective representation of the world and ourselves (we can always turn—unacknowledged—facts about the non-conceptual structure of representings into conceptually represented facts), we can make explicit these conditions of representation, and question their claim to objectivity by revealing their 'objectivating' structure and function (Catren, 2023). Of course, this is a never-ending process: *any* such condition can be potentially challenged though not *all* at once (since some such conditions synchronically function as background enabling constraints of the very intelligibility of our conceptual representations). Yet, the very same non-conceptual representings that condition conceptually represented contents and are synchronically unchallengeable from the latter's point of view can be retrospectively made explicit and challenged from the standpoint of an expanded system of conceptually represented contents operating with novel, more inclusive and more comprehensive concepts and categories. This is how inquiry in its most general sense works: by constant diachronic expansions of its conceptual and categorial space in the face of irresolvable incompatibilities in its systems of conceptual represented concepts, where the resolution of these incompatibilities demands the thematization of the fundamental ideological background assumptions that produced and sustained them and their critical delimitation as objectivating projections in reality that reflect our subjective (evolutionary-psychological-socio-historical) standpoint. This is how objective knowledge is attained from a fractured immanent standpoint.

Importantly, we should not view the above self-correcting process of unearthing the subjective presuppositions of objective representation as sketching a picture of the world in itself by abstracting away from facts about our subjective constitution, evolutionary development or societal position, and conceiving of the in-itself as the remainder of this subtraction. As we saw, in the course of inquiry, many purportedly paradigmatic instances of objective knowledge about the world and ourselves are challenged and replaced by the knowledge of the manner in which the origin, structure and function of our non-conceptual experience (representings) are *objectivating* by actively—albeit obliviously—constructing projections of ourselves-in-the-world. The point here is not to purge that which is comprehended by objective knowledge, namely, the world as it is in itself, from all subjective contribution (e.g. the specific spatiotemporal, causal, sensible and conceptual structure endowed to human organisms by biological and cultural evolution) as if the latter is nothing more than an illusory projection on the in-itself. The point is to *supplement* it with these projective structures (they are *too* a—fractured—way of being of the world in-itself) while simultaneously reconceiving the structure and content of the in-itself such that it becomes possible to understand how this 'pluriverse' of projected worlds constitute intrinsic variations on the same theme (the in-itself) and how the in-itself is such as to intrinsically give rise to this pluriverse. This would be the way of construing the appearance-reality distinction from a fractured immanent—as opposed to a transcendent—viewpoint (see also Chap. 1). In this way, we avoid the spurious equation of objectivity with the elimination of all qualitative and perspectival content while at the same time becoming gradually better in distinguishing what our mind contributes to the world from the world's contribution to our minds.[4]

Let me repeat here that the view that immanence must be fractured directly follows from the rejection of the endogenous Given, namely, the view that we know our minds first and best. If this is not the case, and if, instead, we come to understand the contents of our own minds by modeling them to our knowledge of the external world and to our knowledge of ourselves as social beings interacting at the public sphere through a normatively functioning semantic and categorial framework expressed in language (Sellars, 1956), then it follows that every form of understanding and knowledge of ourselves 'from within' is necessarily mediated by a categorially rich framework that each generation inherits and absorbs through its initiation to our linguistic practices. As I have argued elsewhere (Christias, 2023, pp. 27–62), I take it that this framework has roughly the categorial structure of Husserl's 'lifeworld' (1970) or Sellars' 'manifest image' (1963). This framework revolves around two different yet

[4] Here we agree with Adorno when he insists that 'with the removal of one of its essential moments [i.e. subjectivity] objectivity is falsified, not purified. … The object is not a subjectless residuum' (Adorno, 1998, p. 253).

mutually presupposing and categorially correlated kinds of substances, namely, persons and external objects: persons are substances with normatively individuated mental and bodily capacities and abilities. Persons do not merely causally respond to external circumstances or internal stimulation but are capable of performing actions based on their intentions and desires that can themselves be assessed on the basis of reasons. Standing over against persons we find middle-sized external objects, also considered as substances, individuated by their causal powers and characterized by determinate proper and common sensible properties, indissolubly bound up and experienced as belonging to those objects themselves. The important point for our purposes here is that the basic semantic categories of this framework, through which we understand the relation between persons and worldly objects, thought and reality, namely, the category of intentionality or aboutness which directly links thought and world (to express a thought in the form of an asserted proposition is to use it to say something *about* the world, i.e. something external to the proposition), are socially, public and linguistic, and not inner mental categories. Here I follow Sellars (1956) who argues that thought does not possess 'original intentionality' independently of socially articulated semantic categories of aboutness (meaning, reference, truth). Only by constructing a novel categorial space modeled on linguistic aboutness are we in a position to conceive ourselves as possessors of thoughts, sensations, desires, intentions, etc. construed as semantically significant *inner mental states* to which each of us has direct and privileged access. Of course, this priority of the person-object framework and of the construal of the mind-world connection in semantic terms drawn from our understanding of public language need not be taken as a priority in the 'order of being'. We internalize this framework in our mother's knee by learning to speak and to understand the world and ourselves in terms of this framework. But precisely because categorial frameworks are not Given, no framework, including the framework in terms of which we perceive, think and act in-the-world, has an unchallengeable authenticity.[5] Instead, every framework carries with it the seeds of its own overcoming, as it unfolds and is used in a wider context of immanent

[5] As Sellars eloquently puts this point: 'All classification of objects, however confident and pre-emptory, is a venture, a venture which at no point finds its justification in a pre-symbolic vision of generic and specific hearts on the sleeves of the objects of experience. Classification resembles the grasping tentacles of an octopus, now tentative, now confident, rather than a salesman's selection of a suit for a customer after a glance at his build' (Sellars, 1953, §7).

self-correcting principles of inquiry, which, as was mentioned earlier, gradually de-anthropomorphizes our conception of ourselves-in-the-world by incorporating ever more facts about the non-conceptual structure of our relation to the world (that are projected as objectively situated 'out there' in the world or 'in here' in the subject's consciousness) into conceptually represented facts. In this way, by means of 'immanent critique', that is, by self-correcting principles of inquiry immanently unfolding and expressed in critically controlled conceptual and categorial change in the face of incompatibilities in our conception of ourselves-in-the-world, we gradually disentangle what our mind contributes to the world from the world's contribution to our minds. We do this by revealing the objectivating and reifying character of the categorial structure of the mind while at the same time placing its parochial and subjective elements within a picture of an objective world that *includes* them as its fractured yet immanent world-making projective powers (as opposed to eliminating them from the scheme of things as mere illusions).

3.2 The Rejection of the Given Implies a Non-representational Semantics and Ontology

3.2.1 Non-representational Semantics

An important consequence of the rejection of the myth of the Given as I understand it is that all our central semantic concepts of meaning, reference and truth that at first glance seem to directly connect language or thought to the external world are not at bottom in the business of representing the world but are actually metalinguistic expressive devices that monitor our epistemic and semantic inferences *within* the space of reasons—as opposed to denoting a relation *between* conceptually shaped space-of-reasons contents and non-conceptually worldly events. According to this non-relational view, originally developed by Sellars (1967), semantic terms and descriptions provide functional classifications of linguistic tokens. Sellars' analysis of meaning statements such as 'Rot', in German, means red, or 'Brother' means male sibling, treats them as statements that convey the information that the word mentioned on the left-hand side has a relevantly similar usage to the phrase on the right-hand side. These meaning statements do not explicitly say that the two expressions they each contain have the same usage. Rather, each sentence presupposes that the person it is directed at already understands the background language and uses the expression on the right-hand side as an illustrating sortal to form an indexical predicate (forming a predicate by giving an example bearing the relevant property), which is then applied to the tokens picked out by the expression on the left-hand side of the sentence (de Vries 2005, 2021). From this point of view meaning is a coordination device indexing our moves with respect to the contents of our assertions within the space of reasons; it is not a relation (of 'correspondence', 'representation' or 'intentionality') that directly ties mind and language to the external world. Reference receives an analogous treatment. Extensional contexts are limiting cases of intentional ones, specifying the qualities and properties of objects (Sellars, 1967; Chang, 2022). This is because descriptive extensional vocabulary specifying the identity conditions of objects presupposes modal intentional vocabulary, which makes explicit causal-dispositional properties (Brandom, 2015, pp. 216–235, see also Chap. 6). From this point of view, we can construe reference statements as in fact quantifying implicitly over senses, that is, functions of expressions, and the extensionality of reference can be understood as a relation of material or extensional equivalence between senses (Sellars, 1967, p. 84). Lastly, truth is construed as semantic assertibility; it is conceived as applying univocally to propositions in all domains of discourse characterized by rules of correct assertibility (empirical, mathematical, ethical, aesthetic, etc.), and thus connects language and the world at the semantic level only in a trivial sense in which facts of the form '2 + 2 = 4′, 'Vladimir and Estragon are waiting for Godot', 'it is cruel to inflict pain for no reason', 'Monty Pythons have a keen sense of humor' and 'water is H2O' are the truth-makers to which the respective propositional contents 'correspond' or 'represent' (Sellars, 1967, p. IV). These truth-makers are so diverse, expressing both highly abstract (mathematical) and concrete (empirical) facts, attitude-independent (empirical) and attitude-dependent (aesthetic) facts and even fictional facts, that it makes no sense to say that there is a *substantial* metaphysical correspondence relation that *univocally* ties all these domains with the propositions that are made true by them. And, as we shall see in Chaps. 7 and 8, even empirical truth can be understood as a gradual process of convergence between senses (how things are for us) and referents (how what things are for us ought to be), which yet does not thereby get us beyond the *conceptual* realm. From this point of view, to construe truth-makers as semantically functioning referents that make our beliefs true, is to understand them as objectively and independently existing denizens of an external world that essentially exhibits conceptual form. And, as we shall see, this form, with all that it entails (substance-property categorical structure, the world as comprised of facts, with objects, properties, and relations as constituents), is not itself 'out there' independently of our normative attitudes that institute our semantic practices, but is a projection to the world itself of our (linguistic) *means of representing* that make possible and facilitate our collective coordination within the world.

The above resolutely non-relational, non-representational conception of the function of meaning, reference and truth, which is internally connected to the critique of the Given, has two important consequences: a) Language is not tied to the world by a single 'satisfaction' or 'correspondence' relation, but by a multitude of relationships between tokens of linguistic expressions and the world. b) There is no direct semantic or logical relation between the conceptual and the non-conceptual. Non-conceptual attitude-independent reality is not connected to our beliefs by rational or justificatory relations. Its only direct relation to our beliefs consists in causally evoking or non-epistemically shaping them, for example, by means of evolution or social conditioning. The very intentionality of language and thought, namely, their capacity to point to a semantically isomorphic world outside themselves, can be understood in terms of inferential relations between token linguistic expressions, and not by means of representational or intentional relations holding between linguistic or mental expressions and extra-linguistic, non-conceptually structured reality. Non-conceptual reality poses only causal, evolutionary or, in general, non-epistemic constraints upon our beliefs, thought and actions, not semantic, epistemic or logical ones.

Here, along with Sellars (and his followers, Rorty and Brandom), we reject a widely held view about the way in which the content of our empirical beliefs and thoughts is fixed. According to this orthodox view of content determination, the content of an empirical belief is given by its truth conditions. One understands the content of the empirical proposition 'the streets are wet' if one knows what it is for the streets to be wet, that is, under what conditions the streets are actually wet. In this particular case, the knowledge in question stems from perceptual experience as it seems to be precisely through our sensory modalities that we understand the kinds of things referred to by our empirical beliefs and the conditions under which they are actualized in the world. From this point of view, meaning is understood in terms of propositional contents which function as such by being internal representative signs of objective states of affairs in the external world, independently of their inferential connections to other contents within the space of reasons. Further, in the context of this representational orthodoxy in contemporary analytic philosophy of language, truth is understood in terms of 'representational accuracy' while its inferential transition to other contents is effected by formal rules of inference. This means that our communal normative practices of giving and asking for reasons do not play any *essential* explanatory role in accounting for meaning, reference and truth, but whatever role they are to play must be strictly derived from the truth-conditional relations of linguistic contents as the latter are expressed in formal logic. Yet, from this point of view, it is difficult to explain the epistemic and semantic powers of the 'material' non-logical content of empirical thought, except by appeal either to a self-justifying Given element in experience or to relations of correspondence between the empirical content in question and its truth conditions which are epistemically and semantically invisible (external) to the subject (we can call this 'the thermometer view of content'[6]). And both these solutions seem unsatisfactory (Brandom, 1994; Rorty, 1979).

Here it must be noted that there are more sophisticated representational theories of meaning, truth and reference (McDowell, 1987, 1998), which are not directly affected by the above Sellarsian objections to representational theories, since they construe both parts of the word-world relation as belonging *within* the space of reasons. Moreover, these more refined versions of representational theories of meaning and truth avoid the above-discussed classic objections against more mainstream such theories in contemporary philosophy of language. Yet, they are not without difficulties. Very roughly, this is because relational theories of McDowell's stripe, which place the truth or meaning relation within the space of reasons, thereby conceiving the very externality and independence of the world in conceptual, normative, space-of-reasons terms, *presuppose* that language is always already categorially tailored to represent the structure of the world and cannot independently explain the construction of those very categories themselves through language. Along with Sellars, I think

[6] The 'thermometer model' compares non-inferential knowledge with a good thermometer. The temperature readings of a good thermometer reliably indicate the actual temperature. In a similar manner non-inferential knowledge is a matter of a belief reliably covarying with the environment irrespectively of whether the subject understands its content or has reasons for taking it as true. This externalist view first came under attack in Sellars (1956), as part of a general attack on the whole framework of Givenness.

that there must be some such explanation of the very categorial space connecting language and world, without appeal to space-of-reasons concepts, such as those of linguistic or mental content, or else we succumb to a version of the categorial Given according to which there is a pre-established harmony between our categories and the world. As we shall see later, to achieve this we need to posit an a-categorial, non-normative mode of correspondence between mind, language and world (Sellars' concept of picturing (1967) served exactly this purpose).

3.2.2 Non-representational (Process-Evolutionary) Ontology

How can we combine non-representationalism in semantics which rejects substantial word-world meaning, truth and reference relations, with the (anti-Givenist) realist view to the effect that our thinking does not stop short of the world as it is itself? In my view, this can be best done if we complement non-representationalism in semantics with an ontology that makes this feature explicit in its own structure.

At first glance the combination of realism and non-representationalist semantics might be considered odd, as most contemporary versions of realism adhere to a naturalistic ontology, which, although not directly inconsistent with non-representationalism in semantics, seems to be more naturally combined with representationalist semantics, especially if naturalism is motivated by scientific realist considerations. Doesn't scientific realism, precisely as a self-avowed *realist* position, imply the existence of substantial word-world meaning, truth and reference relations?[7] Not necessarily. I take it that one pressing problem for a view that combines naturalistic scientific realism with normatively laden representational semantics is that, by its own lights, there is a naturalistically unbridgeable explanatory gulf between normative semantic properties and all other non-normative dispositional or categorial properties of the universe to which the scientific realist is ontologically committed (Hattiangati, 2024; Kripke, 1982). What is more, it can be argued that an explanation of the semantic notions of meaning, truth and reference word-world relations in *non*-representational terms is something perfectly natural and only to be expected from a strictly *naturalistic* standpoint which takes empirical

[7] For a thorough investigation of the relations between realism and representationalism about meaning, truth and reference, with special application to scientific realism, see Psillos (1999).

inquiry seriously as the final arbiter of our empirical claims about the structure and content of the world. It is extremely doubtful whether *substantial* (i.e. representationalist) semantic notions of truth and reference can themselves be consistently investigated in a scientific naturalistic spirit, namely, as empirically contingent relations. Instead, in most versions of philosophical naturalism in contemporary analytic philosophy, these semantic properties are treated as (non-naturalistically explicable) brute primitives. But if we focus our attention, as we surely must, on the fact that any scientific naturalist position worthy of the name must ontologically commit only to the entities, relations and facts yielded via the a posteriori descriptive and explanatory methods of empirical inquiry, then it seems that a scientific naturalist has neither need for nor automatic entitlement to the existence of substantial word-world semantic relations—since, by naturalism's own lights, this is at best a broadly empirical matter (Price, 2013, pp. 13–14). If this is correct, and if the best scientific (i.e. naturalistically acceptable) explanation of word-world semantic relations is non-representationalist in character, then we have an argument from *within* empirical inquiry itself, that commits us, if we are realists, to a non-representationalist account of semantic facts. Thus, strangely enough, a *non*-representationalist realism (including scientific realism) is not realist enough unless it leads to an ontology that makes this non-representational dimension of semantics explicit in its own structure. How must the world be like if it is to include human beings whose semantic and epistemic practices orient them to the surrounding world without directly mirroring the latter?

We will examine this issue in more detail in Chap. 6, but for the moment I will merely indicate that to my mind, a most likely candidate for this role would be an immanent Darwinian ontology of *processes of differential adaptation* in which interaction with the environment (other processes) differentially selects (preserves and reproduces) certain elements, while eliminating others. This insight is encapsulated in the concept of habit, which is, roughly, any tendency of a regularity to respond in generally specifiable ways to generally specifiable situations (Peirce, 1931-58, p. 2.148, 5.480). The basic concepts of this Darwinian ontology (random variation, selection, inheritance of adaptive traits and habits of response) are functional and thus broad enough to allow for *specifically different realizations* of evolutionary processes in organic, cultural and even non-organic contexts. That is, they can be used to explain not only evolution at the level of species, but also learning processes at the level of individual, the existence and function of social practices and institutions (including, importantly, semantic social practices) and even the processual dimensions of inorganic structures (including laws of nature—see Chap. 6). For example, from this point of view, the assertion language game can be understood in non-representational terms, as a multipurpose tool—sustained by a special kind of utterance-involving embodied skillful coping—that was selected for due to its capacity to express, align and revise the behavioral dispositions or other variable aspects of speaker's situations in hypersocial niches, where our ancestors increasingly depended on coordination and collaborative action (Price, 2011; Tomasello, 2014). Indeed, semantic content itself, namely, the contents of beliefs and the meanings of sentences, are understood in terms of the roles they play in processes of intelligent reciprocal adaptation between human organisms and the environment. Empirical inquiry and goal-pursuit are the inextricably intertwined aspects of these feedback-loopy processes, ultimately cashed out at the level of know-how. Semantic content is selected at a social level as an adaptive tool that can stabilize, control, align, thematize and criticize our individual verbal and related non-verbal dispositions, thus enabling us to reliably orient (i.e. increase our intelligent responses) at a collective level in these dynamic organism-environment interaction cycles.

Further, it has been recently suggested that the dynamic Darwinian mode of explanation in life sciences goes hand in hand with a process ontology which challenges the substance-based assumptions of traditional ontology and the 'containment' view of the world according to which the latter is a spatiotemporally fixed coordination system containing objects that change properties or location and interact with one another. Processes such as metabolism, development and evolution often underlie and provide the explanation for the persistence of more thing-like or continuant processes. This persistence of objects in space and time is not explained by their substantial form but is a result of more or less stable processes operating at various spatiotemporal scales within the object and between it and its environment. The only reason that we draw a substantial distinction between objects and processes (e.g. melodies) is that the temporal changes characterizing the unfolding of objects can be so minute and slow that they can be practically treated and experienced as immutable (Nicholson & Dupré, 2018). I find this 'processual turn' of life sciences to be on the right track. What is more, I think that we might have reasons to believe that process ontology surpasses substance ontology in its explanatory potential in general, not only in the domain of biology. In what follows I will attempt to briefly cash out this claim.

An important anti-representationalist implication of the rejection of Givenness for ontology is the realization that our whole categorial framework in terms of which we find our way around in our everyday lifeworld, built around the category of an object (and, correlatively, of a subject), conceived as a substance with essential modal (causal-dispositional) properties, such as to make individuals belong to certain kinds characteristic of their nature, does not possess an unchallengeable categorial authenticity as a system of classification. Indeed, we have good reasons to believe that the object-property categorial framework, including its semantic representationalist picture of the world in which the normative order of asserting propositional contents through the use of singular terms, predicates and universally quantified subjunctively robust conditionals mirrors the modal structure of the world, by representing the latter as a realm of causal facts about the properties of objects, is ultimately explanatorily inadequate. And, as we shall see, the reasons for thinking that the process framework is explanatorily preferable to the substance framework arise out of an *immanent critique* of the latter, in which the substance framework proves to be explanatorily inadequate by its own standards.

A general problem here is that the substance-property framework leaves out as contingent and insignificant for conceiving the essence of an individual (be it a person or an object) its occurrent, non-iffy, non-dispositional properties. It can thus account only for the *quidditas*, the 'whatness' ('general essence') of individuals, and not for their *haecceity* or their 'thisness' ('singular essence'): specifically, it can account for their individuality only with the aid of their accidental properties and not on the basis of essential ones, as the latter only characterize the general kinds to which many individuals belong (see Chap. 2).

Moreover, relatedly, the substance-property framework cannot independently explain why under certain circumstances objects possess or lose their causal dispositions (powers). This can only be done by the identification of the intrinsic occurrent processes that accompany the acquiring or losing of these propensities (Russell, 1927; Sellars, 1957). Note that since the causal dispositions of objects are constitutive of the latter's identity, we can independently explain the former only by abandoning the category of object (conceived as substance).

Also, the substance-property framework cannot positively account for observationally unpredicted variation or breakdown of the object's observably expected behavior. For example, in the face of a person's sustainedly incomprehensible deviation from expected verbal or emotional responses in cases of brain-damage the substance-property framework can only classify this behavior as 'anomalous' without any positive explanation of what is actually going on and why in the person's mind[8] (Christias, 2023).

Further, the substance-property framework lacks a categorization of intrinsic qualitative sensory and affective processes in sentient organisms (i.e. an intrinsic characterization of qualia), conceiving them as essential yet not intrinsically occurrent causal powers of independent substances (persons, objects), and thus as essential yet not intrinsically occurrent properties which are effectively subordinated to substances. In this way, it lacks categorial resources for an adequate positive phenomenological characterization of non-veridical experiential phenomena such as illusions or hallucinations. In those cases, the categorial resources of the object-property framework (which expresses what Sellars calls the 'descriptive ontology of ordinary life') cannot positively account for the distinctively *sensory* character of perceptual experience beyond the point of 'assuring us that *something, somehow* a cube of pink in physical space is present in the perception other than merely believed in' (Sellars, 1977, §35). This is connected to the fact that in the context of the substance-property framework cases of seeing and cases of seeming to see, for example, a pink ice cube differ only in the endorsement of the intentional content involved (e.g. 'there is a pink ice cube there'), not with respect to the very content itself. Both seeings and appearings are about the pinkness as a sensible quality exemplified by a physical object. Appearings do not refer to special 'inner' objects, of a different categorial nature than ordinary objects (Sellars, 1956, §16–18). It is exactly for this reason that, for example, I am unable to provide any illuminating explanation about what is happening if it is pointed out to me that although it appears to me that 'there is a pink ice cube in the glass' there is in fact no ice cube, pink or otherwise, in the glass (and, maybe, no glass either). I can only talk about abnormal conditions of observation (and what they do to the perceiver) or I can swear that 'it

[8] Note that to say that behavior is anomalous in the above sense does not amount to saying that it is different from what is expected due to antecedently known abnormal circumstances that hinder the actualization of the object's causal dispositions. The range of circumstances in which the causal dispositions of objects are not actualized due to external obstacles is something that is integral to our understanding of objects, it is antecedently known and, in this sense, it does not constitute a case of observationally unpredicted variation of the object's expected behavior.

was just as if I were seeing a pink ice cube in the glass' but I cannot further determine what is the common experiential element which is phenomenologically indistinguishable in cases of veridical and non-veridical perception alike beyond conceding that 'somehow, in something analogous to physical space, there is actually something that is a cube of pink'.[9] The positive categorial status of a 'cube of pink', of 'physical space' and of the way in which it is 'present in the perception other than merely believed' is left completely undetermined. Their actuality in sensory experience (their non-iffy character) demands an intrinsic characterization based on occurrent, non-dispositional properties that are neither in the external world nor in the subject.[10]

Lastly, the substance-property framework faces a serious challenge in accounting for the relation between substance and property where the former role is played by a subject and the latter by its states of sensory consciousness. This problem emerged as a result of the effort on the part of traditional substance metaphysics to accommodate the modern scientific worldview. It did so by appropriating the scientific 'impact' paradigm of causation (extorting final and formal causes) and by drawing a sharp categorial distinction between two correlated yet incommensurable realms, the mental and the physical. Yet, the existence and causal efficacy of qualitative experience in sentient organisms, including humans, which seems necessary for building a categorial 'bridge' between the mental and the physical, is rendered unintelligible if interpreted according to the above model. If in its bodily dimension, qualitative sensory experience is understood on the model of substances causally interacting by collision, impact or exchange of discreet parcels of information (e.g. particles colliding one another, neurons sending signals to one another), its qualitative dimension can only be salvaged by eliminating the causal difference it makes in the world and vice versa. Reductive materialism saves the causal efficacy of experience only by eliminating its qualitative character, construing it as a property of substances such as atoms and neurons that do not exhibit the homogeneity characteristic of qualitative experience. On the other hand,

[9] Notice that this common experiential element has actual existence—though in an unspecified way and non-apperceived as such—because, even in non-veridical cases, the experience itself actually exists; it is not merely *imagined* to be an actually existing experience.

[10] In the context of the object-property framework, subjects are categorially conceived as substances whose perceptual experience is understood as one of its essential 'iffy' properties—that is, as a dispositionally characterized perceptual capacity. This is why the occurrent, intrinsic character of sensory and affective non-conceptually articulated experience cannot be located in the subject.

substance dualism and emergent materialism preserve the qualitative character of sensory experience at the cost of making experience either non-physical or essentially epiphenomenal[11] (Kim, 1998; Sellars, 1981).

Now, what appears to be a common theme to all the above kinds of criticism against the explanatory adequacy of the substance-property framework is that its basic categories are tailored to account for the individual substances of the framework on the basis of iffy properties (dispositional properties, causal powers, mental capacities) that do not do justice to the *non-iffy* (i.e. non-modal) *occurrently existing, intrinsically characterized and dynamically evolving character* of the spatiotemporal unfolding of the individual in question (be it a sensory-affective experience or a non-experiential inorganic spatiotemporal unfolding). But to do justice to these features of the unfolding of individuals, and thus of their very *haecceity*, we must also challenge several other fundamental structural categories of the framework (the substance framework stands or falls as a whole; its categories are mutually supporting, and cannot be changed in a piecemeal manner): its reliance on kinds which express the nature of individuals based on their iffy properties (capacities, causal dispositions), its understanding of fundamental modal categories (possibility, necessity), its sharp categorial distinctions between correlated yet incommensurable ontological realms (mental vs physical, subject(ive) vs object(ive), real vs illusory), etc.

The process-based categorial framework is precisely what emerges out of an immanent critique of the substance (object-property) framework. The latter raises explanatory demands it is in no position to answer: What is the categorial basis of dispositional properties of objects? What explains the observationally unpredicted variation or breakdown of an object's observably expected behavior? What is the character of sensory qualia when the latter do not correspond to external objects, properties, and do not unfold in public space and time? It is as if our epistemic practices raise

[11] Substance dualism posits an entirely distinct non-physical substance as the locus of sensory (and in general, conscious) experience that runs its course alongside the physical world (the body, including the nervous system), but never interacts with it. Emergent materialism denies that there are any substances that exist other than physical ones, and conceives sensory experience as correlated with, but irreducible to physical brain states. Yet, both substance dualism and emergent materialism fail to explain what the correlation between brain and sensory states consists or why it occurs. Thus, substance dualism makes the causal efficacy of sensory states mysterious (non-physical) while emergent materialism makes it essentially epiphenomenal (since the causal chain at the level of brain states is explanatorily autonomous and sufficient independently of its correlation to sensory states; the correlation itself is an explanatorily superfluous and essentially inexplicable fact) (Landy, 2019; Sellars, 1981).

problems, that is, legitimate explanatory (grounding) demands—about accounting for the *haecceity*, the 'singular essence' or intrinsic character of individual beings—that our proposed 'solutions', namely, the categories of the substance framework, are in no position to answer, and indeed, are obliged to cover over by considering these problems as ultimately unintelligible (see also Chap. 6). This also shows that the process framework is not just another dogmatic metaphysical framework, a mere product of armchair speculation that we have no good reason to accept, or one that purports in the spirit of pre-critical metaphysics to justify itself in an epistemically transcendent manner, *sub specie aeternitatis.*

Another way to make the same point is to see the process categorial framework as the outcome of a transcendental investigation on the conditions under which our conceptual framework is epistemically constrained by an independent external world (as opposed to a world that is a 'shadow of grammar', a projection of our parochial reified linguistic categories). The investigation in question is transcendental because it begins with the Kantian thought that we can only speak meaningfully about the external independent world by construing it as that which affects us through sensibility, and thus by investigating the structure of the latter (as opposed to rushing into an unconstrained metaphysical speculation about the structure of the world in itself). But recall that the Given is a myth. The categorial structure of sensibility and of its ultimate worldly cause does not imprint itself on our minds as a seal imprint itself on wax. Thus, the way in which we are affected by the world—or by our own minds—is not something about which we can have transparent categorial knowledge. While our minds are embedded in the in-itself, they do not categorially mirror the latter. Indeed, the categories through which we conceptually represent the world and ourselves are not God-given and did not evolve for the purpose of ontological insight but under selective evolutionary pressures of survival within a largely unpredictable and hostile environment. But this shows that there is no pre-established harmony between mind and world, or between our everyday lifeworld (based on the substance-property framework) and the structure of the world to which it is a conceptual response. The process framework is immanently constructed as an explanatory attempt to provide this independent worldly and experiential constraint on our conceptual structure by unveiling a deeper ontological *non*-semantic relation between mind and world from the point of which the semantic correlation between mind and world at the level of the substance-property framework is revealed as epistemically indispensable yet ontologically dispensable (see Fig. 3.1). At the same time, the process framework explains why the substance framework *appears* (for all practical purposes) to have real worldly purchase. From its standpoint, the categories of the substance framework are evolutionarily and culturally selected due to their ability to abstract, stabilize and make salient those features of individual processual becomings that facilitate collective coordination (Christias, 2018). From this standpoint, the categories of the substance framework are to be conceived as *special cases* with *restricted validity* (practically optimal under certain evolutionary and cultural contexts and circumstances) of processual categories, in a roughly analogous sense in which the laws and principles of Newtonian physics are special cases of their Einsteinian counterparts.

The fact that the process framework is consciously constructed as a successor to the substance framework is shown by the fact that it eschews the categories of the latter (as philosophically developed in Aristotle and Kant) as a *whole* and not just in a piecemeal manner. In the process framework there are no objects (substances), with (causal-dispositional) properties, belonging to kinds, having natures which distinguish them from their actual causal histories by accounting for their dispositional properties. Also, processes are neither physical nor mental (beyond the physical-mental categorial distinction). They are also subjectless and objectless (beyond the subjective-objective or subject-object distinction), actual (beyond the real-illusory distinction), and endowed with an intrinsic occurrent character (i.e. characterized by intrinsic, non-dispositional properties). The ultimate aim of the process framework, as I understand it, is to provide an immanent critique of the substance framework by explaining its explanatory failures and accounting for its apparent explanatory success and practical utility. From this point of view, to immanently criticize a framework does not merely amount to showing its internal tensions and dead ends. It also essentially involves understanding the *raison d'être* of these tensions by constructing a *new* framework whose categories precisely *instantiate* such an understanding. The issue of whether the process framework in a position to cash out this promissory-noteish dimension will be explored in Chap. 4.

3.3 The Consequences of Non-representational Semantics and Ontology for the Relation between Mind and World

As was mentioned earlier, the rejection of Givenness entails, among other things, that there are no fixed limits to thinking that prevent it from being semantically *about* the world as it is in itself. Yet, the rejection of

Givenness also entails that the semantic structure of language does not mirror the categorial structure of the world, and this fact has consequences about the precise sense in which conceptual thinking is intentionally directed to the world as it is in itself. From the standpoint of a resolutely non-representational semantics and ontology, the realist view that there is no semantic, epistemic or transcendental barrier in thinking being about the world as it is in itself is rehabilitated but is given an important twist: there is indeed no such barrier provided that this direct access of thinking to being in the *semantic* sense of the term is *itself* understood as a *fractured expression* of a *non*-semantic, *non*-representational direct relation between mind and world. Thus understood, the direct semantic access of thought to the world, its semantic aboutness, must be sharply distinguished from the manner of directedness in which thought and world are ontologically related and implicated with one another. Ultimately, the semantic aboutness of thought, namely, its directedness toward a common, public, objective world through reference relations, is revealed as an indispensable yet non-representationally functioning tool for collectively coordinating thought to the world, where both are understood as systems of natural monadic processes and where the relation between them at this level is not referential, logical or causal. This non-semantic expressive relation can be understood in terms of *resonance* between dynamically co-evolving environmental, bodily and brain processes whose degree of adequacy is measured by the degree of empowerment expressed in the conative processes involved. This latter conative empowerment can, in turn, be understood in terms of an expansion of the range of metastable equilibrium of these ecosystemic relations involving our bodily, social and natural environment[12] (Christias, 2023, pp. 212–215).

[12] Roughly, we can distinguish between three different—yet intimately involved—kinds of resonance in the aforementioned non-semantic sense between environmental, bodily and sensory-affective processes. (1) *Topological isomorphisms* (homeomorphisms) established in the qualitative contraction/transformation of intensive worldly information from sensory-affective processes of sentient organisms (this, roughly, corresponds to Sellarsian picturing), (2) *Morphological resonances* between the material and functional characteristics of sentient organisms and their environmental niche (e.g. the hydrodynamic shape and movement of sea organisms resonates better with and in water environments than in land environments), (3) *Functional resonances* between the behavior of a given organism and a given environment (e.g. a human being attempting to swim at the sea, built a house for shelter, telling a tale about the origin of his tribe). Note that the resonances involved in all the above cases are clearly non-representational. They are entrenched by biological and cultural evolution-by-

Yet, it must also be emphasized that these two kinds of immediate connection between mind and world, namely, the semantic and non-semantic (or representational and non-representational) one, are distinguishable yet inseparable (Fig. 3.1). Both are *immediate yet non-transparent* (i.e. not 'clear and distinct') mind-world relations and can be illuminated only by understanding their dependence on one another. In Leibnizian terminology, we can say that the non-representational, non-semantic mind-world immediacy is *distinct yet obscure*, while the representational, semantic mind-world immediacy is *clear yet confused*. The former non-conceptual kind of mind-world immediacy operates at a level that, while being in a sense 'within' us, is best described as a kind of unconscious experience and thinking (this is why it is 'distinct in itself', yet 'confused' from the standpoint of conceptual consciousness). The latter conceptualized mind-world immediacy and coordination can be seen as an evolutionary tool selected and preserved precisely for organizing and making intelligible this distinct-obscure unconscious yet efficacious experience as something (a conceptually articulated content) that can be individually internalized and collectively communicated in a sufficiently reliable manner for enhancing collective survival and well-being. This conceptualized mind-world immediacy thereby turns the 'obscure' non-conceptual mind-world relation into something 'clear' yet only by sacrificing its 'distinctness'. This is done by simplifying, contracting and differentially selecting (simultaneously at several spatiotemporal levels of processing) the amount, variety, source and complexity of information contained in external stimuli for maximizing our allostatic balance. This process was not evolutionarily selected for its ontological insight into the structure of reality at all spatiotemporal levels. In fact, its relative success for the purposes of survival and reproduction depends on its actively contracting this spatiotemporal, intensive and qualitative complexity of the universe—thus avoiding informational overload and energy expenditure levels that would compromise survival. Yet, it is only by way of the conceptual tools provided by this clear-yet-confused level that we can better understand the non-conceptual immediacy between mind and world. We cannot sidestep the conceptual level of thinking about the world and

selection through a largely unconscious process of trial and error. Yet, it is also clear that resonances in the above sense can be better or worse, less or more adequate with respect to the synchronization achieved by the processes involved and the degrees of freedom thereby expressed. See also Fig. 3.1 and Chap. 4 for more on this issue.

ourselves—our reason-giving practices that institute the semantic content of perception, thought and action—and penetrate directly into the non-conceptual mind-world immediacy through some kind of non-conceptual 'method of intuition' (*a la* Bergson or Deleuze). To believe so would be to flirt with mythical Givenness and would amount to a controversial subordination and relativization of conceptual rationality. Yet we are not doomed. The conceptual and categorial tools developed in the course of inquiry and its sophisticated extensions—science and philosophy—ultimately show themselves to be expressions (a subtractive manner of appearance) of a self-fracturing of the non-conceptual level itself which enables it to better regulate itself (see Fig. 3.1 and Chap. 5). The conceptual level thereby asserts its indispensable role for understanding ourselves-in-the-world without ontologically reifying our conceptual representation of the world. By being irreducible to yet complementary and mutually presupposing with the non-conceptual dimension of the mind-world relation, it thereby expresses the way of being of the mind-world relation as one of fractured immanence.

Could a process view of the world or a process language overcome this bifurcation between two levels of mind-world relations (representational-non-representational, conceptual-non-conceptual)? We shall develop a process view of the world in more detail in Chap. 4 but the short answer to this question is: yes and no. Yes, in the sense that the process view of the world and the ensuing language would make salient the non-iffy occurrently existing, intrinsically characterized and dynamically evolving character of the spatiotemporal unfolding of individuals, thereby drawing out attention to each individual's actual causal history for understanding the character and unique particularity of each individual (its *haecceity*). And this would involve a major change not only in our worldview but also in our practical attitudes toward beings: a focus on their unique historically unfolded particularity, as opposed to their being treated as subsumed in universals such as kinds, genuses, in a word, general essences. On the other hand, this does not mean that the complexity and qualitative particularity of the world's dynamic unfolding would be Given ('clear and distinct') in all its exhaustive detail to a mind that possesses a process worldview and a process language. Presumably, a process view of the world would be neither clear-yet-confused (representational level, substance ontology) nor distinct-yet-obscure (non-representational level, monadic process world in all its qualitative complexity), but rather *clear-yet-obscure*. That is, while the process worldview and language would make salient the dynamic character of the individuals and of the relations which account for their unique particularity, if it were to be of any practical use and interest it would still have to informationally compress certain aspects of the spatiotemporal, intensive and qualitative complexity ('distinctness') of the universe so as to make them salient to our semantic reason-giving practices.

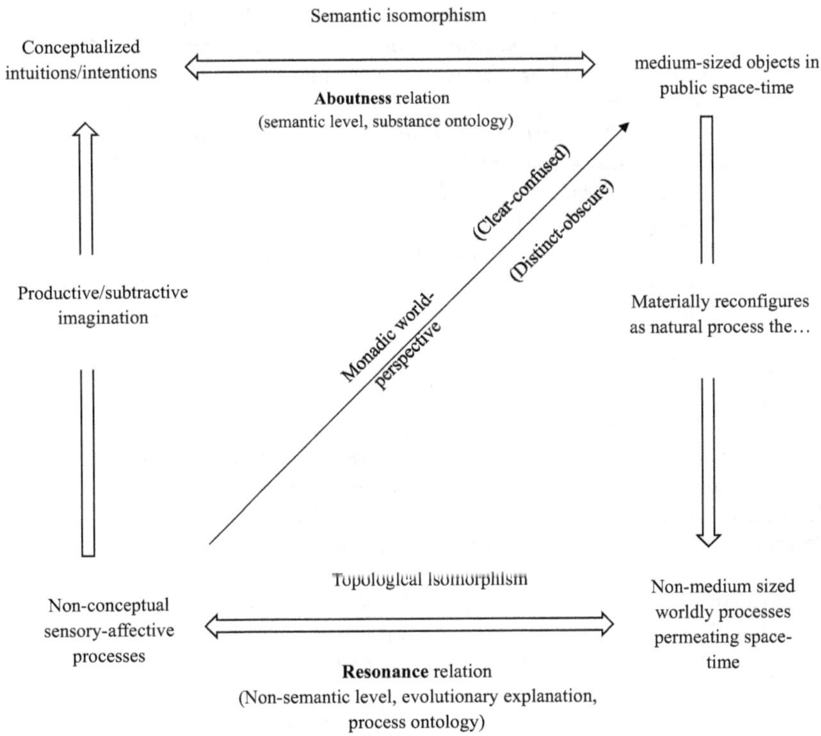

Fig. 3.1 Two distinguishable yet inseparable mind-world relations: (1) Conceptual clear-yet-confused mind-world semantic isomorphism, (2) Non-conceptual distinct-yet-obscure mind-world non-semantic isomorphism. These different kinds of mind-world isomorphisms themselves stand in a fractured immanent relation in that the conceptual mind-world isomorphism is effectively the expression of a self-fracturing of the non-conceptual mind-world isomorphism. This self-fracturing is expressed in the manner of monadic world-perspectives (see Chap. 4)

References

Adorno, T. (1998). *Critical models: Interventions and catchwords*. Columbia University Press.

Brandom, R. (1994). *Making it explicit: Reasoning*. Harvard University Press.

Brandom, R. (2015). *From Empiricism to expressivism: Brandom reads Sellars*. Harvard University Press.

Brassier, R. (2014). Deleveling: Against 'flat ontologies'. In C. van Dijk, E. van der Graaf, M. den Haan, R. de Jong, C. Roodenburg, D. Til, & D. Waal (Eds.), *Under influence - Philosophical festival drift* (pp. 64–80). Omnia.

Brassier, R. (2016). *Transcendental logic and true representings*. Glass bead. http://www.glass-bead.org/article/transcendental-logic-and-true-representings/?lang=enview

Catren, G. (2023). *Pleromatica: Or Elsinore's trance*. Urbanomic.

Cavell, S. (1969). *Must we mean what we say?* Cambridge University Press.

Chang, H. (2022). *Realism for realistic people: A new pragmatist philosophy of science*. Oxford University Press.

Christias, D. (2018). Towards the kantian thing-in-itself: Sellars' and meillassoux's divergent conception of kantian transcendentalism. In F. Gironi (Ed.), *Analytic and continental kantianism: The legacy of Kant in sellars and meillassoux* (pp. 127–153). Routledge.

Christias, D. (2023). *Normativity, lifeworld and science in sellars' synoptic vision*. Palgrave Macmillan.

de Vries, W. (2005). *Wilfrid sellars*. Acumen Publishing.

de Vries, W. (2021). *Wilfrid sellars, the Stanford encyclopedia of philosophy*. https://plato.stanford.edu/archives/fall2021/entries/sellars/

deVries, W., & Triplett, T. (2000). *Knowledge, mind and the given*. Hackett.

Foucault, M. (1970). *The order of things*. Vintage Books.

Hattiangati, A. (2024). Physicalism, intentionality and normativity: The essential explanatory gap. In G. N. Kemp, A. H. Khani, H. S. Rezaee, & H. Amiriara (Eds.), *Naturalism and its challenges* (pp. 69–88). Routledge.

Heidegger, M. (1977). *The question concerning technology and other essays*. Garland Publishing.

Husserl, E. (1970). *The crisis of european sciences and transcendental phenomenology*. Northwestern University Press.

James, W. (1906). *Pragmatism*. Longmans, Green, and.

Kim, J. (1998). *Mind in a physical world: An essay on the mind-body problem and mental causation*. MIT Press.

Kripke, S. (1982). *Wittgenstein on rules and private language*. Harvard University Press.

Kukla, R. (2000). Myth, memory and MISRECOGNITION in sellars' 'empiricism and the philosophy of mind'. *Philosophical Studies, 101*(2-3), 161–211.

Landy, D. (2019). Sellars' argument for an ontology of absolute processes. *Journal for the History of Analytical Philosophy, 7*(1), 1–24.
McDowell, J. (1987). De Re senses. *The Philosophical Quarterly, 34*, 325–363.
McDowell, J. (1998). Having the world in view: Sellars, Kant and intentionality. *Journal of Philosophy, 95*, 431–491.
Nicholson, D., & Dupré, J. (Eds.). (2018). *Everything flows: Towards a processual philosophy of biology*. Oxford University Press.
Peirce, C. S. (1931-58). In C. Hartshorne, P. Weiss, & A. Burks (Eds.), *Collected papers*. Harvard University Press.
Price, H. (2011). *Naturalism without mirrors*. Oxford University Press.
Price, H. (2013). *Expressivism, pragmatism and representationalism*. Cambridge University Press.
Psillos, S. (1999). *Scientific realism: How science tracks truth*. Routledge.
Quine, W. V. O. (1960). *Word and object*. MIT Press.
Rorty, R. (1979). *Philosophy and the mirror of nature*. Princeton University Press.
Russell, B. (1927). *The analysis of matter*. Allen and Unwin.
Sellars, W. (1953). Is there a synthetic a priori? *Philosophical Studies, 20*(2), 121–138.
Sellars, W. (1954). Some reflections on language games. *Philosophy of Science, 21*(3), 204–228.
Sellars, W. (1956). *Empiricism and the philosophy of mind*. Harvard University Press.
Sellars, W. (1957). Counterfactuals, dispositions and the causal modalities. In H. Feigl, M. Scriven, & G. Maxwell (Eds.), *Minnesota studies in the philosophy of science* (Vol. II, pp. 225–308). University of Minnesota Press.
Sellars, W. (1963). Philosophy and the scientific image of man. In W. Sellars (Ed.), *Science, perception and reality* (pp. 1–40). Routledge.
Sellars, W. (1967). *Science and metaphysics: Variations on kantian themes*. Ridgeview Publishing Company.
Sellars, W. (1975). The structure of knowledge. In H. N. Castaneda (Ed.), *Action, knowledge and reality* (pp. 295–347). Bobbs-Merrill.
Sellars, W. (1977). Some reflections of perceptual consciousness. In R. Bruzina & B. Wilshire (Eds.), *Selected studies in phenomenology and existential philosophy* (pp. 169–185). Springer.
Sellars, W. (1981). Foundations for a metaphysics of pure process. *The Monist, 64*(1), 3–90.
Spengler, O. (1926). *The decline of the west*. Oxford University Press.
Tomasello, M. (2014). *A natural history of human thinking*. Harvard University Press.
Wittgenstein, L. (1958). *Philosophical investigations*. Blackwell.

PART II

Outline of a Fractured Immanent Ontology

CHAPTER 4

Toward a Process Nominalist Ontology

In this chapter, I will examine the specifics of a process ontology that attempts to provide successor counterpart categories to the substance-property framework at work in the context of everyday lifeworld.

In what follows, we must always bear in mind that these successor processual categories are not merely alternatives to those of the substance framework, but ultimately aim to provide an immanent critique of this framework by explaining its explanatory failures and accounting for its apparent explanatory success and practical utility. This essentially involves understanding the *raison d'etre* of the tensions of the substance framework by constructing a new framework whose categories instantiate such an understanding. This understanding, in turn, is provided by showing that the categories of the substance property framework are to be conceived as *special cases* of processual categories, in a roughly analogous sense in which the laws and principles of Newtonian physics are special cases of their Einsteinian counterparts. And as such, the categories of the substance framework have *restricted validity*: they are practically adequate under certain—yet not all—evolutionary and cultural contexts and circumstances, where both kinds of contexts and circumstances are specified from the vantage point of the process framework.

We have already seen that the substance framework is a tightly interconnected whole in which the basic substances are particular objects with causal-dispositional properties, standing in relations of material

© The Author(s), under exclusive license to Springer Nature Switzerland AG 2025

D. Christias, *Towards a Hyperstitional Process Nominalism*, Palgrave Perspectives on Process Philosophy,
https://doi.org/10.1007/978-3-032-06350-2_4

incompatibility and consequence. These properties express capacities, abilities and dispositions of objects and are such as to make the objects that have them belong to certain ***kinds***, characteristic of their ***nature***: Socrates (particular object-substance), being a human being (kind), can run, think and sleep (capacity, ability, disposition); gold, being a metal, is a good conductor of heat and electricity, etc. This expresses the central role in ordinary discourse of identifying things as belonging to kinds, which identification carries with it a knowledge of causal—and in general, modal—properties (Sellars, 1949b). Note here that in the substance framework, subjects (persons) and objects are metaphysically on a par at least with respect to the above general categories. They are both substances with causal-dispositional properties standing in relations of material incompatibility and consequence. The difference between them lies in the fact that physical objects, unlike persons, cannot be thought of as acting intentionally, with an end in view, but rather, only out of habit (expressed by their causal-dispositional properties) (Sellars, 1963). Thus, the basic category of object in the substance framework at this level of abstraction also includes subjects or persons since the latter are also essentially conceived as substances with dispositional properties that express their capacities, abilities and dispositions and are such as to make the objects that have them belong to certain kinds, characteristic of their nature.

As is evident from the above, the categorial difference between the process and the substance framework does not merely come down to the fact that in the latter processes and events are derivative from the fundamental category of objects (substances) undergoing change, whereas in the former objects are abstractions of more fundamental relatively stable processes. The difference is far more pervasive: if the basic particulars of the framework are absolute—that is, subjectless and objectless—processes, then just as the category of object (and its corollary, that of a subject) would accordingly be derivative, a shorthand composed of a multiplicity of relatively stable particular processes, so would be the object's causal-dispositional properties, their standing in relations of material incompatibility and consequence, the kind to which an object belongs, the general nature that this kind expresses through which we can understand the object's regular behavior and its deviance from normalcy, etc. The difference between the substance and process framework extends even to the very *form* in which we describe a worldly situation: in the process framework there would be no *facts*, since facts are precisely states of affairs with objects (particulars), properties and relations (universals) as constituents.

And since properties would not be conceived as abstract universals exemplified by objects as particulars (the whole object-property structure would be reconceived in processual verb-adverb terms), the process framework would be an essentially *nominalistic* one. Moreover, the distinction between the fundamental *epistemic* categories of the substance framework, namely, between objective and subjective, real and illusory, or normal and abnormal experiences, would turn out to be derivative (i.e. non-essential) in a process framework. Since these subjective, illusory or abnormal states involve the occurrence of experiential processes, they are *ontologically on a par* with objective, non-illusory and normal ones. They are all equally real, though the latter might well be far more invariant under certain perspectival transformations. Of course, this is consistent with also holding that these categorial distinctions serve important *normative* purposes; for instance, they can be considered as fundamental skeletal coordination principles that facilitate reliable collective coping within a volatile environment.

The above features distinguish the process framework as I understand it from neo-Aristotelian or neo-Hegelian versions of process philosophy (Brandom, 2019; Pippin, 2018), since both these views reject the ontological priority and parity of processes and subordinate the processual dimension of basic particulars to the category of object construed as a propertied substance. The view developed here (inspired by Peirce, Sellars and Deleuze) is closer to the family of more radical process views such as Leibniz's, Bergson's or Whitehead's.[1]

Does this processual metaphysics commit us to ascribe *categorial* structure to the world or experience itself? I do not think so, especially if, following Kant, we understand categories on the model of the forms of judgments as they relate to all objects in general. Processes lack all the characteristic causal, modal and logical structure of Kant's table of

[1] There are also process frameworks that begin at the phenomenological level—that is, at the level of the correlation between thinking and being—and attempt to fuse descriptions based in conceptually undistorted lived experience with a speculative realist metaphysics (Röck, 2022). While I find views of this form interesting, important and close in spirit to process thinkers such as Bergson and Whitehead, I cannot endorse them since I interpret them as a version of 'phenomenological immanence' (Chap. 2) with all the problems that this view entails. My rendering of being as processual goes hand in hand with a conception of fractured immanence, that makes the—essentially 'correlational'—standpoint of lived experience insufficient for a resolutely immanent investigation of worldly, experiential or knowing processes.

judgments. They also completely lack the classic Aristotelian hierarchical categorial structure (genus-species-individual). In this sense, they are not categories—though, of course, they carve up the world in certain ways and preclude others. In my view, if one ultimately endorses a univocal ontology in which being is said in a single and the same sense for all beings (as opposed to it being said in many—mutually incommensurable—ways), one thereby does not think by categories in any philosophically interesting sense of the term. This is because to carve the world into categories just *is* to divide being into mutually exclusive, exhaustive and irreducible metaphysical regions, which in turn amounts to distributing being (existence) to beings in different, mutually exclusive and incommensurable ways. For example, if beings are conceived as particulars (substances) as opposed to universals (properties, relations), as physical as opposed to mental substances, animals as opposed to humans, animate as opposed to inanimate beings, then these types of beings exist in different senses of the term with no common measure between them.[2] Thus, it seems that the very attempt to articulate the world categorially ends up invariably conceiving being in a *non*-univocal manner.

4.1 Basic Features of Process Metaphysics

Let me now sketch the basic features of this alternative process framework in a little more detail.

[2] It could be objected here that at least some of these categorial divisions, such as those between inanimate and animate beings or between animals and humans are not mutually exclusive, but inclusive, in the sense that the animate or the human as layers of being include the inanimate and the animate, respectively, as a proper part: animate beings are composed of inorganic matter and humans are animals, they have animal characteristics (they eat, sleep), instincts, etc. Yet, this view presupposes an additive conception of these categorial features, as if animate beings or humans are emergent layers of being whose fundamental categorial properties, namely, 'life' and 'rationality' are simply *added* to the lower categorial properties of inorganic matter and animal life, respectively. And this additive conception of categories makes the relations between these layers within the same being ultimately unintelligible (Boyle, 2012, 2016). A more coherent view, already held by Aristotle, is the *transformative* conception, according to which each higher category includes the lower ones only by essentially transforming them in accord with its specific difference. Thus, animality exists in human only in rational form, inorganic matter exists in animals only in animate form, etc. This shows that to include nominally one category in another, for example, to say that humans are animals (of the rational type), is fully consistent with *excluding* them with respect to their essence (substantial form).

As we saw, in the process framework, processes as basic individuals can and must be determined without reference to subsuming universals (kinds, counterfactuals, dispositions, laws). More radically, basic processual individuals do not stand in relations of material incompatibility and consequence. This is because these relations define the asymmetrical yet mutually presupposing relation between object (particular) and property (universal) and hence cannot be appealed to in a resolute non-object-bound process ontology. (This is ultimately why Aristotelian or Hegelian versions of process ontology are not resolute enough to count as real alternatives to substance metaphysics as it finds expression in the object-property framework.) A property is the property it is because of its location in a space of compatible determinable families (shape-properties, color-properties) of incompatible determinate properties (spherical-cubical, red-green). These incompatibilities between properties (their standing in the relation of 'determinate negation') are essential to the individuation of the basic particulars of the framework, namely, objects. In fact, objects (particulars) are precisely the substantive unities that must be postulated for accounting for the incompatibilities of properties (universals). For, to be an object just *is* to exclude incompatible properties (to be a table is precisely to be the kind of thing that cannot be red and green, triangular and rectangular at the same time). And, if we recall that these properties, by expressing capacities, abilities and dispositions of objects, are such as to make the objects that have them belong to certain kinds, characteristic of their nature, we see that objects are in the last analysis substantive units of account for the *subjunctively robust exclusions* of properties. Thus, the object-property framework has a very abstract and skeletal yet precise and systematic picture of the conditions of intelligibility that must be in place for anything to have sense or determinate existence: for there to be some determinate way the world is, just *is* for it to be articulated into states of affairs, that is, objects (particulars) possessing properties standing in relations (universals) that include and exclude each other in modally robust ways (Brandom, 2015, 2019).

An important consequence of the above analysis is that non-object-bound processual individuals cannot exhibit the standard metaphysical structure of compatible determin*able* families (polygonal, colored) of incompatible determin*ate* properties (circular-triangular, green-red), which are merely different from determinates falling under other determinables (blue, triangular). This means that nothing of this taxonomic determinable-determinate or genus-species-individual structure of our ordinary lifeworld framework (and of the framework of many current sciences) could be available from the standpoint of a process language or to the corresponding structure of a process world. Does this mean that we have run out of conceptual resources for formulating a framework that does not adhere to the determinate-determinable metaphysical structure? Not necessarily. For example, it is interesting to note that even ordinarily existing objects like shapes are exclusively different from one another only if we restrict ourselves to shapes exhibiting the *same number of dimensions* as the space they inhabit. A three-dimensional pyramid with a rectangular base might be thought to exhibit both triangular and rectangular shapes. The same goes for a cube and a sphere or a square and a circle; they can both be understood as products of a topological deformation via a continuous function, which shows that from this latter point of view they are homeomorphic (topologically equivalent). Also, colors form a family of exclusive different contents only if we restrict ourselves to *monochromatic* regions (e.g. polychromatic—white light envelops all colors). As we saw, in a process language, content determination must be provided in a way that does not make any use of exclusive difference. The mathematical resources of topology (number of dimensions of virtual space, its connectivity, its 'universal singularities', homeomorphisms) and of dynamic systems theory (differential equations, phase space, trajectories, attractors) might be useful to articulate such a radically non-Aristotelian view about content determination. From this point of view, the Aristotelian substance framework would be seen as pragmatically useful (though ontologically non-committal) categorial articulation and taxonomy for coping with ordinary middle-size three-dimensional entities and processes.

The rejection of the determinate-determinable metaphysical structure and of content determination based on subjunctively robust exclusive differences exhibited by an object's essential properties is connected to the fact that at the more fundamental level of the process framework, instead of thinking of a universal genus subsuming a multitude of individuals, we must change our perspective and see qualitied properties as themselves *entirely particularized*. In this case we are more dealing with an individual that is a genus unto itself (Nunziante, 2021), that thereby unfolds its 'singular essence' (see Chap. 3), as opposed to its being determined by a hierarchical determinable-determinate structure. Thus, in the process framework, properties (including qualities) would be *non-sharable* ways of being particulars (individuals). The naming and characterizing function of processes ('this-such' structure) would not be understood along the lines the subject-predicate (one object—many properties) subsumptive model, but on a ***non-subsumptive verb-adverb*** model. From this point of view, what is ostensibly a single particular exemplifying a number of universals (Aristotelian object-property framework) is actually a number of particulars (e.g. qualia episodes), expressed by verbs, exemplifying simple universals (e.g. 'qualia-tropes'[3]), expressed by adverbs (Sellars, 1949a, 1960a). Thus, at bottom, in the process view of the world, the one-many framework is ultimately based on a many-one framework, in which particulars are not subsumed under universals and where the dualism between abstract (universals) and concrete (particulars) ceases to make sense, since we are now operating within a framework in which qualities themselves are entirely particularized and each particular constitutes its own genus (singular essence). This, among other things, shows that in defining the 'architecture of reality' one is not forced to assume that there are (at least) two types of 'building blocks', one corresponding to things and another corresponding to the things' qualitative aspects. Thus, the nominalist ontologist who aims for simplicity and economy is not under any obligation to attempt a complete reduction of everything to either one of these two categories (Morganti, 2012).

As we mentioned above, the process view of reality can be best linguistically elucidated with the use of the verb-adverb model. From this point of view, ordinary sentences of the form 'this table is red' or 'the sea is blue' would be reconceived and transformed into subjectless and objectless sentences of the form 'it tablely reds', 'it bluely seas', etc. Importantly, in the process language the naming function is not performed by objects but by particularized subjectless and objectless properties (a particular 'redding', a particular 'sea-ing'), with the objects being modes (adverbial

[3] Tropes are properties like the particular shape, weight and texture of an individual object. Because tropes are particulars, for two objects to share a property (for them both to exemplify, say, a particular shade of green) is for each to contain a greenness-trope, where those greenness-tropes, although numerically distinct, nevertheless resemble each other, with this relation of resemblance between basic trope-properties considered as ultimate, unanalyzable and capable of independently accounting for the similarity of numerically distinct particulars and for generality without recourse to universals (Williams, 1953). We do not have the space for reviewing the many versions and problems of trope nominalism. I shall only note that, to my mind, trope nominalism becomes stronger if these relations of ultimate resemblance are thought of as a function of the *intrinsic* nature of those basic trope-properties and if this intrinsic nature is conceived as determining intrinsic (basic) *differences* between trope-properties, with the resemblances between them being a surface effect (a derivative function) of those differences.

modifications) of properties, rather than vice versa (as is the case in the object-property framework). In general, the naming function is performed by happenings, doings or acts of unfolding in certain ways (Sellars, 1981). Processes are essentially *adverbial modifications of spatiotemporal occurring*, a way of 'bringing spatiotemporal occurrence about' that is conceptually not separable from the occurring in question which expresses the intrinsic character of the process (Seibt, 2016).

This has pivotal consequences for how processes relate to space and time. Processes *permeate* space and time; they are differing qualitative ways of spatiotemporal occurring. They are not 'in' space and enduring 'through' time, as are the objects of the substance framework where space and time are antecedently understood frames or containers filled by independently understood objects. In this sense, processes are essentially monadic and 'perspectival' in a quasi-Leibnizian sense. By being differing ways of spatiotemporal unfolding, processes constitute their own perspectival and intrinsically differing spatiotemporal microcosm.[4] Of course, this does not mean that processual monads are sealed off from one another and cannot communicate or influence one another in any way. It just means that whatever relations exist between monadic processes will be reconceived in terms of overlapping and interpenetration of differing intrinsic processual characters and the ensuing qualitative fitting of these intrinsic characters into the perspectival structure of the monadic process to which they belong as ingredients.[5]

As we saw, in the process framework time should not be understood as a neutral, homogeneous medium in which self-identical now's indefinitely repeat themselves, differing only in their successive ordering.[6] The same

[4]This entails that there is no one single objective (clock) time, but a multiplicity of asynchronous ontological spatiotemporal perspectives constituted by beings in their monadic becoming. This is in line with Einstein's theory of relativity according to which there is no absolute contemporaneity in the world (see also Whitehead, 1978).

[5]Note that this view is fully compatible with there being irreducible, irreplaceable *relational truths*. It's just that the truth-*makers* for these truths could well turn out to be non-relational ways the world is (Heil, 2021, p. 85).

[6]Interestingly, this entails that, strictly speaking, temporal relations are not in the 'really real' world. The relations of 'before' and 'after' (tenses) can be more perspicuously viewed as temporal connectives, as culturally evolved linguistic devices relating propositions that were, are or will be true (Sellars, 1981). Thus, tense and temporal relations can be viewed as linguistic coordinating devices for our reliable navigation in inferential space, and in this sense, while being practically indispensable tools for prediction, control and collective coordination, they do not directly mirror the unfolding (or the non-conceptual experience) of time.

goes for space. Just as time, space is here conceived as primarily intensive, not extensive or metrical, in its being. It is not thought of as homogeneous and uniform, but as heterogeneous and polarized (think, e.g. of fundamental spatial gradients such as cosmic expansion or gravity). More radically, spacetime itself is conceived as self-differentiated and evolutionarily developed, in the sense that it can produce novel dimensions (change of the space of possibilities). In the process framework, space is not essentially a metric system of external relations of spatial distances between points (this is the Euclidian conception of space), and thus it is not a countable heap of the self-same things, but can be best conceived as a *heterogeneous yet continuous multiplicity* in the Riemannian sense. From this (intensive, non-metric) point of view, the points that form a space can be defined not by rigid lengths from a fixed coordinate system as in the case of metric space, but by the instantaneous rate at which curvature changes at that point (non-metric topological space). Some parts of the space will not be changing at all, other parts changing slowly, and others changing fast. This is a conception of space as essentially non-metric and differential (Deleuze, 1968), in which space in effect becomes *a topological field of rapidities and slownesses*, and via these infinitesimal relations one can specify neighborhoods without having to use rigid lengths (Riemannian conception of space). In their physical application through the apparatus of dynamic systems theory, mathematical multiplicities become the possibility space of a system (its phase space), and its topological invariants become the system's dimensions, singularities or attractors (DeLanda, 2002).

Finally, movement and change receive a similar treatment in the process framework. Movement should not be conceived in terms of mathematical continuity, as a relation between static unchanged entities, for example, between discrete moments in time. *Contra* substance ontology, the identity of a processual individual moving through the spacetime continuum is not independent of its movement. This follows from the rejection of the distinction between the process and the subject or object of movement. Processual movement should not be conceived in terms of a successive passage of an invariant entity through a trajectory that could be conceivably different than what it actually is without affecting the essence of this entity. Processual continuity is qualitative and intensive. As Bergson forcefully argued, becoming cannot be constituted by the mere concatenation of discrete, motionless substances (Bergson, 2002, p. 362). If we spatialize time, construing it as pure succession, composed of indivisible moments or units that succeed each other but without any relation to each other (each unit in the succession is a fully self-present identity), then we risk making movement and change themselves ultimately unintelligible, as shown by Zeno's paradoxes.[7]

4.2 The Individuation of Processes

Now, a crucial question at this juncture is how are we to individuate processes that permeate (rather than exist 'in') space and time? A process ontology must have an account of how ongoing developmental and activity-like processes are situated in time and space, and how they relate to extended temporal and spatial regions. This explanatory challenge looms large if processes are taken to be basic particular entities, which standardly are defined as existing in exactly one location at any time at which they exist. Obviously, a process ontology must provide a different account of individuation for its basic individuals if it is to constitute an alternative to the substance framework. An early modern solution to this problem was presented by Leibniz, who claimed every individual thing (monad) has a corresponding *individual concept* that is sufficient to completely individuate it, and that this concept incorporates all the determinations of the thing, as if it were an infinite list of its universal properties. Leibniz's really challenging claim is that the thing's spatiotemporal location is unnecessary for individuating it, and that it can be determined entirely by the list of its non-spatiotemporal properties.[8] This Leibnizian

[7] This also shows why processes cannot be reduced to series of instantaneous events (*à la* Davidson, 1980). The motivation for this move might be to fit the process framework to the classic interpretation of quantified statements (of the form 'there is an *f* such that *x* is *f*') which implies the existence of thing-like, countable particulars. Yet, if we adopt the formula 'there is *something* which…' as a reading of quantified statements, then we can make room for variables ranging not only to thing-like countable particulars, but also to stuffs, properties, processes, etc. (Sellars, 1960a).

[8] By contrast, as is well known, Kant *contra* Leibniz held that some grasp of spatiotemporal location is necessary to completely individuate a thing, but that this is nevertheless insufficient without the addition of a general concept. This ties back to the distinction between sortals and qualities insofar as not just any general concept will do here. Sortal concepts are just those general concepts whose combination with some grasp of location is sufficient to completely individuate something. This is precisely because they are tied to procedures for counting the number of instances of a universal in a given location (i.e. what Kant would call 'schemata'). All we need to completely individuate a thing is enough information about its location to get the relevant counting procedure to give you the answer '1'. This idea is devel-

view might be in the right direction in that processual individuals can be usefully seen as particularized perspectives of the whole world (which they confusedly include as a result of their actual evolutionary development and interaction with all other relevant worldly processes). Yet, Leibniz's view about spacetime and its role in individuation is both problematic (since, as Kant showed, spatiotemporal polarities—e.g. between left and right—cannot be captured even by Leibnizian individual concepts) and unnecessarily radical. To do justice to the insight that monadic processes permeate spacetime we do not need to view the latter as epiphenomenal. All we have to do is to conceive spacetime as *inseparable* from the quantities (differences in degree) and qualities (differences in kind) that deal with it. From this point of view, we can conceive of spatial dimensions as emerging simultaneously with systems of quantitative and qualitative dimensions. Given the above account of processes as essentially adverbial modifications of spatiotemporal occurring, we can now see that it is precisely this adverbially particularized occurrent spatiotemporal character that would differentiate a process *intrinsically* from other processes, independently of relations of material exclusion (determinate negation) or consequence (inferential mediation) it might bear, as a contentful state, to other worldly and linguistic contentful states.

This has serious consequences for the fundamental logico-metaphysical notions of identity and difference. Difference is primitive and is not mediated by exclusive difference or self-same identity. This means that difference is not conceived as difference between already existing identities (e.g. of the form 'x is different from y', where x and y are already self-same objects). Difference is primitively defined by means of a differential self-relation (dx/dy). From this point of view, identity is conceived as an effect, a product (an equalization, a covering over) of a pre-existing infinitesimal self-differentiation (e.g. of a force-field).[9] And if we take it that the ontological cash value of differentials can be provided in terms of dynamic systems theory, we see that difference can be also viewed as morphogenetic engine. It is not conceived negatively, as lack of resemblance (mere difference), or in terms of exclusion, based on a pre-existing notion of

oped by Frege, Russell and Quine as the claim that there is nothing more to existence than instantiating such a sortal concept. Obviously, this solution, however intuitive it might sound, cannot be of immediate use within a resolutely process framework as it presupposes essential modal and spatiotemporal categories of the object-property framework.

[9] This suggests that even the logico-mathematical concepts of identity and necessity can be grounded in a metaphysics of difference. For an attempt in this direction see Ardoline, 2024.

negation and identity (exclusive difference), but positively or productively, as that which—in its dimension as *intensive difference* or difference of potential—drives a dynamical process. Relatedly, the identity of basic processual individuals is based on a notion of repeatability which is not to be confused with exact repetition of the same under rigidly fixed laws. Instead, it is based on a notion of re-identification without generality, or need for an original self-identical entity, and in which every identifying act is simultaneously an act of (non-interchangeable) variation.[10]

Here it could be objected that a language of processes with the above characteristics would be insufficiently determinate for individuating processes in a stable manner. But I do not think that this is necessarily the case. Processes can be nameable and identifiable—and thus sortable—in terms of the causal roles which their intrinsic (qualitative or intensive) character plays within a system of other processes.[11] The causal roles

[10] Note that this view of becoming allows for the simultaneity of opposites. For example, something that 'becomes' and never 'is' (with regard to its size) can be viewed as 'bigger than what it was and smaller that what it becomes' at the same time (Deleuze, 1969). Does this violate the principle of non-contradiction (PNC)? Not necessarily. PNC governs predication, namely, how we ascribe properties to self-identical *substances* (a ball cannot be both red and not-red at the same time and in the same respect). PNC also excludes the co-instantiation of contrary properties, even in the absence of underlying substances, provided that these properties are conceived as self-identical fixed essences (*a la* Plato) and thus as mutually exclusive. But, in both these versions, PNC does not directly apply to the categories of becoming as such since the latter are a-substantial and bereft of a self-identical essence. Indeed, the categories of becoming account for the derivative nature of substances and self-identical essences by way of the more fundamental concept of a differential intensive field. From this latter point of view, where the simultaneity of opposites is understood as a differential and not as a contradiction, the logic of becoming exceeds the limits of PNC-based logic without necessarily negating it.

[11] Here it must be noted that since in a process view of the world, the category of causality is also abandoned, strictly speaking, the causal-functional role in terms of which processes are individuated must be reconceived in non-causal terms, while preserving the informational content of the former. As we shall see in Chap. 7, instead of construing efficient causation in terms of a local interaction between self-standing objects, in a process framework such interactions would be understood as wavelike ripples in the texture of spacetime, with ordinary objects reconceptualized as processual *modes* of the unfolding of spacetime, while efficient causation would be reconceived in terms of patterns of interference in the unfoldings of these processual modes. If we want to retain the notion of causation to describe these patterns of qualitative spatiotemporal alterations between interfering processual fields, we could perhaps call the relation between these interfering processes one of *expressive* causality (Chap. 2): Here what is expressed (the 'cause') is not an independently existing and prior condition of the expression (the 'effect'), for example, in the form of an antecedently existing possibility

performed by the intrinsic character of processes in effect individuate processes by what they *do*, by the *difference* their continuous activity makes to other processual doings—those that they presuppose and those they engender. As this kind of differentiation of processes is due to their *intrinsic* character as occurrings arranged in a certain way, it individuates and sorts processes independently of their instantiating general properties, abstract contents or abstract relations to those properties or contents. From this standpoint, the basic particulars of the object-property framework (e.g. specific physical objects or human beings) would be understood as 'logical constructions' out of a multiplicity of more basic complex processual individuals. For example, instead of being individuated in terms of its differing in 'matter' among members of the same essential 'form' (species), where the latter is conceived as a natural kind defined by characteristic 'iffy' dispositional properties, a particular human being, for example, Socrates, would be reconceived as the locus or dynamic field of historically unfolding and synchronically retained (across different space-time scales) co-occurring and co-selected patterns of bodily-brain processes. The collective resonance (allostatic balance) of these nested processes sustains, in relative stability through change, what we might call the 'Socrates-event'. And, as was mentioned above, this event would be individuated by the difference it synchronically and diachronically makes to the processes it presupposes, engenders and encounters, where this difference is encapsulated in the processes' *power to affect and be affected*. This would be a conception of Socrates as a complex particular (Sellars, 1949a), namely, as a complex of particular episodes of its occurrently retained and evolving causal history, as expressed in the above multiplicity of episodic co-selected co-occurring processes that generate and maintain the Socrates-event (by constantly transforming it) through its worldly encounters.[12]

or power in the cause waiting to be expressed in the effect. Rather, it is one of progressively unfolding differentiation which constantly changes (even if minutely) the very spatiotemporal qualitative space which occurrently expresses what we ordinarily call 'the object's power to affect and to be affected'. Transposed to current discussions of causality, expressive causality is closer to models of non-mechanistic, non-linear causality, as the latter is applied to encompassing dynamic organizations, expressing deep context-sensitivities and interdependencies, including loops and networks of non-linear control structures with feedback and thresholds.

[12] Simonelli argues for a similar point in explicating Sellars' distinctive brand of process nominalism. As he puts it: 'Something's being a lion is not its instantiating the general kind

Yet, at this point one might ask how we can understand, in this dynamic processual model, the conditions under which a processual individual comes and ceases to be, as opposed to its undergoing changes in its property-episodes (which do not affect its identity), without subsuming it to kinds, expressive of its general essence? The short answer here is that this difference is *itself* accounted for in processual-dynamic terms, being ultimately a function of the ongoing relative stability (allostatic balance or imbalance) of the interdependent field of forces and the related enabling constraints[13] that dynamically sustain (or fail to do so) the processual individual through its spatiotemporal unfolding. There is no underlying essence or substratum to account for this 'identity through change' of a processual individual. Its identity is something that itself becomes and is expressed by the ongoing activity of the dynamic multiscalar interdependent field of forces around it.

Here, it should be noted that the above introduced new notion of processual identity must be *non-transitive*. In this way a process can preserve and update the heterogeneity of its interdependent 'parts' while holding them together in their becoming—without introducing a new 'object' for representing this 'togetherness'. Perhaps Priest's 'gluonic' theory of composition can serve as a model for this novel processual conception of identity. The gluon is non-transitively identical with each of its parts or forces involved (thus binding them together), but the parts themselves are not equal to one another (Priest, 2014). In this way, a fundamental

lionhood. Rather, it is its doing what lions do, being the locus of the patterns of processes that unfold in the world whenever you have a lion. So, to be a lion is to eat gazelles, to bear live young, to reflect light with a certain frequency, and so on. In this context, the criteria for the application of the sortal term "lion" to some particular lion, conceived of in terms of the pattern of processes that unfolds whenever you have a lion, can be articulated directly in terms of these processes. Sentences that articulate the conceptual content of the expression "lion" such as "Lions are tawny", "Lions are carnivorous", and so on, can be reconceived as functioning to regiment the language such that the patterns of the use of these expressions track the patterns of processes that unfold whenever you have a lion' (Simonell, 2022, p. 1057).

[13] Enabling spatiotemporal constraints are context-dependent constraints that irreversibly link and couple previously separate processes at the same scale as the constraints. By lowering barriers to energy, matter, and information flows such that independent entities become conditional on each other, enabling constraints drive phase transitions to metastable emergent coordination patterns, realized as mutually dependent relations. Examples of these processes range from convection cells to homeostatic regulation in organisms (Juarrero, 2023, pp. 73–74).

assumption of mereology needs to be abandoned, namely, the view that a part of a part of a whole is itself a part of that whole (transitivity).[14] But, in any case, mereology does not seem to fit well with a nominalistic process view of the world in the first place. For example, mereology allows for the formation of new particulars from old ones, independently of spatiotemporal or causal connections, and it has no account of alteration except adding or subtracting parts (moving any of the parts does not change the mereological whole).

It is worth here to be reminded of the radicality of this nominalistically oriented process model of individuation, which fuses the abstract and the concrete, the particular and the universal. Not only is this view directly opposed to determination in terms of the substance framework, on the basis of something's (Felix, Socrates) being a contingently differing instance of the same 'essential form' (cat life-form, human life-form), characterized by specific counterfactual modal properties (capacities, powers, dispositions) standing in relations of material incompatibility and consequence. Even if in a Hegelian manner, these essential forms are dynamically understood as themselves subject to change, gradually 'becoming what they are', the logic of this change would still be bound to a normative unity or identity retrospectively rationalizing the progressive development of the units of change. This is not the case in a resolutely process view of the world in which process is prior to products (e.g. goals, ends) and not just a means for achieving an antecedently determined goal. Normative unities are at best ways of stabilizing, controlling and coordinating ceaseless processual becoming; they are not the latter's ontologico-teleological *raison d'être*.

Another potentially serious objection to the process framework is that it might seem difficult to see how, lacking modal categories altogether, it can accommodate without loss the dispositionally rich properties of ordinary lifeworldly objects understood in terms of the substance framework. In the process framework, ordinary counterfactual propositions about the dispositional properties of objects would be replaced by propositions expressing uniformities in the co-occurrences of particular interconnected processes comprising the object (those that are—synchronically or

[14] See also Seibt (2008) whose process ontology is based on the logically primitive relation 'is part of' which is asymmetric, irreflexive, extensional, but *not* transitive, since a process can be wholly present in *all* of its parts. In Chap. 5 we shall elaborate on why a process ontology ought to reject a classic mereological part-whole logic.

diachronically—implicated in its occurrent dynamic force-field). The substance framework of objects whose natures, by accounting for their dispositional properties, distinguish them from their actual causal history would be replaced by a process framework where objects would be reconceived as *loci* of bundles of (micro-macrophysical) co-present overlapping processes whose 'natures' would be such as to make these dynamic processual individuals *identical* to their (occurrently retained) *actual* causal histories. Processual individuals are essentially path dependent and 'carry their causal histories on their back' as it were. Their very dynamic architecture and characteristics embody the conditions in which they were created and evolved. But it is precisely here that the question arises: How can our dispositionally infused experience of ordinary objects be retained in a process framework in which objects are reconceived as bundles of enveloped co-occurring processes without any dispositional features whatsoever?

The short answer here is that the dispositionally rich part of our experience of the object would be informationally contained in a transposed form in its occurrent dynamic unfolding, which would be now experienced as directly and occurrently expressing its power to affect and to be affected. But, again, how can this notion of 'power'—and our experience of it—be conceived in the context of a process view that eschews ordinary 'iffy' properties and modal categories? Here, process ontology could benefit much from introducing Deleuze's concept of the *virtual* as a successor concept that replaces our ordinary concept—and associated experience—of possibility, as well as the related 'iffy' notions of disposition, capacity and tendency.[15] The notion of the virtual, adapted for our purposes, enables us to make intelligible the view that the powers that our ordinary notions of disposition, capacity and tendency express in an 'iffy' manner (i.e. in the form of an essentially counterfactual if-then inference) can be also expressed in a non-reified, *non*-counterfactual manner as enveloped

[15] One could argue here that the notion of a possible world as is used in possible world semantics is all we need for elucidating the modal concepts of necessity, possibility and impossibility. I do not share this widespread view. Indeed, I think that easy talk of possible worlds in effect obscures the extreme difficulty of understanding what makes it the case that something is possible or impossible, what makes it true that, for example, this glass would have broken had I dropped it. As Heil puts this point: 'To declare that something is possible because it occurs at some world is really just to assert in a pointlessly indirect way that it is possible. If the question is, what is the source of possibility and necessity, what makes claims about possibilities, impossibilities, and necessities true, appeals to possible worlds simply reiterate what is at issue' (Heil, 2021, p. 11).

and dynamically evolving in the *occurrent* spatiotemporal unfolding of a monadic processual individual. The trick here is to disentangle this kind of dynamic *occurrent* unfolding of a process from its mode of existing as a series of *actual* successive states within a metricized spatiotemporal field. This latter view is shared by many Humean actualists and if the process framework is to be understood in this manner, it is difficult to see how powers can be salvaged and not altogether disappear from view; in which case, it is difficult to see how change itself, let alone 'identity through change' (necessary for individuating things even in this minimal Humean 'one damn thing after another' view of the world), would be so much as possible. The Deleuzian virtual thereby avoids the Scylla of collapsing to an actualist Humean ontology which eschews modality altogether, and the Charybdis of a Platonic reification of possibilities as abstract universals (laws of nature, forms, essences) that are merely exemplified in actual events, states and objects in the world.

The virtual, as opposed to the classic notion of possibility, is not a domain of already individuated entities inhering in some abstract realm of potentialities waiting to be realized. Moreover, virtual entities are not conceived on the model of actual entities by abstracting the latter's essential properties from its accidental properties. This way of thinking presupposes that everything that is real must also be possible, and it is unable to explain why that which is possible has not already come into being. By contrast, the virtual makes possible a conception of a power of becoming itself which a) is *occurrently* expressed, b) it is real yet not actual or abstract and c) is dynamic, generative and morphogenetic. An example of a virtual function here might be that of biological evolution that immanently produces the infinite heterogeneity of life-forms. Indeed, at a more general level, evolutionary principles can be said to be virtually operating even at the level of becoming itself, providing a dynamic account of the plane of possibilities itself: each throw of the 'cosmic dice' (i.e. the spatiotemporal unfolding of becoming) is at once a repetition and variation upon the past. The past is 'repeated', 'inherited' and in this sense retained and entrenched in each present moment while at the same time the 'rules', the very field of possibilities that can be actualized, are constantly (even if minutely) changing. In this way, the world is understood as something that can constantly produce novelty which is yet constrained by the course of its past causal history as the latter is occurrently embodied (informationally contained in the form of sedimented and entrenched constraints[16]) in the present. This constrains the constantly variable field of novel future possibilities, since these can be actualized only if they become part of an evolutionary line which is 'convergent' with the virtual past contained in a compressed form in each present moment in the unfolding of each monad (see also Deleuze, 1968, 1969). To use another image, the past history of the world, in the form of differing sedimented and entrenched constraints informationally contained in actual beings, differentiates the attractors and thus the possibility space in which the latter are embedded. These monadic possibility spaces and the likelihood of their being expressed in actuality are not defined or explained in terms of static forms or essences but are themselves under constant—even if usually minute and imperceptible—change and evolution. At the same time, their constant reproduction as information-containing dynamic regimes through repetition from each past to each present moment confers long-term processual stability to the extent to which in the long run the features and dynamics that support resilient-cum-evolvable structures gradually become entrenched.[17] This is a very different, dynamic, image of possibility or power from the traditional one in which everything that is real must also be possible, where possible contents differ from actual ones only in modal status, while being identical in their essence, and where capacities, abilities and dispositions are conceived as static atemporal general essences exemplified in particular things.

[16] For example, bound information in the human genome can be considered as a sedimented record and memory of phenotypes that were successful in earlier contexts. The same goes for inorganic processes (geological traces in volcanoes that continue to produce significant effects on today's climate) as well as for social organizations and institutions (Juarrero, 2023, pp. 143–148).

[17] This is just an application to a cosmic level of evolutionary principles: resilience and evolvability confer selective advantage thanks to their capacity to persist as a coherent unit across a wide range of circumstances and to evolve over longer periods of time. Non-resilient features, dynamics and regimes are selectively extinguished over time. Consider, for example, robustness, namely, the ability of a system to maintain its function despite mutations and to open up vast networks of genotypes that produce the same phenotype giving the opportunity for populations to explore huge areas of genotype space without changing what works (Wagner, 2014). Although natural selection at the organic level presupposes the existence and function of robust organic systems, robustness is itself an object of selection at the inorganic level (e.g. in the context of prebiotic autocatalytic sets) precisely due to its favoring the ability to adapt.

4.3 Is a Nominalist Process Language Possible?

A nominalistic process metaphysics must be accompanied by an equally nominalistic process account of *language* which dispenses with abstract entities such as 'senses', 'contents', 'objects', 'properties', 'facts', etc. Note that senses and propositional contents are things that exist only in actual and obtainable conceptual representings. In other words, they exist in the world only insofar as we conceptually represent it in experience (Kant, 1999; Sellars, 1967), and thus only to the extent to which we represent the world (and ourselves) as *conceptually* articulated, which, in turn, essentially involves the modal categories of the object-property substance framework. This means that a process philosophy which aims to articulate an alternative to the object-property categorial framework, bereft of the latter's modal categories and the conceptual-judgmental articulations of its propositional contents, must sooner or later confront the issue of how a radically nominalistic process language is so much as possible as a coherent alternative to our familiar conceptually articulated language which represents states of affairs by propositional contents articulated in the forms of judgment. In other words, it seems that a serious challenge for a nominalistic process framework is to make intelligible a *contentless language*, namely, a language for saying that 'things are thus and so' without using properties and propositional contents. Since properties are, after all, things that we *say* of things, and propositions are things that we *say*, how can a process nominalism which rejects the existence of both properties and propositional contents (as universals) avoid the paradoxical conclusion that nothing can be said? If the world is a collection of processual individuals, how can we make sense of what we are doing in saying that 'things are thus and so'? (Brandom, 2015, p. 270).

In response, we can say that while we are certainly saying *something*, for example, when we say of a table that it is red, we are not necessarily committed to the view that there is some *thing* that we are saying of it (Sellars, 1980, 2018). 'That is red' does not pick out some *thing* (the property of 'redness' construed as an abstract universal) but functions to characterize what we are *doing* in saying this: making certain sorts of move in the space of reasons—committing us to saying that the table is colored, precluding us from saying that it is green, etc. (Simonelli, 2022). In general outline, a nominalistic process language would have the following form: propositions would not be conceived as expressing contents referring to objects, properties and relations, as constituents of facts, with the latter conceived as truth-makers of these propositions. Rather, they would be understood as 'languagings', as particular (not universal) repeatable (not general) happenings (not objects), identifiable in terms of the difference they make among other linguistic doings—de facto precluding or mandating them in linguistic practice. These linguistic acts would be processual doings of a certain *sort*, namely, *sayings*, due to their functional role in *the game of giving and asking for reasons*, where this functional role would be played by *intrinsic* features of these languagings as processual acts. From this nominalistic point of view, there is no thing ('content' as abstract universal) that is said over and above what is done. Sayings really are just doings of a certain sort, moves in the game of giving and asking for reasons, sustained by a complex collective network of normative attitudes (see Chap. 8).

We can illustrate how this language might function by bringing to mind Sellars' fictional language of Jumblese (Sellars, 1962), which is meant to make perceptually perspicuous the fact that the function of predicates is *not* representational. Predicates (relations, properties) do not stand for special kinds of abstract entities (universals) but are *ways* of saying something *about* what is nameable (referred to), and this is perspicuously shown at the very perceptual level by this language by various different ways of spatiotemporal arrangement of the perceptible qualities of Jumblese 'sentences'. Thus, in this language, the compatibilities and incompatibilities of semantically significant styles are reflected in the compatibilities and incompatibilities of quality spaces. The 'basic singular statements' of this language are thus non-predicative (i.e. not of the subject-predicate form). This language opens up a world of *nameable processual individuals*, not facts, dynamically pictured by natural-linguistic events and the multifarious empirical arrangements and relations in which they stand (Sellars, 1967, pp. 120–125; Brandom, 2015, pp. 269–270). The natural-linguistic events in those arrangements are intelligible as singular terms (names) picturing non-linguistic events in virtue of the holistic functional role they play (by way of intrinsic features of the arrangement of the natural-linguistic events in question) in the inferences relating the sentential sign-designs formed from these natural-linguistic events.

Consider, for example, the basic singular statement '*A*'—written in Sellars' nominalistic Jumblese language. Recall that in this language predicates are dispensable, their functional role being that of helping bringing it about that two (or more) names stand in a certain dyadic (or polyadic) matter-of-factual relation. '*A*' could represent (picture) a certain non-linguistic particular a by virtue of being a-shaped and represent a as being red and triangular by virtue of being italicized and capitalized. In this way, the complex state of affairs that we might represent in a logical calculus by an extended conjunction—'Ra & Ta & Bb & Sb & aLb'—could be represented in a Jumblese-style representational system by a *single composite sign* whose mode of composition is a matter of its parts *simultaneously instantiating* a multiplicity of (representationally relevant) properties and relationships. In this way, it becomes possible to conceive, for example, of our own total visual field at a given time as a single composite state functioning as a pictorially complex patterning resonating with the then and there visually available environmental information.[18] And insofar as they are appropriately caught up in dispositions to inferences (expressing mastery of the game of giving and asking for reasons) and certain courses of action, such sensory or affective states could be said to function as highly complex Jumblese-style 'sentences' (Sellars, 1967, p. 120; Sellars, 1980, p. 57; Rosenberg, 2007, pp. 111–113). Indeed, certain complex neurophysiological brain-patternings, which are now understood analogically with respect to their qualitative structure on the model of inner speech (which transposes in the inner mental realm the semantic properties of overt public language), could be ultimately revealed to be the locus of (intrinsic qualitative properties that *directly* function as) nominalistic-processual Jumblese *thoughts*. Again, here we must always bear in mind that since this nominalistic language must reflect processual categories if it is to be perspicuous, its articulation must be of the verb-adverb variety, where the naming function is performed by verbs expressing occurrences instead of subjects referring to objects ('it reds') while the characterizing function is performed by adverbs expressing spatiotemporal ways of occurring instead of predicates representing properties ('tablely'). Recall that this language is not subsumptive: it does not subsume particular objects (substances) to general properties and relations (universals) through the use of the copula (is), which is another way of saying that it does not objectify predication by attributing to it a special semantic referential function. Thus, although the statements of this language have a this-such

[18] The resonance in question is 'clear yet confused' if environmental information is conceived as affordance-like, and 'distinct yet obscure' if it is conceived as preserving the structure while sensorily transforming the intrinsic qualities of its external source (see Fig. 3.1). In what follows, I use the Sellarsian term 'picturing' to refer to *both* these dimensions of resonance relations between mind and world.

articulation, they do not directly function as judgments if for no other reason than that they lack the subsumptive function of the copula.

The way in which these process-like nominalistic 'statements' are more-than-causally yet non-judgmentally related to the world can be elucidated if we recall that picturing, in Sellars' technical sense, is a non-intentional, material correlation between minds, considered as natural processes and their physical and social environment, also considered as natural processes (in abstraction from their conceptual form). Picturing relations, being resolutely non-normative and non-intentional, are *not* reference relations. Reference is a semantic concept with normative connotations: how things are set standards for the normative assessment of the truth of our judgments, the material correctness of our inferences and the success of our actions. On Sellars' account, referential correctness presupposes pictorial adequacy, where the latter is conceived in non-normative terms, for example, as an evolutionarily developed 'mapping', itself regulated by relations of dynamic homeostatic equilibrium between brain processes and the environment (Sellars, 1960b). If we model nominalistic language after Sellarsian Jumblese, we can say that this language would be 'about' the world in the sense of dynamically *picturing* it, not in the intentional sense in which a propositional content (sense) semantically represents a fact (referent).

Here it must be stressed that since the statements of this nominalistic language would not directly function as ordinary descriptions or explanations, as they would be bereft of 'iffy' structure.[19] Yet, since the process language in question would possess counterpart 'non-iffy' modal categories (modeled off the Deleuzian virtual), it could also be said to function on the basis of counterpart (monadic) notions of description and explanation: the latter would be dynamically embodied in languagings in the mode of monadic world-stories as expressed by the language's method of projection (see below). The statements of this language would also not directly function as judgments, claims or assertions about the truth or falsity of states of affairs, and they would not represent the world through negation. Nominalistic assenting or dissenting acts can be in error only in

[19] According to this Sellarsian view, which I take to be in the right direction, offering any empirical description, attributing any empirical property, involves modal commitments, that is, commitments as to what *would* happen to what is so described under various circumstances, what would be true of it if various other claims were true (Sellars, 1957). For example, the ordinary descriptive sentence 'That lion is sleeping lightly' has as necessary conditions that some moderate stimulus (e.g. a sufficiently loud noise, bright light) would wake the lion.

the sense that the defeater of atomic this-such languagings is not the *fact* that things are *not* as the sign depicts them, but that the *pictured complex of processes* are *other* than as the sign as complex languaging pictures it (an otherness which is itself a matter of degree). The linguistic functions of description and explanation can be nominalistically cashed out only in terms of the adequacy of the *pictorial method of projection*—that is, of a whole system of patternings of perceivings, inferrings, intendings and actings that these functions holistically instantiate at the non-conceptual level of brain-body-environmental resonating processes—and of our success in navigating the environment with the aid of this method of projection and the world-story it embodies. Note here again that unlike the relation between senses and referents which is *normative* and can be decided on a case-by-case basis in an absolute manner (where sentences are considered as judgments which can be true or false depending on whether the facts asserted in the propositional content involved obtain or do not obtain) the crucial notion of pictorial adequacy is *non-normative*, comes in degrees involving intensive gradients (as opposed to binary oppositions) and is essentially *holistic*: it would be cashed out by the practical success of the method of projection instantiated by the patternings of linguistic doings involved in the relevant linguistic assenting and dissenting acts in orienting our brain-body complexes in—that is, in maintaining dynamic homeostatic equilibrium with—the surrounding physical and social environment.

Here, it must be stressed that, as we shall see in Chap. 8, the dynamicity of the picturing embodied in this nominalistic language (which, importantly, includes the shared picturing embodied in our scorekeeping inferential practices) does not consist in its passively registering or 'chasing after' the dynamic structure and changes of its targets but is essentially *projective*. It maps a relatively independent territory (the world, our body, social deontic attitudes) needed for satisfying the fundamental needs of an organism only by occurrently and dynamically *anticipating* the imminent future course of itself in relation to the physical, bodily and social environment, acting on this anticipation and correcting its method of projection accordingly. In this way, picturing has an essentially active, performative and world-transforming dimension which proves necessary for its ultimate *raison d'être*, namely, the striking of ever-expanding cycles of allostatic balance with the surrounding physical, bodily and social environment. Thus, a nominalistic language that dynamically pictures the world has an essentially *hyperstitional* character (Chap. 8). It also has an equally essential *affective* dimension: its abovementioned *raison d'être* is best served by our achieving ever-increasing degrees of freedom in the exercise of our sensory and intellectual powers, which at the affective level are experienced as joyful ends-in-themselves. Recall that the complex patternings of languagings whose method of projection dynamically embodies and anticipates worldly processual becoming are understood as *doings* of a certain character, where this character is cashed out in terms of the languagings' power to affect and be affected (e.g. by inducing and forging specific arrays of associative connections and salience patterns with other languagings, perceivings, intendings, etc.). Thus, from this point of view, the affective dimension of languagings, though distinguishable from their projective picturing function, is inseparable from it, since it is an essential component of their very identity and individuation and of the very *raison d'être* of the projective picturing function of language itself.

4.4 Processual Becoming as Self-Differentiating and Self-Fracturing

As we saw, the process view of the world is an essentially dynamic one which overcomes the particulate image of the world of discrete substances (e.g. particles) bumping into each other on the model of billiard-balls. This is eloquently shown if we construe basic processual individuals on the model of (electromagnetic, gravitational, vector, tensor) fields. Spacetime is itself a structural quality of (gravitational) fields, while physical particles are quantized excitations of a field (Birkhard, 2003). Moreover, fields, in virtue of their intrinsic nature, provide the categorical basis for the dispositional nature of forces. For instance, a particle placed in a field will be subjected to a certain force because there is an electromagnetic field present in that region. With this model in mind, we might reconceive reality as a tissue of fields which can interact and hang together in ways that free us from the substance paradigm. For example, we can model interaction among absolute processes on the sorts of interference that results when different fields are overlapped (Rubinstein, 1997; Heil, 2021). Note that in this process view of the world, matter is not 'dead', passive, inert or undifferentiated, waiting to be formed (determined) by something other than itself (active intelligible forms), but is understood in terms of fields with *differences in potential* that can intrinsically induce movement and change.

This view of the world as a ceaseless becoming in which objects as discreet particulars with fixed properties are derivative practically useful abstractions, relatively stable—yet not substantial or eternal—axial points of reference for our successful collective coordination, has important consequences for our conception on individuation. As Simondon has argued in this connection, individuation must itself be viewed as a *process* contemporaneous with that which it individuates and not as something that characterizes a completed product. From this point of view, particular beings *individuate* as they *become*. They are contemporaneous with their becoming in the sense that becoming is not the becoming of an already individuated being but the becoming of the being's individuation. Since it is not determined by a pre-existing essence that just unfolds itself in becoming, individuation itself becomes together with that which it individuates (Simondon, 2020, pp. 364–365, DeLanda, 2002). This is connected to the fact that from this processual point of view, being itself has an essential pre-individual dimension in an energetic state of overtension the release of which comes in the form of its immanently self-differentiating (self-individuating) in beings. Thus understood, individuation is not static, but itself becomes and never exhausts the pre-individual's potential. As Simondon himself puts this point: 'The individual is grasped as a relative reality, a certain phase of being which supposes a pre-individual reality prior to it and which, even after individuation, does not fully exist by itself, for individuation does not exhaust in a single stroke the potentials of pre-individual reality' (Simondon, 2020, p. 7). Pre-individual reality is essentially self-differentiated. This is because being as pre-individual is conceived in terms of *supersaturated potential charges* within a metastable system. It is precisely the degradation of energy sequential to a state of overtension of the system that progressively self-differentiates and self-individuates being into beings. The phase-shifting of a metastable system charged with a supersaturated energetic potential individualizes while also simultaneously spouting—from its not-yet-individualized internal tensions—a profusion of individualizing forms, which are then capable of being structured into further systems and reforming into new metastable equilibria.

This progressive self-differentiation and self-determination of pre-individual being into individuated beings can be understood as the progressive determination of a problematic field into certain (always partial) (re)solutions, namely, individuated actual beings. This problem-solution dialectic and terminology comes from Deleuze (1968, 1969) and Simondon (2020), and it is to be understood ontologically as inhering in the very fabric of being, expressing its evolutionary-selectional dimension at all levels of description and spatiotemporal scales. *Contra* pragmatism, problems and solutions in this sense should not be understood anthropomorphically: they are not essentially relative to human interests, and do not merely express subjective epistemic limitations of the evolutionary and cultural development of our species. Human interests themselves emerge as (provisional) evolutionary solutions to problematic situations (evolutionary pressures) whose resolution *is* the human organism and its distinctive way of life. This is not to say that problems in this non-anthropomorphic sense are not intimately and internally *related* (without being *relative*) to human interests (this is the grain of truth in pragmatism). Our explicitly formulated questions in inchoately encountering a recalcitrant experiential situation which cannot directly be captured by our ready-made categories of thought, namely, questions of the form 'what is going on with this?', 'why is this happening?', 'how is this going to evolve?', are conceptually articulated responses (inaugurating a line of investigation and further conceptual and categorial articulation for finding an answer to the question) precisely to worldly problematic situations as they non-conceptually impinge on and affect us through experience.

To understand the problem-solution dialectic in its non-anthropomorphic dimension it is instructive to focus on evolutionary adaptive processes. Think, for example, of how a growth of a tree involves a maximization problem, namely, how to maximize exposure to sunlight. The tree's growth traces its way through the shifting environment, but each moment of growth, occurring at an imperceptible (for us) scale of time, is changing the very environment in which the sunlight is maximizable by hoisting its branches. Of course, the tree does not deliberately settle on this strategy. The tree *is*, in a sense, this strategy (also in response to other problems that make up its becoming), progressively determining the conditions of the problem (the possible differential interconnections between cells, water and sunlight) whose provisional solution it *is*, while producing its own milieu in the process (a world—an *umwelt*—composed of sunlight, air, water, dirt and growth).

Moreover, there is no reason for thinking that the problem-solution dialectic in this non-anthropomorphic sense is limited to the domain of sentient or organic beings. In this more commonsensical interpretation of this dialectic, we would conceive of problems non-anthropomorphically yet essentially involving the tendency to satisfy a certain need activated by the registering of a disturbance in the homeostatic regulation of an organism. Although here we are in a highly speculative territory, I think that provisionally at least, we must resist the tendency to completely dislodge the realm of the organic from the realm of inorganic, viewing the latter as inert matter, with no internal self-organizing dynamics and intrinsic properties and as comprising passive objects with no powers of their own that are individuated exclusively by external spatiotemporal features and by their conforming to external forces (natural laws). Indeed, we can conceive of the essential differences of the organic from the inorganic level, epitomized in the former's built-in homeostatic regulatory mechanisms for maintaining a complex self-organizational dynamics, and the different qualitative forms it takes (sensory organs, affective states), as *themselves* being cases of *novel emerging solutions* to supersaturated energetic tensions (differences of potential) occurring at the inorganic level itself.[20]

I take this move to be justified by the theoretical benefits that a monistic univocal ontology confers over an essentially equivocal ontology characterized by incommensurable ontological regions and categories, as in the latter case the relation between these two categories of being (in this case between organic and non-organic) becomes incomprehensible. This is not to say that the organic level, especially in its sentient dimension (organisms endowed with sense organs), cannot be literally an emergent level in relation to the inorganic. As we shall see in Chap. 5, this might well be correct, and not only in the case of the relation between the organic and the inorganic. The real issue is to understand how and in what sense emergence is possible within a resolutely univocal (and in our case nominalistic) ontology, which accounts for the plurality of beings and the distinctive character of different regions of being (inorganic, organic, sentient, sapient) without making their emergence a brute fact and the relations between them unintelligible.

A word here is in order about a major theme running through this book, namely, the self-fracturing of being. This self-fracturing takes the

[20] For instance, Simondon argues that individuation at the level of organic beings retains and expands the most precocious phase of individuation at the level of inorganic beings (e.g. instability conditions in crystal growth), such that the organic level would be the inorganic level in suspense, slowed down in its process and indefinitely expanded (Simondon, 2020, pp. 163–164). Obviously, my purpose here is not to commit myself to the literal truth of this particular explanation of the emergence of the organic from the inorganic, but to give an example of a possible form of explanation of this kind of emergence within a resolutely immanent, univocal process ontology. See also Chap. 5.

form of self-differentiated being immanently expressed in a plurality of beings. The theme of self-fracturing is intimately connected to that of emergence, and even more fundamentally, to the very *possibility* of processual becoming and evolution. For only if being is conceived as essentially self-relating and self-affecting can we view it as becoming and as essentially evolving. As we shall see in Chap. 5, at the most basic level, emergence—and even becoming—itself can be usefully understood in terms of self-fracturing, that is, of a *folding* of being upon itself, a continuous deformation of its supersaturated potential as an intensive multiplicity, immanently expressed in the form of monadic actual *umwelten* beings, that is, microcosms enacting their own spacetimes and confusedly expressing each in their own way 'the whole world'. Folding and unfolding—in the modes of tension and release (Simondon, 2020), contraction and relaxation (Bergson, 1907), slowing down and speeding up of a spatiotemporal process (altering the processes' rate of change), involution and evolution (Deleuze, 1968; Deleuze & Guattari, 1980)—express the manner in which being affects itself, thereby spouting beings in a discontinuous manner (emergence) through a continuous self-differentiating process (folding-unfolding). From this standpoint, the fold is 'a continuous labyrinth [which] is not a line dissolving into independent points, as flowing sand might dissolve into grains, but resembles a sheet of paper divided into infinite folds or separated into bending movements, each one determined by the consistent or conspiring surrounding' (Deleuze, 1993, p. 18). Folding is essentially a doubling of a single surface from which its two opposing sides continuously emerge (e.g. interior-exterior, past-present). This self-fracturing (folding, self-contraction) of being initiates the very topological distinction between 'inside' and 'outside' placing it at the very heart of beings. An 'inside' that 'is not something other than the outside, but precisely the inside of the outside' (Deleuze, 2000, pp. 96–97). In a word, the inside is none other than a fold of the outside, and not a self-sufficient enclosed subjective space. From this point of view, the very opening (unfolding) of a dimension of inorganic nature is to be itself viewed in terms of emergence of monadic *umwelten* beings, initiating a primitive topological distinction between 'interior' and 'exterior', that is, between individual and *milieu*. Interestingly, according to this line of thought, the organism-environment relation is a *further development* of this primitive relation (overcoming its inherent limitations) and not a brute fact emerging from merely passive inorganic matter bereft of internal structure, intrinsic character and intensive differentiation. This move, which lies at the heart of a univocal process ontology, in the spirit of Whitehead (1967, 1978) and Deleuze (1968), aims to overcome the division between a mechanistic and an organicist vitalist worldview.

Finally, it must be stressed that the above account of processual becoming in terms of a folding-cum-unfolding immanent relation between being and beings does not bake experience or thought into the clay of reality. What is baked into the clay of reality is only the distinction between interior and exterior or monadic individual and external *milieu*. These structural features of experience are indeed taken to be intrinsic to being-as-unfolded-in-beings. But these distinctions and the notion of perspective involved are more primitive than and do not presuppose the distinction between *experiential* perspective and the world or between *thinking* and the world. In fact, it is precisely these latter perspectival phenomena operative in sentient organisms (qualitative experience) and human beings (symbolic thinking) that presuppose the primitive inorganic monadic spatiotemporal perspective and not the other way around. Being's immanent self-fracturing is something that *itself develops* as a self-relation, taking different forms over time, with sentience (qualitative experience) and sapience (thought) being relatively recent novel ways of self-fracturing, of perspectival ways of world-making.

References

Ardoline, M. (2024). *Deleuze, mathematics, metaphysics: Difference and necessity*. Edinburgh University Press.

Bergson, H. (1907). *Creative evolution*. Dover Publications.

Bergson, H. (2002). *Henri Bergson: Key writings*. Bloomsbury.

Birkhard, M. (2003). Variations in variation and selection: The ubiquity of the variation-and- selective retention ratchet in emergent organizational complexity, part II: Quantum field theory. *Foundations of Science, 8*(3), 283–293.

Boyle, M. (2012). Essentially rational animals. In A. Günter & J. Conant (Eds.), *Rethinking epistemology* (pp. 395–428). Walter de Gruyter Verlag.

Boyle, M. (2016). Additive theories of rationality: A critique. *European Journal of Philosophy, 24*(3), 527–555.

Brandom, R. (2015). *From empiricism to expressivism: Brandom reads Sellars*. Harvard University Press.

Brandom, R. (2019). *A spirit of trust: A reading of hegel's phenomenology*. Harvard University Press.

Davidson, D. (1980). The logical form of action sentences. In D. Davidson (Ed.), *Essays on actions and events*. Clarendon Press.

DeLanda, M. (2002). *Intensive science and virtual philosophy.* Bloomsbury Publishing.
Deleuze, G. (1968). *Difference and repetition.* Columbia University Press.
Deleuze, G. (1969). *The logic of sense.* Columbia University Press.
Deleuze, G. (1993). *The fold: Leibniz and the baroque.* The Athlone Press.
Deleuze, G. (2000). *Foucault.* University of Minnesota Press.
Deleuze, G., & Guattari, F. (1980). *A thousand plateaus: Capitalism and schizophrenia.* University of Minnesota Press.
Heil, J. (2021). *Appearance in reality.* Oxford University Press.
Juarrero, A. (2023). *Context changes everything: How constraints create coherence.* MIT Press.
Kant, I. (1999). *Critique of pure reason.* Cambridge University Press.
Morganti, M. (2012). Sellarsian particulars. *Acta Analytica, 27*(3), 293–306.
Nunziante, A. (2021). Tropes variations: The topic of particulars beyond Sellars' myth of the given. *Synthese, 199*(5-6), 12019–12043.
Pippin, R. (2018). *Hegel's realm of shadows: Logic as metaphysics in the science of logic.* The University of Chicago Press.
Priest, G. (2014). *One: Being an investigation into the unity of reality and of its parts, including the singular object which is nothingness.* Oxford University Press.
Röck, T. (2022). *Dynamic realism: Uncovering the reality of becoming through phenomenology and process philosophy.* Edinburgh University Press.
Rosenberg, J. (2007). *Wilfrid Sellars: Fusing the images.* Oxford University Press.
Rubinstein, E. (1997). Absolute processes: A nominalist alternative. *Southern Journal of Philosophy, 35*(4), 539–556.
Seibt, J. (2008). Beyond endurance and perdurance: Recurrent dynamics. In C. Kanzian (Ed.), *Persistence* (pp. 133–164). Ontos Verlag.
Seibt, J. (2016). How to naturalize sensory consciousness and intentionality within a process monism with a normative gradient: A reading of Sellars. In J. Oshea (Ed.), *Sellars and his legacy* (pp. 187–221). Oxford University Press.
Sellars, W. (1949a). On the logic of complex particulars. *Mind, 58*(231), 306–338.
Sellars, W. (1949b). Aristotelian philosophies of mind. In R. W. Sellars, V. G. McGill, & M. Farber (Eds.), *Philosophy for the future: The quest for modern materialism* (pp. 544–570). Macmillan.
Sellars, W. (1957). Counterfactuals, dispositions and the causal modalities. In H. Feigl, M. Scriven, & G. Maxwell (Eds.), *Minnesota studies in the philosophy of science* (Vol. II, pp. 225–308). University of Minnesota Press.
Sellars, W. (1960a). Grammar and existence: A preface to ontology. *Mind, 69*(276), 499–533.
Sellars, W. (1960b). Being and being known. *Proceedings of the American Catholic Philosophical Association, 34*, 28–49.
Sellars, W. (1962). Naming and saying. *Philosophy of Science, 29*(1), 7–26.

Sellars, W. (1963). Philosophy and the scientific image of man. In W. Sellars (Ed.), *Science, perception and reality* (pp. 1–40). Routledge.

Sellars, W. (1967). *Science and metaphysics: Variations on kantian themes*. Ridgeview Publishing Company.

Sellars, W. (1980). *Naturalism and ontology*. Ridgeview Publishing Company.

Sellars, W. (1981). Foundations for a metaphysics of pure process. *The Monist, 64*(1), 3–90.

Sellars, W. (2018). In P. V. Amaral (Ed.), *Wilfrid Sellars notre dame lectures 1969–1986*. Ridgeview Publishing Company.

Simondon, G. (2020). *Individuation in light of notions of form and information*. University of Minnesota Press.

Simonell, R. (2022). "Sellars's Ontological Nominalism". *European Journal of Philosophy, 30*(3), 1041–1061.

Wagner, A. (2014). *The arrival of the fittest: Solving evolution's greatest puzzle*. Penguin.

Whitehead, A. N. (1967). *Science and the modern world*. Free Press.

Whitehead, A. N. (1978). *Process and reality*. Macmillan.

Williams, D. C. (1953). On the elements of being, part I and II. *Review of Metaphysics, 7*(3-18), 171–192.

CHAPTER 5

Nominalism, Emergence and Process

In Chap. 4, we mentioned that at the most basic level, being expresses itself in the form of a ceaseless becoming of actual beings, and that this self-expression of being can be usefully understood in terms of self-fracturing, that is, of a folding of being upon itself. We saw that this folding is a continuous deformation of the supersaturated potential of being as an intensive multiplicity and is immanently expressed in the form of monadic actual *umwelten* beings, that is, microcosms enacting their own spacetimes and confusedly expressing, each in their own way, the whole world. We also briefly mentioned certain modes of folding and unfolding—namely, tension and release, contraction and dilution, slowing down and speeding up of a spatiotemporal process (altering the processes' rate of change), involution and evolution. These express the way in which being affects itself, spouting beings in a discontinuous manner through a continuous self-differentiating process (folding-unfolding). This way of understanding emergence, namely, in terms of a discontinuity resulting from underlying modes of continuity, can be also helpful for making sense of the emergence of different strata *within* the realm of actual beings: the emergence of living sentient beings out of inorganic ones, the emergence of symbolic thinking (sapience) out of sentient living beings, etc. In this chapter, I shall attempt to show how emergence of this (diachronic) kind is both real (causally efficacious) and non-mysterious if it is understood in topological terms, that is, as folding, contraction, relaxation, slowing

© The Author(s), under exclusive license to Springer Nature Switzerland AG 2025

D. Christias, *Towards a Hyperstitional Process Nominalism*, Palgrave Perspectives on Process Philosophy,
https://doi.org/10.1007/978-3-032-06350-2_5

down or speeding up of certain spatiotemporal developmental processes. Correlatively, it will be argued that any viable model of emergence must be such as to avoid the shortcomings of most such accounts on offer in the contemporary scene. These are (1) the nearly inevitable epiphenomenalist implications of most non-reductive naturalist views and (2) an appeal to non-explanatory relations of supervenience between different ontological levels (Heil, 2021; Kim, 1993). I take it that the topological understanding of emergence successfully avoids these problems. Moreover, I examine whether and in what sense can process nominalism account for levels-emergence. On the face of it, strict nominalism seems to allow only for the emergence of non-universal structures and properties based on combinations of a finite stock of actual basic particulars. How can these limited resources account for emergence and novelty? Here I use Deleuze's notion of intensity (Deleuze, 1968), stressing its topological properties, as a way of developing a non-traditional intensive processual nominalism which can account for emergent levels of being in a nominalistically kosher way. In this connection, I attempt to show that Deleuzian intensity is the key for understanding the monadic, intrinsic and perspectival character of inorganic processes and the possibility of its being contracted and qualitatively transformed in its interaction with sentient beings, while at the same time avoiding panexperientialism (endowing inorganic processes with qualitative experiences). Further, I highlight an important implication of my view about physicalism and panpsychism. Both physicalism and panpsychism are reductive positions that rightly reject overly simplistic (e.g. non-reductive physicalist, supervenience-based) accounts of emergence, yet wrongly concluding that this amounts to a rejection of emergence *tout court*. It is suggested that both physicalism and panpsychism are blind to the possibility of emerging discontinuity from underlying continuity—for example, the emergence of discontinuous actual levels (strata) from topologically continuous deformations of intensive spatiotemporal processes. Lastly, I situate my position on emergence with respect to a view very close in spirit to it, namely, Juarrero's (2023) recent account of emergence as resulting not from *sui generis* forceful causes (efficient causes transferring kinetic energy through impact from cause to effect) but from the imposition of limiting or enabling spatiotemporal constraints on the hitherto existing possibility space.

5.1 Is Nominalism Compatible with Emergence?

As we saw in Chap. 4, process nominalism, while being a resolutely univocal ontology, allows for real—causally efficacious—emergence in the world. Different levels of self-organizing processes, namely, inorganic, organic, sentient and sapient ones, can all be thought of *both* as characterized by univocal ontological principles, and as emergent levels with the power to act as such on the real world. Yet, how can *nominalism*, even if it goes hand in hand with a process ontology, account for levels-emergence? On the face of it, strict nominalism seems to allow only for the emergence of non-universal structures and properties based on combinations of a finite stock of actual basic particulars. How can these limited resources account for emergence and novelty?

The short answer to this question is threefold:

1. The nominalism endorsed in this book is a non-traditional form of nominalism that, unlike classic empiricist nominalism, does not eliminate universals altogether and thus does not hold that only particulars exist. Rather, it *reconceives* universals and their relation to concrete particulars in a way that (a) universals cease to be abstract or general entities inhering or exemplified in particulars, and (b) universals do not stand in a relation of logical mediation to concrete particulars while particulars are not conceived as actual instances logically subsumed under universal essences. As we shall see, it is precisely the topological properties of these non-traditional universals (tension-release, contraction-relaxation, spatiotemporal slowing down or speeding up of morphogenetic processes) and the ensuing deformation of the whole spatiotemporal field in the context of which actual processes unfold, that accounts for the emergence of novel properties (and possibility spaces) in actuality in a strictly immanent manner, that is, without invoking generalities or abstractions operating as *sui generis* top-down efficient causes which would be transcendent with respect to their field of application and whose causal intervention in the world would violate conservation laws of physics.
2. Correlatively, nominalism in our non-traditional sense is not committed to—and indeed is committed to rejecting—the view that at the most fundamental ontological level what exists are ready-made unchangeable actual entities (actual particulars) which explain

change in terms of combinations of unchangeable parts. Specifically, nominalism of this sort rejects a simplistic 'cooking model' for understanding emergence and novelty. According to this model, (a) novel properties are combinations of already existing properties, the emergent element consisting in the novel mixture produced as a new whole or unity out of these properties, where the latter are now conceived as differently organized parts of this new whole, or (b) novel properties are completely new externally added features over and above those already existing properties, in which case the new emergent whole does not have these (however 'differently organized') properties preceding its existence as proper parts and neither do these latter properties have any intelligible relation or connection to the emergent ones beyond being their antecedent causal conditions of emergence. Option (b) is meant to capture in imagistic terms the view that emergent wholes operate as *sui generis* top-down efficient causes, while option (a) captures the intuition that they arise by altering the organization of their parts. As we shall see in Sect. 5.3, this simplistic picture in which existing and emergent properties are understood on the model of distinct ingredients of a cooking recipe, the implementation of which produces novel outcomes by mixing already existing ingredients or by externally adding new ones, can be said to constitute a common background presupposition of both physicalist and panpsychist positions that make them blind to the possibility of accounting for the radical discontinuity introduced by emergent properties in actuality by appeal to immanent as opposed to transcendent causes, continuous deformations as opposed to discontinuous ruptures.

3. The nominalistic rejection of any form of final, formal or efficient causes on the model of human agency and the reconceptualization of efficient causes in mechanistic and non-agentive terms is historically connected with the rise of the 'impact' paradigm of causation and mechanistic explanation in which causal influence is understood on the 'push-and-pull' model of physical interaction. However, nominalism is not metaphysically committed to the existence of efficient causation or to any specific view about its nature. Indeed, we can very well reject this restrictive understanding of causation and even reject the causal closure of the physical *to the extent to which* it is understood as the logical consequence of projecting this mechanistic understanding of causal explanation to everything that exists,

while *remaining* nominalists. Recall (Chap. 3) that the impact paradigm of causation is one of the main reasons why the qualitative (sensory-affective) features of consciousness cannot be considered as *both* emergent and perfectly naturalistically respectable properties within a scientific worldview (Sellars, 1981). The only causal constraint put on emergent properties by nominalism is that they make a causal difference to the world as concrete processual individuals and not as abstract types, generals or kinds standing in real relations of 'exemplification' to actual particulars (their 'tokens'). The precise nature of nominalistic causation is left open and is even consistent with models that reject efficient causation (transfering kinetic energy through impact from cause to effect) altogether as an account of emergence. Also, it must be stressed that nominalism can accept that there are perfectly legitimate normatively induced logical relations (type-token, exemplification relation, genus-species-individual) between universals and particulars. It's just that these logical relations must not be ontologically reified, since then we risk attributing causal force to all kinds of normatively constructed abstract universals. Note moreover that to reject the causal closure of the physical as a principle operating under an impact-based paradigm of causation, where efficient causes are understood as transferring kinetic energy through impact from cause to effect, does not imply a corresponding rejection of the conservation laws of physics, since the validity of the latter does not essentially depend on a specific paradigm of understanding causation and causal interaction.

For the time being, the above three points might suffice for avoiding the temptation to equate nominalism with a reductionism (of the form 'only particulars and combinations of particulars exist') hostile to emergence. I shall attempt to show how nominalism can provide a plausible understanding of emergence in more detail in Sect. 5.4. Let me just briefly point out that the rejection of the traditional nominalist model of ready-made basic particulars and of the cooking model of emergence goes hand in hand with a rejection of the mereological and supervenient conception of physical and mental causation (supervenience presupposes classic mereology and a traditional substance ontology). As we shall see in Sect. 5.2, mereological and supervenient-based accounts of emergence can only conceive the latter either as epiphenomenal or as a *sui generis* causal power that mysteriously inverts the logic of mereology and supervenience at the level of the mental.

5.2 Problems of Mereological and Supervenient-Based Accounts of Emergence

In this section I shall briefly comment on why mereological and supervenient-based accounts of physical and mental causation make emergence inexplicable and mysterious and must thus be rejected if one wants to understand emergence. These accounts have been highly influential, especially in analytic philosophy, because they seem to provide a way for articulating a scientifically informed naturalistic view which nevertheless makes room for the reality of sensory states and mental phenomena characteristic of sentient and sapient beings. This non-reductive naturalistic view offers itself as a corrective to an extreme physicalism which reduces mental phenomena and sensory states to scientifically respectable properties bereft of the qualitative structure of sensory states and the normative dimension of thinking. This is perhaps why it is still one of the dominant ideological trends in contemporary analytic philosophy of mind.[1]

Now, the problems of mereological and supervenient accounts of emergence are at bottom problems about the causal efficacy of emergent properties (e.g. mental states) over the lower-level properties of their component parts (e.g. neurophysiological states). How can emergent properties be causally efficacious *as such*, as opposed to being surface effects of the interactive dynamic arrangements of their propertied parts? On the current model of understanding causation, namely, as efficient causation that works by transferring kinetic energy through impact from cause to effect, for an emergent whole to be causally efficacious over its component parts is for it to be able to exert downward causation over them, that is, top-down whole-to-part causal influence which outstrips the causal powers and relations of those parts.

[1] I will not discuss non-reductive so-called liberal naturalist views (De Caro & Macarthur, 2010; McDowell, 1994) that attempt to dissolve the problem of accounting for the irreducibility and *sui generis* nature of the causality involved in mental properties by construing this *sui generis*, conceptually and normatively articulated dimension of mental states as an aspect of nature itself (see e.g. McDowell's 'second nature'), where the latter is conceived as the 'material isomorph' of the logical powers of thinking. To my mind, these views solve the problem of the emergence of mind in nature only verbally since they effectively transpose this problem *within* nature (as an irreducible emergence of second nature out of first nature).

5.2.1 Platitudinous Vs Robust Mereological Emergence

Here we must immediately pause to explain why certain more commonsensical conceptions of emergence would be ontologically unilluminating as accounts of emergence. For example, a property is not emergent in a philosophically interesting sense just because there are truths about wholes not applicable to, or derivable from truths about their parts, or predicates that apply truly to wholes but not to their parts. Consider, for instance, the fact that a tomato has the power to roll owing to its spherical shape. The tomato's parts, however, need not be spherical and need not themselves have the power to roll. Yet it does not seem that the tomato, or the tomato's shape, is, in any interesting sense, ontologically emergent. The tomato and its shape are just what you get when you put the tomato's parts together in a particular way, in a dynamic network of connections and—chemical, electromagnetic, gravitational—interrelations (Heil, 2021, p. 136). Yet, to be emergent in a philosophically interesting sense, characteristics of wholes must be something *in addition* to interactive dynamic arrangements of their propertied parts.

It seems that if we are to exclude this platitudinous sense in which the tomato's shape is 'emergent' upon the dynamic network of interrelations of its parts, the only thing we might mean when we say that the novel characteristics of emergent wholes are something *over and above* the interactive dynamic arrangements of the properties and powers of their parts, is that this emergent whole must be such as to exert *efficient downward causation* to their lower-level component parts. Yet it is difficult to see how they could be effective causes in the above manner unless these novel emergent properties introduce novel (mental) powers and laws into the universe that break or alter the course of their component parts (neurophysiological states). But since this implies the existence of an inexplicable transcendent causation that violates fundamental conservation laws, it seems an exceedingly high price to pay; the costs clearly outweigh the benefits of this transcendent top-down emergentism.

On the other hand, as we saw earlier, to say that a property is emergent just because there are truths about wholes not applicable to or derivable from truths about their parts (the power of the tomato to roll due to its spherical shape) is to use the concept of emergence in a philosophically unilluminating manner since the novel properties of these mereological wholes seem nothing more than a shorthand for the interactive dynamic arrangements of their propertied parts, and do not exert any downward causation to their component parts. This platitudinous mereological emergentism does not seem to have direct ontologically emergentist consequences. Parts of the whole can well interact causally with one another and with entities outside the whole, and the whole can undergo vast changes as a result of these interactions. The same parts, differently organized, might constitute very different kinds of wholes. But this by itself does not show that the wholes themselves could somehow act independently of the parts that make them up (Heil, 2021, pp. 46–47).

Thus, mereological emergentism seems to come in two equally implausible forms: (1) Platitudinous mereological emergentism which as we saw does not have direct ontologically emergentist consequences, and (2) robust mereological emergentism according to which emergent properties exert downward causation and introduce novel powers and laws in the universe (e.g. mental over physical ones), but which entails a highly controversial form of transcendent causation that violates the causal closure of the physical, breaks conservation laws and makes mysterious and inexplicable as a matter of principle the interaction between emergent wholes (e.g. mental states) and its low-level component parts (physical states). This view is effectively committed to a kind of property dualism in which it becomes unintelligible how emergent mental properties *can so much as be* the right kind of thing as to affect and be affected by physical (bodily, neurophysiological) properties.

5.2.2 Process and Mereological Emergence

Before turning to supervenience-based emergence and its problems, it is important here to stress how differently we would understand emergence in the context of a process ontology. This is because our conception of mereological relations largely takes for granted a substance-based ontology. But a change of paradigm in ontology entails a change in our conception of mereological relations. Think, for example, that what we regard as complex wholes turns out to be, for example, process-like disturbances in fields. In this case, such wholes could be described as having parts, but these parts would not make up wholes as particles might be thought to make up an ordinary object such as a rabbit or a tomato. Their relation to wholes would be analogous to the relation regions of space or spatial points bear to space: the former depend on the latter, not the other way round. Thus, parts of complex objects, and indeed the objects themselves, would turn out to be aspects or modifications of processes, just as the crest of a wave might be thought as an aspect of the wave, and not as some kind of independently individuated part out of which the wave is materially composed or constructed. Moreover, a process as such could be said to involve or be 'composed' of many sub- or supra-processes at multiple spatiotemporal scales, where the latter are best conceived not as 'parts' but as *ingredients* of the process in question. And ingredients relations, *pace* part-whole ones, are *not* necessarily spatially or temporally homogeneous (as points are ingredients but not part of a line, lines are ingredients but not part of a surface, spatial points are ingredients but not parts of space). It is thus obvious that emergence would be treated very differently in a process framework. For example, it would exclude *both* platitudinous and robust mereological emergentism as described above.

5.2.3 *Supervenience and Emergence*

I now turn to supervenient-based accounts of emergence and their problems. As is well known, notions of supervenience are attempts to distill the intuition that higher-level (e.g. mental) properties depend on lower-level (e.g. bodily, neurophysiological) properties, while at the same time respecting the intuition that higher-level properties are irreducible to lower-level ones. The motto of supervenience is: 'no change at the higher level without a concomitant change at the lower level'. Supervenience seems especially fit for non-reductive naturalists since it is supposed to respect both our view that science, and ultimately physics, provides the most reliable knowledge about the world and our place in it and our intuition that mental phenomena, such as intentionality, meaning and morality, are palpably real yet irreducible to lower-level physical, neurophysiological or biological processes occurring in human organisms. Thus, supervenient relations between higher- and lower-level properties seem to go hand in hand with and account for our hierarchical, layered ontological view of the world in which higher-level entities correspond to higher-level taxonomies (e.g. middle-size perceptible objects, mental states) and are taken to be dependent on, necessitated by, yet distinct from and multiply realizable in entities corresponding to lower-level (scientific) taxonomies all the way down to the lowest level described by physics.

Yet, a supervenience-based account of emergence seems to offer only a semblance of an explanation of the relation between lower-level and higher-level emergent properties and not 'the real thing'. At bottom, to say that the dependence relation of a higher-level state or property to a lower-level state or property consists in the fact that there can be no change at the higher level without a concomitant change at the lower level is to give a purely formal specification with no ontological purchase and thus no explanatory value. To say that a lower-level state or property asymmetrically necessitates and is synchronically correlated with a higher-level state or property does not amount to making the latter intelligible as causal actors of their own and is thus compatible with rendering them essentially epiphenomenal. There is no explanation whatsoever for understanding why higher-level properties are correlated with lower-level properties, in what this correlation consists in and why it occurs (Kim, 1993). And to say, for example, that higher-level properties are token-identical but not type-identical with the lower-level ones (Davidson, 1970) does not remove the air of mystery regarding the explanatory connection between levels and neither does it offer any help on the issue of the causal efficacy of higher-level properties, since the latter are causally efficacious not as such—as types—but as tokens, that is, only in the guise of their material realizers and by virtue of their lower-level properties. The supervenience dependence relation of higher-level properties to lower-level properties thus proves to be too weak to be explanatory and leaves us with an explanatorily idle conception of higher-level properties. From the standpoint of lower-level physical properties, higher-level properties are explanatorily dispensable, since to explain the behavior of the latter one need not appeal to anything over and above physical laws. If we are to take higher-level properties as real and irreducible to their lower-level properties, while at the same time supervenient on them, we cannot but conceive them as epiphenomenal. But then they cease to be emergent in an ontologically interesting sense, since they lack causal powers of their own.

5.3 Physicalism and Panpsychism as Inadequate Monistic Solutions

The failure of mereological and supervenience-based accounts of emergence, and thus of the non-reductive naturalist's aspirations to eat his cake (adhere to a scientifically respectable naturalistic picture of the world) and have it too (admitting mental properties as real and irreducible to physical ones), revived interest in more unabashedly reductive accounts that rejected emergence altogether, such as reductive physicalism and panpsychism. Since the above models of emergence were unable to explain the relation between mental properties and their physical realizers in the brain, in a way that accommodates the reality and causal efficacy of qualitative features of sensory and affective consciousness, why not reject emergence altogether along with the hierarchical multi-level picture of the world as populated with lower-level physical and higher-level mental properties connected by relations of supervenience? Why not settle for a reductive and resolutely monistic view according to which, since emergence of conscious qualities from non-conscious physical processes is no longer an option, we must either (1) understand mental properties on the model of physical properties, for example, by holding that qualia are just dynamic arrangements of physical properties with no special non-physical character (reductive physicalism) or, conversely, (2) understand physical properties on the model of mental properties, holding that conscious qualities in some—perhaps attenuated—form must be present in what we had previously thought were inanimate bits of matter that make up everything, including stones and particles (panpsychism).

Unfortunately, the initial attraction of these reductive monistic views is immediately counterbalanced by their highly implausible conception of the world and our place in it.

5.3.1 *Objections to Reductive Physicalism*

The first alternative, reductive physicalism, holds that all organized qualitative wholes over and above the fundamental particles, powers and laws of physics are ultimately epiphenomenal. But, by construing sensory and affective experience in terms of non-qualitative and non-homogeneous parts (e.g. neural firings, interrelations and combinations thereof), it effectively eliminates from our worldview our very experience of the world, and in particular its essentially extensive qualitative features (e.g. shape and color properties). We cannot make qualitative and homogeneous wholes out of combinations and interrelations of non-qualitative and discreet, particulate parts. However we might combine or interrelate discreet particulate and colorless neurons, molecules or atoms, the resulting complex physical state will not amount to a homogeneously colored[2] experiential quale. Experiential qualia (whether they are taken to be features of external physical objects, states of human perceivers or qualitative

[2] To say that color is essentially homogeneous is to say that every part of it is also colored. The same goes with all other experiential qualities (Sellars, 1981).

body-brain-environmental co-evolving processes) are not made up of imperceptible qualities in the way in which a tomato's shape is made up of imperceptible parts that need not themselves be spherical (with sphericity being the result of the parts held together in a particular way, in a dynamic network of connections and interrelations).[3] Experiential qualities are essentially homogeneous not only because we *experience* them as thusly ultimately homogeneous but also for the more philosophically compelling reason that the ability to represent the (arbitrarily selected) parts of any object of representation presupposes the ability to represent that object as having some sensible quality (Landy, 2019). More generally, for something to be represented as an object *at all* (whether it is an external object or an inner experiential one) we must delimit its spatiotemporal boundaries, and doing so requires that the object be represented as having some quality or another, be it color, sound, resistance to pressure, etc. (Kant, 1998). This point stands whether the object in question is perceptible or imperceptible: objects, including imperceptible physical objects postulated by theoretical physics, cannot have primary qualities only; structural and mathematical properties presuppose what might be called 'content qualities'.[4] It does not follow from this that physical objects posited by theoretical science must *themselves* have secondary qualities (as Berkeley thought) but it does follow that they must have content *counterpart* qualities to secondary ones that retain the second-order structural feature of homogeneity (matched by a counterpart conception of extension, since, arguably, extension is inconceivable without color).[5] Thus, by identifying experiential qualities with non-qualitative and particulate complex physical structures that lack ultimate homogeneity, the reductive physicalist does not only eliminate qualia from our worldview, but also inadvertently eliminates the world *as such* to the extent that the attribution of 'content qualities' is necessary for making sense of the physical reality (causal efficacy, power to affect and be affected) of the imperceptible objects, properties and processes posited by theoretical science itself (their mathematical and structural properties do not suffice for rendering them efficacious in physical reality).

[3] Thus, platitudinous mereological emergence cannot work in the case of sensible and affective experiential qualities.

[4] Kant, Berkeley, Whitehead and Sellars have pointed this out, though each of them draws different conclusions from this observation.

[5] We shall see in the next section how Deleuze's concept of intensity might provide such a counterpart conception of extended objects with secondary qualities.

But what if reductive physicalism somehow retained the 'content properties' of physical objects and aimed only at eliminating experiential *qualia* from our worldview? Unfortunately, it would still be an untenable position. For the identification of experiential qualities with complex cerebral processes lacking experiential qualities amounts to taking our experiential and affective life out of our world-picture altogether, thus making it unintelligible how things could even *appear* to have experiential qualities (e.g. color) (Sellars 1963). This makes reductive physicalism a nearly self-contradictory position. If neural structures cannot really be colored, and colors really exist in or as modifications of the mind, as the reductive physicalist has it, then we lack the conceptual resources to explain why things can so much as *appear* to our minds to have color, except perhaps if it is argued that things 'appear to appear' to have qualities, but they really do not appear to have them. But obviously the is-seems distinction cannot be applied to appearances as such. Even if the qualitative content of external objects is ultimately phenomenal, belonging only to a world of appearances, sensations themselves must be *real* if they are to explain how the world can so much as appear in the way it does to perceivers.

5.3.2 *Objections to Panpsychism*

This explanatory dead-end of reductive naturalism regarding sensory and affective appearances, coupled with skepticism about emergence induced by the failure of mereological and supervenience-based accounts of emergence, have increasingly drawn many philosophers who take consciousness seriously toward panpsychism. For, given the above premises, it seems that the only way to explain how conscious qualities came to exist in the world is to hold that experiential qualities, in some attenuated form, must be present in what we had previously thought were inanimate bits of matter. Just as we could not hope to produce something extended from parts none of which were extended, so we could not produce conscious experiences from parts, for example, neurons, and the molecules and particles they are made of, wholly lacking in consciousness (Strawson, 2008, pp. 60–67, Goff, 2017). Experiential qualia must be present from the start in every bit of matter, or else they could not be features of humans or sentient beings at all. In this way the problem of the causal efficacy of experiential qualities is solved directly, without recourse to emergence of any sort, by taking it that macro-experiential properties are causally efficacious by being grounded to micro-experiential properties situated at the level of fundamental physical entities.

Yet, notoriously, panpsychism is subject to a major challenge, namely, the combination problem: How do the experiences of fundamental physical entities such as quarks and photons combine to yield the familiar sort of human conscious experience? Just as private human minds do not agglomerate into a higher compound mind (collective spirit, collective will), so single atomic micro-experiences or 'feelings' of inanimate or imperceptible bits of matter (particles, molecules) do not seem to be the sort of thing that can aggregate into macro-experiences (James, 1895). While we clearly understand what it is for a house to be formed out of bricks, or of a car engine formed by mechanical parts, the idea of many minds forming some other mind is considerably less clear and does not seem to make much sense. This problem concerns both the subjective and the qualitative dimension of our experience: given any group of subjects and any further subject, it seems possible in principle for the first group of subjects to exist without the further subject. If so, then no group of micro-subjects necessitates the existence of a macro-subject (James, 1895). Moreover, it seems that each subject's perspective *excludes* the perspective of all other subjects. A conjunction of the experiences that a subject is having and of those that they are not having constitutes that subject's unique point of view, so that if those experiences were combined with the experiences of another subject, the new subject would have to instantiate a conjunction of these two sets of experiences, which is contradictory (Coleman, 2014). Combination problems of a different sort arise in our attempt to understand the relation between the micro- and macro-experiential *qualities*: given that there is a vast array of macro-qualities, including many different phenomenal colors, shapes, sounds, smells and tastes, and given that there is presumably only a limited palette of micro-qualities, how can this limited palette of micro-qualities combine to yield the vast array of macro-qualities? The initial issue here is whether qualities can combine to yield other qualities at all. We understand how this can work when perceived qualities are co-instantiated. If the same object simultaneously instantiated a degree of redness and a degree of whiteness at the same location, it will instantiate pinkness. Something similar goes for co-instantiated phenomenal qualities. If the same entity is simultaneously aware of a degree of redness and a degree of whiteness at the same location, it is thereby aware of pinkness at that location. But separately instantiated qualities—for example, the redness and whiteness of distinct objects—do not yield a combined quality nor do separately instantiated phenomenal qualities. Perhaps we can avoid this problem if we take it that the full range of macro-qualities is included among the micro-qualities of matter-energy. Yet, in this way we sacrifice the simplicity of our worldview at the level of fundamental physics without being able to explain why this fundamental level *appears* to be characterized by simplicity, and without gaining anything at the level of really understanding how conscious experience can have a foothold in the world (Chalmers, 2016). All we do is project the qualitative character and structure of macro-experiences to micro-experiences without having any grip on how this can be so much as possible.

5.3.3 The 'Cooking Model' of Emergence

I contend that an important common underlying presupposition of both reductive physicalism and combinatorial panpsychism, that explains why they find reductive solutions to the problem of emergence appealing, is a simplistic 'cooking model' for understanding emergence and novelty. As was mentioned above, according to this model, (a) novel properties are combinations of already existing primary (not further decomposable) properties, the emergent element consisting in the novel mixture produced as a new whole or unity out of these properties, where the latter are now conceived as differently organized parts of this new whole, or (b) novel properties are completely new externally added features over and above those already existing primary properties, in which case the new emergent whole does not have these (however differently organized) properties preceding its existence as proper parts and neither do these primary properties have any intelligible relation or connection to the emergent ones beyond being their antecedent causal conditions of emergence. At bottom this simplistic picture is one in which existing and emergent properties are understood on the model of distinct ingredients of a cooking recipe the implementation of which produces novel outcomes by mixing these already existing ingredients or by externally adding new ones.[6] Physicalistic material micro-particles and panpsychist micro-

[6] I take it that the main reason for being under the influence of this model of emergence is the need to understand part-whole relations as transitive so as to abide by classic extensional mereology. Obviously, this commitment to extensional mereology is not at all naïve and usually goes hand in hand with a respective commitment to representational semantics. This latter representationalist view is challenged in this book (3.2), though the proper formula-

experiences (along with their 'primary properties') are the distinct ingredients, laws of nature and the combinatorial possibilities based on the intrinsic natures and interrelations between micro-physical or micro-experiential atomic building-blocks play the role of the cooking recipe and its implementation, qualitative experiential wholes are the novel outcomes of the implementation of the recipe, experiential qualia as mereological compositions of micro-physical or micro-psychical parts correspond to the mixing of already existing ingredients, and experiential qualia as *sui generis* emergent whole-causes correspond to ingredients externally added to the recipe. I take it that it is largely due to the seductive influence of this picture for understanding part-whole relations and novelty that makes both reductive physicalism and panpsychism blind to the following possibility: accounting for the radical discontinuity introduced by emergent properties in actuality by appeal to immanent causes in the form of *continuous* deformations whose (*discontinuous*) outcomes are processes with a novel intrinsic character, the 'parts' of which are not conceived as building blocks arranged or mixed in a certain manner. As we have seen (Sect. 5.2.2), they can best be conceived as same-level *aspects* or a process, or as its spatiotemporally heterogeneous yet organizationally coupled *co-ingredients* (i.e. its sub- or supra-processes co-existing and co-evolving at multiple spatiotemporal scales in a multidimensional phase space).

5.3.4 *Monadic Panpsychism*

Lastly, let us here mention the interesting case of monadic panpsychism, which rejects both emergence *and* combinatorial panpsychism (Kadić, 2024). According to this view, the basic building blocks of consciousness are micro-subjects with empty points of view without experiences, mere conduits for experiences as it were, situated at the basic level of matter-energy. Our dynamic and ever-changing experience is relationally constituted out of a multitude of interrelations between these atomic perspectives. There is no 'emergent' subject here: just as it might be thought that there are no tables but just particles arranged table-wise (van Inwagen, 1990), accordingly one might hold that there are no emergent unified selves, but only consciousness-involving ultimates arranged 'human consciousness-wise'. This proposal indeed avoids the pitfalls of

tion and defense of the non-representationalist alternative would require a book-length treatment.

combinatorial panpsychism, but only at the cost of invoking a barely intelligible notion of a subjective point of view without experience, which, though bereft of qualitative experience or self-awareness, is nevertheless intrinsic, sharp and determinate (while our dynamic and ever-changing experience at the human level is relationally constituted out of a multitude of interrelations between these atomic perspectives, ending up being vague and indeterminate) (Kadić, 2024, p. 5). It is very difficult to understand how a notion of perspective, of a point of view, which is supposed to be 'intrinsic, sharp and determinate', but which is explicitly said to be a mere *condition* for complex qualitative experience (subsequently coming on the scene to fill in this experiential placeholder), can *actually* (not merely potentially) produce, by virtue of its multifarious interrelations to other such micro-perspectives, anything remotely akin to the internal experiential viewpoint of even the simplest organic beings, without any appeal to emergence. Also, it is difficult to see how interrelations between discreet, discontinuous monadic perspectives could account for the continuity and qualitative homogeneity of experiential qualia.[7]

This is not to denigrate this monadic version of panpsychism. Indeed, I take the notion of a monadic perspective to be an important first step in articulating a coherent and comprehensive univocal process ontology that rejects the sharp categorial distinctions and the ensuing unbridgeable ontological divisions between mind and body or between living beings and 'dead' matter. In Chap. 4, I attempted to sketch the notion of a monadic perspective by construing objects as intrinsically differing ways of spatiotemporal unfolding. As we shall see in Sect. 5.4, given that worldly processes are characterized by ultimate homogeneity and by continuity of unfolding which discontinuously produces (through self-contraction) intrinsic perspectival actual structures, qualitative experiences can be thought of as emergent in a topological sense, without the need to presuppose some kind of minimal experiential qualitative structure gradually filling in an 'empty' perspective inhering at the atomic level of matter-energy. Indeed, given the above fundamental features of process ontology,

[7] Here, it can be argued that the homogeneity of experiential qualia is a feature of what experience represents (e.g. expanses of red objects) and not of the experience itself (Stoljar, 2001). But this seems to make experiences as such phenomenologically undetectable in principle, which is surely counterintuitive. Moreover, it does not seem to explain cases of hallucinations, illusions and dreams, where our experiences remain homogeneous without representing anything external to themselves.

emergence becomes only something to be expected even in its ontologically demanding, non-platitudinous sense.

5.4 The Emergence of Sentience and Sapience in a Nominalist Process World

In this section, I shall argue that the notion of intensity (Deleuze, 1968; Simondon, 2020), and its topological properties, provides a way of developing a non-traditional intensive processual nominalism which can account for emergent levels of being in an immanent and nominalistically kosher way, while at the same time avoiding both physicalism and panpsychism. With the aid of the concept of intensity, we can provide a via *media* between mechanistic reductive physicalism and panpsychism: *contra* reductive physicalism, we can thereby reject the view of an inert, 'dead' inorganic realm in which objects are passively pushed and pulled around by externally determined natural laws and explain how inorganic processes can be monadic, perspectival and replete with spatiotemporal dynamisms which make possible their self-differentiation, self-organization and transformation. What is more, if we view inorganic processes as essentially intensive, we can render intelligible the possibility of their contraction and qualitative transformation in their interaction with sentient beings, while at the same time (*contra* panpsychism) avoiding endowing the inorganic world with qualitative experiences. Importantly, by viewing processes as essentially intensive, we can satisfy the demand (Sect. 5.3) that the reality and causal efficacy of experiential qualities can be accommodated only if their emergent 'bases' are characterized by homogeneous properties (both synchronically and diachronically). These counterpart properties must be ultimately homogeneous *without* themselves having experiential or qualitative, for only then could they independently account for sensible qualities while avoiding relapsing to an implausible panqualityism or panexperientialism. We shall see that the notion of intensity is precisely what is needed to play this theoretical-explanatory role.

5.4.1 *Making Intelligible the Emergence of Sentience*

The above issues boil down to the question of how exactly we are to conceive of the ultimate homogeneity that according to a process ontology must characterize all levels of processes, including inorganic ones, without ending up to panpsychism. How can intensity throw some light on this issue?

As we saw earlier, since the essential second-order property of all sensations, their ultimate homogeneity, cannot be rendered intelligible by combining and interrelating discreet component parts that are not already qualitative in nature (particles, neural signals, digital information processing in the brain), it must itself be a feature of whatever is thought to be the sensations' 'component parts'. Moreover, the process view of the world endorsed in this book, along with Sellars (1965) and Whitehead (1978), accepts the view that since, as Berkeley convincingly argued, experiential qualities (e.g. as expanses of external middle-size objects) are unintelligible without extension and vice versa, the notion of 'structure (mathematical, formal, metrical) properties' without 'content (qualitative) properties' makes no sense *across the board*. Thus, all processes are to be understood as having an intrinsic character, and as ultimately homogeneous in nature (Sellars, 1963a, 1965). That includes both sensory-affective processes, such as 'reddings', 'buzzings', 'itchings', 'languagings', *and* micro-physical processes, such as 'electronings' and 'quarkings'.

Yet, if we are to avoid panpsychism or Whitehead's panexperientialism, we must hold that the qualitative structure of micro-physical processes is not intrinsically *experiential* (for a view of this sort see Sellars, 1963b). Every process, even rocks and mountains, differentially responds to features of the external environment and incorporates part of the latter into itself. But this does not amount to saying that these processes have qualitative experiences. In the absence of sense organs, the differentially responsive behavior of inorganic processes does not suffice for calling their viewpoint 'experiential', though it is the unfolding of an intrinsic qualitative character. Only if we reach the level of sentient organisms does it start to seem plausible to equate qualitative and experiential structure.

However, it is difficult to distinguish this 'panqualityist' view from panexperientialism. How can the intrinsic *qualitative* dimension of a process be such as to involve no *subjective* and *experiential* dimension? Think, for example, of the notion of qualia which paradigmatically combines those dimensions. How could a process have intrinsic qualitative character without at the same time being a (however minimally) perspectivally experienced quale? It seems that panqualityism is committed to the existence of intrinsic qualities without qualia, a view that is very hard to make sense of.[8]

It is precisely at this juncture that Deleuze's notion of intensity might be used for giving at least the form of a solution to that problem. This is because Deleuzian intensity, among other things, provides a way of accounting for the fact that processes cannot just have mathematical, formal and metrical structure, yet without necessarily construing the contentual dimension of all processes in proto-experiential terms (Whitehead's prehensions) or non-experiential yet qualitative terms (Sellars' panqualityist world of particulars). Intensive processes are essentially continuous, which is not to say that they are indivisible (*à la* Bergson), but that they cannot be divided without changing in kind. They are thus essentially non-linear and non-metrical. They are analogous to qualitative properties in that they are continuous and non-metrical, yet, unlike qualia, they can be divided, and unlike extensive quantities they cannot do so without changing in kind. For example, the temperature of a volume of water can be divided by heating the container from below, causing a temperature difference between the top and the bottom. Yet, if the temperature differences reach a certain threshold (if they attain a certain intensity), the system will undergo a phase transition, losing symmetry and changing its dynamics, entering into a periodic pattern of motion (convection) which displays extensive properties of size. More importantly, it can be argued that intensive differences (differences in potential) constitute the genetic condition of actual objects with extension and qualities. Roughly, this involves topological alterations in the rates of change of spatiotemporal dynamisms that progressively determine the internal and external *milieu* of objects (and thus objects themselves) and proceeds through processes of *integration* (compression, stabilization, simplification) and *resolution* (scale, granularity, accuracy), resulting in the production of actual objects with extension and qualities, with an internal and external *milieu*. The distinction between internal and external *milieu*, expressive of the essential self-fracturing of being in its morphogenetic self-differentiation into

[8] These highly controversial features of panqualityism and panexperientialism give rise to cryptic views (Deleuze, 1968; Whitehead, 1978) according to which even inorganic beings like rocks, solely by existing as such, that is, merely by contracting energy forces in certain ways, have a primitive sense of 'self-enjoyment', which yet is not sensed in the manner of ordinary affective qualia.

beings, is defined at the most general level by the selective capture—in the form of spatiotemporal-cum-informational contraction—of energy forces and not by the presence of sentience or qualitative experience.[9] This general morphogenetic process model propagated by regimes of intensive differences is multiply realizable and takes place differentially and simultaneously, progressively determining inorganic, organic, sentient and even linguistic bodies (Deleuze, 1968; Smith et al., 2022). Note that none of the fundamental properties of intensive spatiotemporal dynamics, namely, continuity, non-linear dynamics, non-metrical structure and divisibility with change in kind is (necessarily) experiential or qualitative in character. By being non-formal, non-discreet, non-digital and non-metrical, intensive dynamics preserve essential properties of the *ultimate homogeneity* of sentient experience, without being literally experiential or qualitative in nature. By forming the intrinsic character of all beings, including inorganic ones, intensive properties can thus serve as the *theoretical counterparts* of the qualitative properties of sentience, on the basis of which we can understand how the emergence of sentience is so much as possible in the first place.

The (synchronic) emergence of experiential qualia, of sentience, in organic processes, would then be philosophically unproblematic since it could be understood as a result of *selective contraction*, *synchronization*, *integration* and structure-preserving *homeomorphic transformation* of (intensively understood) information effected by the dynamic energetic *coupling* and the *internal resonance* established between inorganic processes and 'shielded' organic bodies.[10] As for the diachronic emergence of sentience, it might be suggested that these processes of contraction, synchronization, integration and homeomorphic transformation can

[9] All inorganic processes (e.g. the formation of rocks) differentially respond in a lawlike manner to environmental energy forces. These reliable differential responsive dispositions characteristic of every inorganic process stable enough to count as an object of reference, and thus their selective absorption, appropriation and response to environmental energy sources, along with the fact that at bottom all processes (including inorganic ones) are ways of spatiotemporal occurring, draw within every such process a topological limit between an interior and exterior environment (between what it selectively absorbs and what it selectively responds to). A candle flame that actively maintains its own spatiotemporal integrity might be a better example of inducing this limit between interior and exterior. This limit, the primal self-fracturing of being in its expression in beings, itself undergoes self-differentiation and development in organic, sentient and sapient processes.

[10] This coupling is itself manifested in the perspectival form of a self-constituting topological division and polarization between an 'inner' and 'outer' sensed *umwelt*.

themselves be understood as evolutionarily novel adaptations intelligible in the wider context of ***continuous topological deformations*** in the timing and rhythms of the developmental paths of processes belonging to previous evolutionary stages: such timing deformations of the developmental paths of inorganic processes could result in the production of polarized membranes characteristic of living beings and analogous timing alterations of the developmental paths of non-sentient living beings exhibiting primal forms of decentralized sensitivity to the environment (chemically induced motility, receptivity, reaction), could result in the emergence of a centralized sensory-affective expression of environmental sensitivity (nervous system). For instance, Simondon has proposed that the emergence of the organic level of self-organization (a novel and more autonomous form of self-organization through homeostatic self-maintenance far from thermodynamic equilibrium) can be understood in terms of neotenic[11] processes of 'slowing down' of the developmental process of inorganic crystals.[12] In this way, the organic could emerge out of the inorganic by 'repeating' the initial unstable phases in the latter's pattern of individuation (before it reaches its stable equilibrium) while at the same time slowing it down in a process akin to neoteny. This suspension would make possible the conversion of instability into metastability and the exploitation of the flexibility involved in these unstable conditions to increase the degrees of freedom of the novel emerging structures (Simondon, 2020, pp. 163–164). This increase in degrees of freedom goes hand in hand with an increase in the dependence of the complex emerging structures on the external environmental conditions. The emergence of synergies between autocatalytic and self-assembly processes needed to create simple proto-membrane

[11] Neoteny or developmental retardation is a radical way of generating evolutionary novelty through shifts in relative timing of developmental 'programming'. It is the slowing or delaying of body development, compared to non-human primates, resulting in features such as a large head, a flat face and relatively short arms. Neoteny has been proposed as a crucial mechanism that contributed to the rise of many human-specific features, including an increase in brain size and the emergence of human-specific cognitive traits. Simondon widens the application of these neotenic processes to explain the emergence not only of animal species but of different stratifications of being itself (organic out of the inorganic, animal sentience out of plant non-sentient sensitivity, sapience out of sentience, etc.)

[12] When the modification of the energetic conditions (temperature, pressure) of a system of amorphous substances passes a certain threshold, a transition takes place from an amorphous state to a crystalline structure. The potential energy contained in the amorphous substance is actualized by effectuating a series of phase-shifts from a stable equilibrium to a metastable equilibrium and the beginning of crystallization (Voss, 2018, p. 97).

structures[13] and their increasing complexification, for example, in the form of recursively self-maintaining organisms such as bacteria might be understood as novel evolutionary solutions for maintaining the integrity of the complex systems involved in the face of increasing evolutionary pressures (dependence on an increasing number of external conditions with increasing volatility). A similar model can be suggested in the case of the emergence of perceptual and affective modalities from non-sentient living beings: The emergence and coupling of sensory and affective modalities to motor responses and action (nervous system) express a novel resolution of evolutionary problems situated at the level of non-sentient living beings (e.g. limited and slow-paced control over frequent and speedy alterations of environmental circumstances crucial for the survival of the organism). This resolution involves the dilation and premature expansion of already existing non-sentient living regimes (motivity, receptivity, reaction through decentralized chemical forms of information processing and propagation) and comes in the form of a centralized nervous system.

This is not, of course, to say that we have an even remotely adequate understanding of the specifics of these transformational processes. This is a matter of scientific inquiry and philosophy can go only so far as to sketch the most general categories that might liberate science from problematic background presuppositions that block the road to inquiry or delay it by diverging it to dead ends or unnecessary detours. The purpose of the above speculative examples is only to illustrate how the emergence of radically novel objects, properties, kinds, forms, capacities, and indeed of novel *categories* of being, can be readily intelligible in a resolutely *immanent* manner if we see space and time not as an inert, neutral and metricized stage-setting inhabiting objects, but as non-metric, polarized and dynamically evolving modes of being internal to the very being and individuation of particular objects (properties, kinds, forms, capacities) themselves. Under this light, the dynamicity of spacetime itself, expressed in its various modes of topological alteration, can perfectly account for novelty and emergence both synchronically and diachronically without any need for violating conservation laws of physics or for positing events that break the course of nature, inducing to it external, transcendent—and thus ultimately unintelligible—determinations. Thus, we come to see that immanence is not only compatible with emergence but internally related to it at

[13] Deacon (2013) shows how synergies of this form might be crucial for the emergence of DNA-based life.

least to the extent to which we have come to see why and how discontinuous actual structures can be understood as the outcome of the continuous variation of intensive morphogenetic factors.

5.4.2 *The Emergence of Sapience as a Self-Fracturing of Sentience*

A classic example of transcendent emergent causation that 'breaks the course of nature' is Kantian free will construed as a *sui generis* causality of freedom that violates natural causality and is imposed on it from outside. And since free will and rationality are the main traits that traditionally distinguish humanity from nature and other organisms, we owe here a brief account of how the emergence of *sapience* is so much as possible and *immanently* so in a nominalistic process world. Again, here, our purpose is to enlarge the space of possibilities so that—*pace* reductionist and transcendent top-down accounts of emergence—we can make intelligible and non-mysterious the immanent emergence of the mind's abilities for symbolic displacement. It is not to substitute scientific research with armchair metaphysical speculation.

As we have seen in Chap. 1 and shall see in more detail in Chap. 7, to say that our distinctive characteristic as human beings is sapience, as opposed to mere sentient sensitivity to the environment, is not to say that our minds are special in the sense of inhabiting a distinct and separate ontological realm over and above our physical bodies or the minds of other animals. It is to say that our mental states are understood *normatively*, as *loci* of normative authority and responsibility. Construing humans as essentially normative beings has serious consequences about rationality and free will: persons are not identified and individuated based on considerations having to do with the material constitution of their bodies but rather by their functionally characterized capacities and abilities of these bodies. If a complex physical system is bound up in a network of reciprocally recognized semantic and epistemic rights and duties and can be functionally characterized as having intentions (thoughts that motivate one to realize their content), then this complex physical organism—whatever its ultimate material constitution—*is* a person (Sellars, 1963a). Among other things, this means that the very distinction—crucial for the possibility of the freedom of the will—between our being disposed or impelled to act in a certain way (e.g. running away from danger) and our having compelling *reasons* for acting or not acting in this way, which can be potentially in conflict with our desires, dispositions or impulses, is itself a *normative functional* one, and not an ontological distinction between different incommensurable levels of reality. If a complex physical system exhibits sensory-affective patterned regularities such that the latter can be saliently redescribed in normative-functional terms, as expressing the ability to choose between alternatives and act on the basis of reasons that can overtake desire, then this complex physical system *does* have free will. Thus, the naturalistic and immanent process view of the world presented in this work is compatible with the existence of free will even in a robustly normative sense of the term.

Yet, this line of thought offers at most the form of the solution to this problem, by making some important distinctions that can untangle knots which can create philosophical confusion. I take it that a fully comprehensive process naturalism which aspires to be resolutely immanent must in the end attempt not only to make the ontological level of occurrent-intrinsic properties of our sentient dimension *compatible* with the normative capacities of our sapient dimension, but to show their essential *inseparability* at the very level of ontology. Moreover, if we do not move in that direction, we only ascribe a platitudinous emergence to sapience: predicates that apply to sapient wholes are not applicable or derivable from truths about their (sentient) parts. Yet, as we saw, this does not amount to a robust notion of ontological emergence. Here it might be argued that this is just as well: why do we need to ascribe a non-platitudinous notion of emergence to sapience, especially if our goal is to accommodate sapience into a broadly naturalistic worldview? I am in agreement with this deflationary line of thought that takes sapience to be logically irreducible yet causally reducible to sentience (O'Shea, 2007; Sellars, 1953). Yet, I take it that this ontologically deflationary strategy is plausible only if we want to avoid hypostatizing the mind as a causally efficacious universal—and thus, only if we conceive of emergence as the emergence of *universals* (kinds, types, generalities). Yet, if we work with a nominalistically constrained conception of emergence, we can be faithful to the materialistic spirit of this ontologically deflationary strategy about sapience while at the same time recognizing a sense in which sapience, nominalistically reconceived, makes an ontological difference to the world.

A very rough first suggestion in this direction might be the following (see also Christias, 2023, pp. 252–253): the specifically human highly symbolically displaced abilities to will otherwise, or to deliberate and act on the basis of reasons that can overtake desire, are based as ways of experiencing ourselves-in-the-world on a specific *distinctive* manner in which we non-conceptually represent ourselves-in-the-world: the ability of our senses to picture part of their *own intrinsic* activity, namely, the continuous occurrence of particular (visual, oral, tactile) sensory impressions, affects and images, as occurrently instantiated *schemata* that selectively arrange (spatiotemporally enhance, mute, bring to foreground, withdraw to background, highlight as relevant, play down as irrelevant) these sensory and affective states into a *digitally* organized variety of discrete salient patterns of similarities and differences. This marks the emergence of compositional structure (e.g. iteration and recursion), that is, the syntactic infrastructure upon which semantic inferential relations (incompatibility, consequence) articulating conceptual contents are built when this compositional structure is put into use in practical contexts. It can be suggested that the above self-synthesizing sign-making function is based on the imaginative ability to sense *in* what it is (a 'complex of impressions') something *other* than it (an 'impression of a complex'),[14] and I contend that it is precisely this ability that *immanently fractures* this self-picturing consciousness into a normative and a non-normative dimension. This immanent self-fracturing might be effected through evolutionarily explicable natural and cultural processes of compression, stabilization, separation and integration of information in the manifold of sensuous and affective particularity, thus yielding sapient self-consciousness through both *productive* imagination (integrative processes) and *subtractive* imagination (compressive, stabilizing, separating-abstractive processes) (see Fig. 3.1). In this way, it is as if self-consciousness at the nominalistic, non-conceptual level becomes self-dirempted in its striving to better regulate itself. This does not mean that the sapient space-of-reasons level has independent or emergent causal powers of its own. Its productive powers (e.g. the subsumption of associatively connected schemata to inferentially articulated concepts-rules, the distinction between subjective viewpoints and a public objective world, etc.) are, as it were, a *subtractive manner of appearance* of a (non-transparent) relation of non-conceptual self-consciousness

[14] Note that this process of meaning-making is more primordial than seeing something *as* something else, as belonging to a certain *kind*. The capacity we are envisaging here is antecedent to the distinctions between universals-particulars, meaning-reference, type-token, kinds-species-individuals, etc. These latter distinctions are sophistications upon this more primordial process of understanding the world.

with *itself*, namely, its *self-displacement*[15] at the non-conceptual level of sign-instituting we-attitudes.[16] These we-attitudes, in turn, can be seen as unproblematically emergent in the same sense in which sensory qualitative features of sentient organisms are emergent. In the case of sign-instituting we-attitudes what emerges is *affective* or *orectic* qualitative experience (a primitive form of the 'desire for recognition') along with a kind of *transindividual* group-resonance or attunement (mimetically mirroring the attitudes of others in the group)[17] which enables the skillful coping with signs. I take it that this collective yet non-conceptual attunement along with the 'desire for recognition' of desiring organisms as expressed in recognitive attitudes of *treating* each other as *authoritative* or *holding* each other as *accountable* for the *inferences* they *draw* from each other's individual picturing perspective needs to be recognized as another fundamental kind of emergent structure, along with sensory experience, in complex sign-emitting desiring organisms like us. For we must not forget that qualitative experience is not exclusively sensory but also has a—perhaps even more fundamental—affective, orectic, and transindividual dimension.

[15] One way of understanding this self-displacement of the primordial mimetic-cum-imaginative ability to 'see in what it is something other than it' is in terms of a change of orientation from imitating the object so as to control it by getting 'inside' it (magical thinking) to taking a distance from it in order to control its course based on a representation of its invariant features (Horkheimer & Adorno, 2002). Note that *both* these orientations, including the latter which lies at the heart of conceptual articulation, are features of non-conceptual consciousness.

[16] As we shall see in Chap. 8, this 'productive-through-subtraction' manner of existence of the space-of-reasons level is necessary—in the form of positing impersonally and timelessly valid normative statuses—for better calibrating and collectively coordinating divergent private perspectival picturings.

[17] For a detailed analysis of the evolutionary anthropological underpinnings of normativity see Stovall (2022), and especially his notions of 'shared practical picturing' (a physiological capacity for environment-responsive sensorimotor attunement between humans involved in transperspectival planning), and its variant, 'deontic picturing' (involving the transperspectival exercise of non-discursive emotional and reactive attitudes). See also Sachs' discussion of 'shared' or 'collective' picturing (Sachs, 2019). I take it that shared picturing in the above sense is directly implicated in the fundamental social phenomenon of being attuned to what 'everyone' is supposed to do under certain circumstances (Heidegger's anonymous *das Man*). Note also, importantly, that a grasp of the perspective of others in the we-picturing mode need not be understood as a grasp of the other's internal subjective standpoint. Instead, the other's perspective can be construed as an externally and directly observable feature of the other.

Through the abovementioned self-fracturing of non-conceptual self-consciousness and by way of productive-through-subtraction procedures, a normative space is opened—sustained by recognitive attitudes—for distinguishing between a relatively stable self, its expression in actuality (perception, volition, action) and the conditions under which it is within its power to think, will and act, thus making meaningful a sense in which one can be entitled, committed, have authority or be held responsible for one's beliefs and actions. In effect, this self-fracturing of self-consciousness enables us to collectively and individually transform an *actual inability* to actually survey and control what is beyond our perceptual powers, into an *imaginative ability*: the symbolically laden mind is a transposed world-model, a world 'within our powers', whose 'conceptual world-making' structural features can be used as models for anticipating—and thus reliably guiding us through—the perceptually inaccessible but potentially affectively relevant part of the real world. Yet this self-fracturing also comes at a cost: it is not an epistemically transparent process, and indeed misconstrues its own imaginative nature projecting it in a 'direct realist' manner as a feature of the world as it is in itself, by—ironically and insidiously—*applying its own productive and subtractive imaginative tools to its own self-understanding*[18]: Our imaginative ability of seeing in what it is something other than it is not transparent to itself in that it sees *in* what it *itself* is something *other* than it, thus opening the space that make possible the dimensions of *reification* (the sense of an unmediated cognitively transparent access to the world), *projection* (of our subjective activity to the objective world) and *alienation* (estrangement from our collective normatively instituting world-making activity).[19] Yet, at the same time and by the same token, this imaginative ability makes possible the *liberation* and *creative*

[18] This direct realist self-misrecognition is only to be expected from an evolutionary standpoint: for most practical purposes, and in human-adjusted time scales, not only would it not jeopardize our survival, physical or cultural reproduction, but it would significantly compress and simplify external and internal informational influx in ways that are necessary and conducive to survival in a changing and hostile environment.

[19] Thus, this sign-making function achieves semantic self-consciousness only gradually and only as a cultural achievement. Initially, in animistic hunter-gatherer societies, although the ability 'to see in what it is something other than it' does of course exist, it is not apprehended in *semantical* terms, as implying a distinction between type and token, act and content or a relation of representation or intentionality between mind and world, but is construed in mimetic, quasi-magical terms, as something that enables one to 'take the place' of the worldly phenomenon to which the word refers and thus to directly influence and control it (Cassirer, 1957). But the seeds of the above distinctions, as well as of their direct realist reification (i.e.

self-expression of ourselves-in-the-world by opening novel possibilities and ways of perceiving, thinking and acting, by constructing new purposes, interests and desires, and by creating the necessary semantic and epistemic distance from which it can *self-correct*, that is, critically assess its first-order (empirical) application in reality.

A last note is here in order: since the imaginative capacity 'to see in what it is something other than it' is a self-relation at the level of what we might call 'nominalistic self-consciousness', we must be more precise about what kind of relation it is, since it must be such that it does not commit us to causally efficacious universals (types, kinds, etc.). Perhaps we could call it a *virtual-differential* self-relation (where the terms are understood along Deleuzian lines), in the sense that it is the expression of a capacity of occurrent imagistic consciousness to occurrently (non-dispositionally) produce—at the level of we-attitudes—a *differential intensive field* which at once virtually differentiates qualities and fuses them in a *single* concrete actual qualitative experience. This capacity is antecedent to the object-property-relation, essence-appearance, type-token, act-content distinctions. It is also antecedent to intentional, representational and referential semantic relations between mind (language) and world. Indeed, it is the condition of possibility of *all* the above distinctions and relations. Thus, this self-diremptive non-conceptual self-relation does not function as a universal (with particular instances), and it is nominalistically kosher. Being antecedent to the object-property framework as well as to the appearance-essence, type-token, act-content distinctions, and to the classical Aristotelian notion of potentiality (capacity, disposition, etc.), this self-diremptive self-relation can be considered as an emergent qualitative regime without losing its nominalistic credentials.[20]

5.5 Emergence as a Form of Constraint

We will end this chapter with a brief review of Juarrero's (2023) recent account of emergence which is very close in spirit to the view presented in this book. Juarrero presents a broadly processual and univocal picture of the world using the conceptual tools of complex systems dynamics.

their unmediated ascription to objective reality), are already present in and presuppose the ability 'to see in what it is something other than it'.

[20] We shall have more to say about normativity as an emergent intensive regime with a distinctive hyperstitional character in Chap. 8.

Processes, inorganic and organic alike, are conceived as dissipative structures, namely, as open systems far from equilibrium that exchange matter, energy and information with their environment, and which self-organize and act as dynamically stable totalities in response to constraints. This view challenges classical essentialist assumptions about identity (which ignore the role of present context and past circumstances in the coherence and identity of a process) and causation (which construes causal interaction as resulting from transferring kinetic energy through impact from cause to effect). This enables a novel understanding of emergence as a result of the imposition of enabling and constitutive constraints on the hitherto existing possibility space of the unfolding of a process. Constraints bring about effects by making available, structuring, channeling, facilitating or impeding energy flow, rather than from energy-transferring forceful impacts. From this point of view, nature self-organizes and sorts itself into real but contextual coherent dynamics, emerging from contextually constrained interactions among individuals, which, in turn, as coherent dynamics, constrain the individuals and circumstances from which they emerge. Bottom-up enabling constraints such as positive feedback, catalysts and autocatalysts, recursion and iteration take systems to a threshold of instability and precipitate non-linear phase transitions to novel and coherent interdependencies, thereby explaining the emergence of resilient, robust and causally effective interdependencies displaying novel properties. Once these interdependencies have coalesced, order parameters of that coordination dynamic, acting as top-down constitutive constraints through negative feedback loops, stabilize and sustain this dynamic by laying down the coordinates of a new topology, a new possibility space in which this dynamic unfolds. Examples of this emergent dynamics based on enabling and constitutive constraints are hurricanes, laser beams, convection cells and superconductivity (in the inorganic realm), symbiosis, homeostasis, ecosystem stability (in the organic realm), and even human symbolic thought, norms, values and social practices such as economics (Juarrero, 2023, pp. 23–25). From this point of view, the higher-order properties of these coherent dynamic processes, that is, the phase locking, synchronization and resonance they establish in hitherto independent processes, do not operate as top-down efficient causes, but as top-down *constraints* on the possibility space of the lower-level properties. Note that since constraints are not directly involved in energy transfer, they can exert a form of downward causation ('causality of constraints') that can yield all these emergent dynamic regimes without violating causal closure or conservation laws, and thus without introducing a philosophically problematic transcendent conception of causation that breaks the course of nature from outside.

I am very sympathetic to the spirit of Juarrero's monistic processual account of emergence, and I fully agree that the notion of enabling constraint is important for making sense of the reality and causal efficacy of emergent properties and processes at all levels of reality in a resolutely immanent manner, without introducing transcendent efficient causes to nature. Moreover, I grant that the notion of constraint is a key element in determining the dynamic identity of processes over time (see Chap. 4), and my view of emergence as characterized by topological properties and dynamics is also to be found (though not in the same sense or as having an equally central role) in Juarrero's work as part of her conception of spatial and temporal context-dependent and context-independent constraints (polarities, attractors, symmetry breaks, temporal constraints on genome's expression and epigenetic profiles, on childhood development, etc.).

However, I have the following reservations about her view (which are not meant to downplay the overwhelming convergences with the project developed in this book). Sometimes Juarrero operates with a conception of emergence that comes close to platitudinous emergence. In many places she seems willing to equate emergent properties with properties that are true about wholes but are not derivable from truths about their parts, or with predicates that apply truly to wholes but not to their parts. Presumably, this is why she is willing to count as a novel emergent property virtually every property of wholes that is semantically irreducible to properties of their parts, such as those of traffic jams, cities, corporations and nations. This tendency on Juarrero's part may be due to her using a very broad notion of emergence that includes many related yet importantly different uses of the term. To say that (a) the actions of a social institution emerges from the actions of individual humans, (b) the various stages of the lifecycle of a silkworm emerge, (c) the self-maintaining behavior of a bacterium emerges from its internal and external interactions or (d) mental states emerge from brain processes is to use the term 'emerge' in different—though not totally unrelated—ways (Campbell, 2015) that hover between platitudinous emergence (in the first two cases) and ontologically robust emergence (in the latter two cases). By not distinguishing between different senses in which the term is used, and by effectively treating the term 'emergence' as a natural kind, Juarrero's account sometimes seems to treat certain cases of platitudinous (synchronic and diachronic) emergence as cases of robust emergence. Here we might get carried away by the fact that in the context of ordinary language it makes perfect sense to say that wholes are 'over and above' their parts if there are truths about these wholes that are not derivable from their parts. But as we saw (Sect. 5.2.1) this is not enough to make a property emergent in a robust *mereological* sense that implies downward causation exerted by emergent wholes over their parts. As the example of the tomato rolling due to its spherical shape reminds us, putting the parts of a whole together in a dynamic network of connections and interrelations does not by itself show that the whole exerts top-down causation over its parts. Instead, we might say that in these cases, the parts, due to their being arranged in novel interactive ways and by thereby involving novel dynamics, are characterized by properties that by *themselves* suffice to account for the causal properties of the whole without appeal to a top-down novel form of causality from wholes to parts that affects the latter's behavior. Of course, it can be convenient to use this causal-hierarchical 'top-down' talk to describe these properties of wholes that are semantically irreducible to their parts. For example, we can say that congestion shockwaves in traffic jams have the qualitatively novel property of propagating backward, a property which its parts (vehicles) lack (Juarrero, 2023, pp. 22–23). But this does not commit us to an ontologically interesting kind of emergent properties in addition to the patterns of interactive dynamics and thus of novel arrangements between the parts themselves that thereby acquire new relational properties.

I do not mean to imply here that Juarrero's 'causality of constraints' is unilluminating as an account of emergence of novel properties. Constraints are indeed crucial for understanding the formation of novel, coherent dynamics and properties, even if we disentangle cases of platitudinous emergence from those of robust emergence. That is, the causality of constraints is important for understanding cases of robust emergence even if it can also be used (unperspicuously in my view) to describe cases of platitudinous emergence. And surely Juarrero would concur that constraints are not emergent properties *in addition* to the interactive dynamic arrangements of the propertied parts of wholes. What is not clear is why novel properties originating from enabling constraints must be described as exerting a kind of *downward top-down* causation directed from wholes to parts. This remains unclear even in the cases of robust, non-platitudinous emergence. It seems to me that in many such cases (e.g. homeostatic regulatory mechanisms of animals) the situation is more perspicuously described by saying that the novel properties induced by enabling constraints are an outcome of the novel interactive dynamics (e.g. of a chemical and/or ecological nature) within a system of elements ('parts' of the organism-environment complex) that thereby acquire new intrinsic and relational properties *themselves*, thereby constituting a *novel process*, without the need to postulate a novel emergent 'whole' that exerts top-down causation to its parts. Perhaps the situation is even more perspicuously described by dropping mereological talk of wholes, parts and their relations entirely, and conceiving of the 'parts' of 'wholes' characterized by emergent properties that originate from enabling constraints as *aspects*, *modifications* or *ingredients* of the (novel) *processes* involved (see Sect. 5.2.2). The 'parts' of this novel process should not be conceived on the model of building blocks arranged in a certain manner. They are to be regarded as aspects of the process, just as the crest of a wave might be thought as an aspect of the wave, or they might be construed as its spatiotemporally heterogeneous yet organizationally coupled ingredients (i.e. its sub-processes or supra-processes co-existing and co-evolving at multiple spatiotemporal scales—in a multidimensional phase space). In any case, the ingredients or aspects of a process are not independently individuated parts out of which the whole is materially composed or constructed. If the 'parts' in question are understood as aspects or ingredients of processes in the sense defined above, then talk of top-down or bottom-up causation ceases to make clear ontological sense, and different spatiotemporal dimensions and levels of organization, with their constraint dynamics and interpenetrating patterns of interference, can now *themselves* be seen as aspects or ingredients of the unfolding of a process. Thus, what emerges is the process *itself* (along with it non-mereological structure), by being the *topological knot* of organizational couplings situated at all these different spatiotemporal dimensions and levels. And this is not perspicuously described as an emergence of a higher-order whole over and above its lower-level parts. Thus, the process framework enables us to conceive of a *robust non-mereological emergence*: the ontological reality and efficacy of robustly emergent properties and of the related enabling and constitutive constraints can be accounted for without putting them in a mereological straitjacket and thus without positing causal downward or upward paths, causally interacting levels and hierarchies.

References

Campbell, R. (2015). *The metaphysics of emergence*. Palgrave Macmillan.

Cassirer, E. (1957). *Philosophy of symbolic forms. Volume three: The phenomenology of knowledge*. Yale University Press.

Chalmers, D. (2016). The combination problem for panpsychism. In G. Bruntrup & L. Jaskolla (Eds.), *Panpsychism: Contemporary perspectives* (pp. 179–214). Oxford University Press.

Christias, D. (2023). *Normativity, lifeworld and science in Sellars' synoptic vision*. Palgrave Macmillan.

Coleman, S. (2014). The real combination problem: Panpsychism, micro-SUBJECTS, and emergence. *Erkenntnis, 79*(1), 19–44.

Davidson, D. (1970). Mental events. In L. Foster & J. W. Swanson (Eds.), *Experience and theory* (pp. 79–101). Duckworth.

De Caro, M., & Macarthur, D. (Eds.). (2010). *Naturalism and normativity*. Columbia University Press.

Deacon, T. (2013). *Incomplete nature: How mind emerged from matter*. Norton.

Deleuze, G. (1968). *Difference and repetition*. Columbia University Press.

Goff, P. (2017). *Consciousness and fundamental reality*. Oxford University Press.

Heil, J. (2021). *Appearance in reality*. Oxford University Press.

Horkheimer, M., & Adorno, T. (2002). *Dialectic of enlightenment*. Stanford University Press.

James, W. (1895). *The principles of psychology*. Henry Holt.

Juarrero, A. (2023). *Context changes everything: How constraints create coherence*. MIT Press.

Kadić, N. (2024). Monadic panpsychism. *Synthese, 203*(2), 1–18.

Kant, I. (1998). *Critique of pure reason*. Cambridge University Press.

Kim, J. (1993). *Supervenience and mind*. Cambridge University Press.

Landy, D. (2019). Sellars' argument for an ontology of absolute processes. *Journal for the History of Analytical Philosophy, 7*(1), 1–24.

McDowell, J. (1994). *Mind and world*. Harvard University Press.

O'Shea, J. (2007). *Wilfrid Sellars: Naturalism with a normative turn*. Polity Press.

Sachs, C. (2019). In defense of picturing: Sellars' philosophy of mind and cognitive neuroscience. *Phenomenology and the Cognitive Sciences, 18*(4), 669–689.

Sellars, W. (1953). A semantical solution to the mind-body problem. *Methodos, 5*(3), 54–84.

Sellars, W. (1963a). Philosophy and the scientific image of man. In W. Sellars (Ed.), *Science, perception and reality* (pp. 1–40). Routledge.

Sellars, W. (1963b). Theoretical explanation. In *Philosophy of science: The Delaware seminar* (Vol. II, pp. 61–78). John Wiley.

Sellars, W. (1965). Scientific realism or irenic instrumentalism: A critique of nagel and feyerabend on theoretical explanation. In R. Cohen & M. Wartofski (Eds.), *Boston studies in the philosophy of science* (pp. 171–204). Humanities Press.

Sellars, W. (1981). Foundations for a metaphysics of pure process. *The Monist, 64*(1), 3–90.

Simondon, G. (2020). *Individuation in light of notions of form and information.* University of Minnesota Press.

Smith, D., Protevi, J., & Voss, D. (2022). *Gilles deleuze, the Stanford encyclopedia of philosophy.* Stanford University. https://plato.stanford.edu/archives/sum2022/entries/deleuze/

Stoljar, D. (2001). Two conceptions of the physical. *Philosophy and Phenomenological Research, 62*(2), 253–281.

Stovall, P. (2022). *The single-minded animal: Shared intentionality, normativity, and the foundations of discursive cognition.* Routledge.

Strawson, G. (2008). *Real materialism and other essays.* Clarendon Press.

van Inwagen, P. (1990). *Material beings.* Cornell University Press.

Voss, D. (2018). Simondon on the notion of the problem: A genetic schema of individuation. *Angelaki, 23*(2), 94–112.

Whitehead, A. N. (1978). *Process and reality.* Macmillan.

CHAPTER 6

A Novel Account of Grounding

In this chapter, I apply the basic modal principles of our proposed process ontology (Chap. 4) to contemporary discussions about grounding in analytic metaphysics. Based on a (Deleuze-Sellars-Peirce inspired) evolutionary rendering of the principle of sufficient reason (PSR), and a reconceptualization of the relation between metaphysical possibility and necessity (where, *contra* Meillassoux (2008), metaphysical necessity implies nomological contingency and vice versa), I offer a novel account of grounding, against and beyond neo-Humeanism, neo-Aristotelianism and neo-Hegelian absolute idealist readings. It will be suggested that grounding relations are immanent modalities, which yet do not explain in terms of hierarchical metaphysical dependence, but via co-expression within morphogenetic fields. From this point of view, it turns out that truths are not passive mirrors of an already structured world (their truth-makers). Truth-makers are internal articulations of the world that emerge as the world self-organizes through evolutionary processes of expression and differentiation, while truths also emerge in the same manner as we, embedded in those processes, develop conceptual tools that express and resonate with this organization. As we shall see (6.1.3), this whole view is based on a conception of being as an evolutionarily developed intensive multiplicity. Yet, far from being transcendent metaphysical substances or essences, intensive multiplicities and their evolutionary principles of becoming just *are* the *mode* in which being self-differentiates. To say that

© The Author(s), under exclusive license to Springer Nature Switzerland AG 2025

D. Christias, *Towards a Hyperstitional Process Nominalism*, Palgrave Perspectives on Process Philosophy,
https://doi.org/10.1007/978-3-032-06350-2_6

being is an evolutionarily developed intensive multiplicity is just what it *means* for being to be unconditional yet internally self-fracturing. Thus, this grounding principle—itself an outcome of PSR—does not function as an external independent ground that explains being and its relation to beings from outside.

6.1 Being as an Intensive Darwinian Multiplicity: A Heterodox Account of PSR Through Sellars, Peirce and Deleuze

What is PSR and how exactly our account thereof differs from its more orthodox understanding?

Traditionally, PSR implies that for every thing, fact or occurrence there must be a *ground* (reason, cause, essence, law of nature) that explains *why* it is thus and so as opposed to registering *that* it is thus and so. Reasons can be thought of as grounds of worldly phenomena to the extent to which they account for them by appeal to logical or 'conceptual' necessity while causes are grounds of actual phenomena if they explain them by appeal to *real* necessity. The same goes for laws of nature, to the extent to which, unlike accidentally true generalizations, they are counterfactually robust—and hence physically necessary. Explanations by appeal to laws of nature differ from more traditional (e.g. Leibnizian) versions of PSR, in that they offer non-teleological (yet still necessitating) reasons for why things are thus and so. Moreover, a thing's essence has been traditionally thought of as a ground of its properties (and sometimes—though not always—of its existence in the actual world), where the necessity involved in this case is metaphysical ('water is H_2O'). Necessity, whether it be logical (whose negation is a contradiction), metaphysical (whose negation is false in all possible worlds) or nomological (whose necessity ranges over all possible histories of the actual world), can be—and historically has been—used in all these forms to ground the occurrence of contingent worldly facts or experiences.

In what follows, I shall present my own version of PSR, building on the work of Sellars, Peirce and Deleuze who are all united in rejecting PSR in its traditional form. Importantly, although these philosophers end up holding very different views over the ontological status of PSR, I take it that we can extract key aspects from their views that can contribute to the formulation of a novel position on the issue, which can throw new light on contemporary discussions about grounding in analytic metaphysics.

6.1.1 PSR as a Non-metaphysical Framework Principle of Inquiry

As I have suggested elsewhere (Christias, 2023), based on Sellars' analysis of causation (Sellars, 1957), we can conceive of the *causal* modalities—for example, of counterfactually robust laws of nature—as not directly ontologically committal, as not committing us to laws, powers or capacities existing *in re*, but as indispensable semantic tools of our categorial framework which make possible the self-correction of our descriptive and explanatory practices and epistemically orient this self-correcting process of inquiry toward the identification and occasional revision of (our conception of) what is physically necessary, possible or impossible. From this standpoint, causal necessity, that is, the necessity of the causal laws of nature, is not 'in the world'. What exists at the level of the 'in-itself' is, at most, relatively stable regularities or, at best, unrestricted constant conjunctions (Sellars, 1957). Yet this stability does not entail any kind of real necessity: the necessity of the causal laws of nature is, at best, *conceptual* (the shadow of alethic norms) not real necessity. Stable regularities, even *unrestricted* constant conjunctions, can well be physically *contingent*.

We can better understand this view of the causal modalities if we focus on the function of the 'causal principle' ('every change has a cause') (Sellars, 1957). The causal principle cannot be justified by induction (empiricism) or rational intuition (rationalism). Yet, it is a priori in that it functions as a *regulative meta-principle* of empirical inquiry. Specifically, along with the logical and deontic modalities, it provides the *framework* for the thinking by which we reason our way into the making of *new* commitments and the abandoning of the old. For instance, the proposition 'there are causal connections which have not yet been discovered' is not to be understood as a descriptive statement but rather as embodying the idea—and ideal—of the descriptive predicates it will be reasonable to *introduce* or to *discard* in our effort to develop and improve our language in accordance with rational explanatory procedures (Sellars, 1957, pp. 302–307). I take it that in the case of the 'causal principle' this regulative status finds concrete expression in our practice of *revising* parts of a conceptual system (e.g. a particular scientific theory or 'paradigm') or even the latter as a whole, by offering *causal explanations* of apparent deviations of hitherto lawlike behavior of observed phenomena, in terms of an underlying reality whose (unobservable) entities, laws and principles explain why the deviations in question are only apparent, thereby restoring the lawful character of reality (Sellars, 1961). Moreover, it can be argued that the way in which causal laws prove indispensable in this process of revision, or for the self-correcting character of empirical inquiry in general, is by functioning as rules which determine the space of physical necessities, possibilities and impossibilities. To see this note that, as can be seen from a close inspection of our scientific practices, every change in our conception of the *form* of our fundamental causal explanations of the phenomena effects a corresponding change (and, in a sense, expansion) of the space of physical possibilities and impossibilities, that is, of what inferences we take to be counterfactually robust (distinguishing lawlike generalizations from merely accidentally true generalizations). Specifically, by changing the very *space* of physical necessities, possibilities and impossibilities, explanations on the basis of causal laws 'convert' what from the point of view of the preceding framework may have seemed physically impossible into a genuine physical possibility (or even, into a physical necessity), and conversely, seemingly genuine physical possibilities or necessities into physical impossibilities (i.e. possibilities or necessities which turn out to be merely illusory ones). Moreover, they thereby make possible the identification of novel phenomena and possibilities which were conceptually invisible in the former conceptual framework. Finally, and most importantly, in this way they bring to light and challenge certain fundamental yet tacit/unthematized presuppositions of the predecessor theory, which, precisely due to their centrality in the predecessor theory, were constitutively invisible by the system's own lights.[1]

In sum, it can be argued that the 'causal principle', its concrete expression in the practice of giving and revising explanations of observed phenomena on the basis of underlying entities, causal laws and properties, as well as the notions of physical necessity and possibility involved (expressed by the counterfactual robustness of natural laws) are metalinguistic *expressive* devices whose ultimate function is *normative* and *practical*: they enable the user of a conceptual framework to represent to himself (make explicit) the range of available epistemic alternatives that are open to him and that demarcate his choices for improving his epistemic position. In this way, they enable the user of a conceptual framework to be *consciously self-critical* toward his own past, present and future tokenings of propositions licensed by the rules (natural laws) of a given conceptual framework in the face of explanatory anomalies (Sicha, 1980). From this point of

[1] Think, for example, of Newton's three laws of motion as seen from the standpoint of Einstein's causal laws which, unlike Newtonian laws, are invariant in all inertial frames.

view, the causal principle expresses a framework regulative principle which provides empirical inquiry with a general epistemic orientation toward a very schematic (and only diachronically approximated) ideal of rationality, while the specific (less—yet still—schematic) forms of causal explanations which are taken for granted in each empirical-scientific theory function in effect as *synchronic paradigms of rationality*, as concrete expressions or momentary crystallizations of what we take a rational procedure to *be* as regards the discovery of new facts and the explanation of old ones, in a certain cross section of the history of inquiry. And the causal principle here is nothing other than the expression of PSR, where the 'sufficient reasons' in question are construed as 'causal-explanatory grounds' that make intelligible (due to their internal connection with counterfactually robust laws of nature) why things are as they are.

6.1.2 *Darwinian PSR as a Metaphysical Ground that Ungrounds Traditional PSR*

We can now make a further move, which goes beyond the quasi-Sellarsian conception of the PSR discussed above, and argue that it is precisely the *relentless application* of PSR *itself*—as expressed in our constructing and revising causal schemata for explaining apparent deviations of hitherto lawlike behavior of observed phenomena—that ultimately shows that this principle has non-metaphysical, normative, practical and critical import. It is only through the provisional understanding and application of PSR as having ontological/explanatory purport (i.e. of the causal explanations as ontologically committal) that we can ultimately see it for what it really is, namely, as something that while not being 'in the world' has an indispensable practical, normative and critical import in coordinating empirical inquiry. For sooner or later we arrive at a point where causal necessity is *itself* put into question as to whether it is the ultimate 'groundless ground' of empirical phenomena or has a metaphysically arbitrary character: What grounds real necessity itself? Why not settle for unrestricted constant conjunctions? That is, in a Humean spirit, we can use the inner logic of PSR to *question* specific grounding principles (in this case, the causal principle) that purport to express PSR. And this further means that the relentless application of PSR *does* yield ontological implications: at the deepest level of explanation, real connections, far from being fundamental 'unexplained explainers', prove, on philosophical-cum-scientific grounds, to be *themselves* in need of explanation by a-categorial structures that *lack* causal-dispositional, logical and deontic properties altogether. For example, if causal explanations have an essentially counterfactual character and causal properties have an essentially dispositional character, what accounts for this latter non-factual dimension of their being? Moreover, if the dispositional character of causal properties is conceived as explanatorily ultimate, then all such properties (e.g. flammability) can only be rendered intelligible by understanding the nature of their manifestations (e.g. burning), which in turn requires understanding the nature of the ways in which these latter phenomena are *themselves* manifested (e.g. fire) and so on ad infinitum (Heil, 2003; Lowe, 2006). Does it not seem that PSR itself pushes toward a direction in which dispositional (relational, 'iffy') properties must be explained in terms of categorical (intrinsic, non-iffy) properties? And does this not mean that the theoretical explanation of the very being of causal properties must ultimately involve non-causal, non-dispositional terms if it is to be really explanatory?[2] Similar considerations can be applied to the metaphysical explanatory sufficiency of logical or deontic grounds (Brandom, 1994; Carroll, 1895).

This way of thinking about PSR leads to a picture in which being remains unsubordinated to fixed essences, forms or causal laws of nature.

[2] Indeed, Sellars himself pointed toward this direction yet without connecting it to PSR. As we mentioned in Chap. 3, he draws our attention to the fact that our object-property categorial framework, which expresses our ordinary lifeworld metaphysics, gives rise to questions and explanatory demands about the categorial basis of dispositional properties it is not in a position to answer. Sellars takes it that the abandonment of the object-property framework might be necessary for explaining the dispositional dimension of causal properties that are constitutive of the very identity of the *objects* of the framework (Sellars, 1957, p. 263). As he puts this point: 'The soluble object has some character such that it is a general fact that anything having this character which is also put in water, dissolves. But this character cannot be another causal property on pain of circularity. … More, indeed, must be known of an object than the mere fact that it was put in water, in order to infer that it dissolved. But such thing-kind generalizations as "Salt dissolves in water" include this more not by specifying additional part causes, but by restricting their scope to identifiable kinds of thing, in identifiable kinds of circumstance. But the philosophically more exciting part of the solution consists in distinguishing between the causal properties of a certain kind of thing, ***and the theoretical explanation of the fact that it has these causal properties.*** For while causal generalizations about thing-kinds provide perfectly sound explanations, in spite of the fact that thing-kinds are not part-causes, it is no accident that philosophers have been tempted to think that such a phenomenon as salt dissolving in water must "at bottom" or "in principle" be a "lawfully evolving process" ***describable in purely episodic terms.*** Such an "ideal" description would no longer, *in the ordinary sense*, be in *causal* terms, nor the laws be *causal* laws' (Sellars, 1957, pp. 262–263).

From this point of view, the non-metaphysical (normative, practical and critical) function of PSR presupposes the *ungrounding* of the traditional version of PSR which itself follows from the above *non*-traditional yet thoroughly metaphysical interpretation of PSR according to which the causal modalities are not deep enough to ground *why* things are as they are (as opposed to the fact *that* they are as they are).

Note, by the way, that in rejecting traditional PSR, the Humean does not provide an alternative account thereof, but just takes it that, at bottom, metaphysically there are only regularities, a certain spatiotemporal arrangement of externally related 'atomic' qualities, without reason, necessity, 'causal glue', etc. Thus, the Humean metaphysical picture of the world does not provide any deeper grounding for why things are as they are. I think this is because the Humean metaphysical picture of the world is resolutely *actualist*: there are no potentialities in nature; only actual states exist. As we shall see later, we can construct an account of grounding, which is Humean in spirit (as its basic grounding principles lack causal, logical or deontic structure) though not in letter (since these principles are morphogenetic, thus not strictly speaking actualist). Interestingly, this way of viewing grounding can reconcile categorical intrinsic properties and powers (potentialities) by viewing powers not as iffy-like counterfactual properties but as an intrinsic occurrent feature of processual becoming (and thus as akin in important respects to categorical qualitative properties). The trick here is *not* to equate the occurrent with the actual: the *occurrent* continuous spatiotemporal unfolding of processual becoming must be distinguished from the *actual* extensive and qualitative states or objects existing within space and enduring through time. Actual states and objects are temporary stabilizations or 'resolutions' of occurrent processual becoming and must not be confused with becoming itself. In this way, at one stroke, we can avoid both Humeanist actualism and neo-Aristotelian dispositionalism about grounding.

Now, in the beginning of this chapter we stated that grounding principles have an essential evolutionary-selectional dimension. It is now time to discuss this issue in more detail. We shall see that an evolutionary rendering of PSR can account for the 'form' and 'content' of actual beings, relations and laws without presupposing hylomorphically construed powers, capacities, dispositions or essential forms. Perhaps the philosopher closer to the spirit of this Darwinian way of understanding PSR is Peirce who rejected the necessity of the laws of nature *by appeal to PSR*, arguing that it remains arbitrary why a law of nature has this specific content rather than another, whereas if we construe it as a patterned regularity that evolves from a process of random variation and selection at the inorganic level we are in a position to understand *why* it is the way it is (Peirce, 1887, 1892). As Peirce (1887) characteristically puts this point: 'Since law in general cannot be explained by any law in particular, the explanation must consist in showing how law is developed out of pure chance, irregularity, and indeterminacy' (I.407) [...] 'all laws are results of evolution; underlying all other laws is the only tendency which can grow by its own virtue, the tendency of all things to take habits' (6.101) [...] 'nothing but a principle of habit, itself due to the growth by habit of an infinitesimal chance tendency toward habit-taking, is the only bridge that can span the chasm between the chance medley of chaos and the cosmos of order and law' (6.262). Peirce adds, plausibly to my mind, that for this to work we must presuppose the virtual existence of a continuum of an indefinite multitude of dimensions, something very much akin to Deleuze's virtual continuous multiplicity. As we shall see later (1.3), this means that 'chaotic' becoming, as the self-relating 'movement' of (virtual) being to (actual) beings and vice versa, must itself be structured, in order for this movement to be intelligibly described and expressed in evolutionary-selectional terms.[3]

Note here, importantly, that a globally applied evolutionary model of explanation based on random variation and selection through inheritance of adaptive traits can provide an *independent* ground of the causal modalities (power, laws, capacities, dispositions), without itself being capable of further grounding. Thus, global evolutionary grounds can be seen as the ultimate satisfiers of the demands of PSR itself. For the origin, function and scope of the causal modalities can be understood as derivative from an ontology whose ultimate explanatory posits are *processes of differential adaptation* in which interaction with the environment (other processes) differentially selects (preserves and reproduces) certain elements, while eliminating others. All processes (inorganic, organic, psychological, social) can be understood as the evolution-by-selection of a population of habits—where habit is understood as the tendency of a process to repeat any activity that has been performed before (Peirce, 1989). In this way, evolutionary explanations make intelligible the contingent emergence of collective order from individual randomness by providing concrete, situated historical narratives of local, contingent, mutable practical reciprocal

[3] This means that, paradoxically, chaotic becoming has its distinctive kind of order. The concept of virtual continuous multiplicity is precisely a suggestion about what that order might amount to at this ground level of being.

accommodations of particular ordered structures and their habitats (Peirce, 1887, Brandom, 2011, pp. 36–37). From this global evolutionary standpoint, what (under the spell of Newtonian models of explanation) we take to be eternal, immutable, necessary, universal laws of nature, can be understood as themselves, in the broadest sense 'habits' (sufficiently stable patterns/regularities) of the universe—a kind of order that has arisen contingently, but ultimately statistically explicably, by a selectional-adaptational process operating on a population of such regularities, which in turn provides the dynamic habitat to which all must collectively adapt.

We can understand the grounding potential of evolutionary principles if we realize that the 'survival of the fittest' is just a special case of the more general principle of the *survival of the stable* (Dawkins, 1976). Indeed, the earliest form of natural selection, already at work in the inorganic realm, was simply a selection of stable patterns and a rejection of unstable ones. The crucial explanatory element here is the fact that the mere preservation and 'repetition' of a minimally stable pattern (that can always arise as a chance event as a result of a phase transition of an inhomogeneous continuous n-dimensional multitude) is *itself* a selectional factor that infinitesimally contributes to its gradual entrenchment by *shifting the range and mean of variation* for subsequent selections[4] (Peirce, 1887). Importantly,

[4] Arguably, this is a primitive form of hyperstitional process (see Chap. 8) imprinted in the very fabric of being. This view is expressed in another characteristic passage from Peirce: 'All things have a tendency to take habits. For atoms and their parts, molecules and groups of molecules, and, in short, every conceivable real object, there is a greater probability of acting as on a former like occasion than otherwise. This tendency itself constitutes a regularity and is continually on the increase. In looking back into the past, we are looking toward periods when it was a less and less decided tendency. But its own essential nature is to grow. It is a generalizing tendency; it causes actions in the future to follow some generalization of past actions; and this tendency is itself something capable of similar generalizations; and thus, it is self-generative' (Peirce, 1887 I: 245). For Peirce, even the spatiotemporal continuum itself is ultimately an evolutionarily entrenched 'habit' of the universe. As he puts it: 'The substances [monadic processes] carrying their habits with them in their motions through space will tend to render the different parts of space alike. Thus, the dimensionality of space will tend gradually to uniformity. At the outset, the connections of space were probably different for one substance from what they were for another; points adjacent or near one another for the motion of one body would not be so for another. But the mutual actions of bodies would have tended to reduce their habits to uniformity in this respect; and besides there must have arisen conflicts between the habits of bodies and the habits of parts of space, which would never have ceased till they were brought into conformity' (I: 416). From this point of view, the very establishment of resonances between different monadic durational rhythms is itself accounted for in evolutionary terms.

this shows that chance variation and directive selection are not (always) two independent temporally sequenced factors of evolution: selection is not only variation-eliminating but also variation-preserving (it preserves previous variations that do not negatively affect stability and adaptation), and even creates the variation that it acts upon (it builds adaptation in stages by preserving, generation after generation, the favorable part of a random spectrum of variation, thereby increasing the probability of novel combinations of rare patterns or mutations if the latter turn out to be beneficial within a certain environment) (Beatty, 2016; Gould, 1977).

Obviously, at this level of explanation, this understanding of evolution based on random variation and selection through inheritance of adaptive traits, as well as the corresponding concept of 'habit'-tendency, is *functional*, and thus broad enough to allow for *specifically different realizations* of evolutionary processes in organic, cultural and even non-organic contexts. Each of these evolutionarily developed contexts can be characterized by different kinds of organization, dynamics and material constitution, and it is the task of scientific inquiry (aided by a scientifically informed metaphysics) to determine which of these differences must (or need not) be accounted for by the presence of emergent dynamic regimes in the respective organizational contexts. Thus, natural selection in the most general sense is not defined in terms of replication of genetic information (Deacon, 2013, p. 315). This latter mechanism is a specific *expression* of the functional principles of natural selection. Interestingly, this suggests that evolution *itself evolves* with respect to the complexity, intricacy and refinement of the specific material structures that express the basic functional evolutionary principles (chance variation, competition for energy resources, differential selection through inheritance of adaptive traits).[5]

Note also that these evolutionary grounds provide ultimate why-explanations *not* in the Leibnizian sense of showing why something is necessarily what it is as opposed to being otherwise, but in the sense of making intelligible the unfolding of increasing complexity and self-organizational novelty of being. The emerging actual structures—including natural laws—remain ultimately contingent, yet not in the sense that 'they could be otherwise' (as if there is a ready-made stock or totality of

[5] In this sense, we are not just the products of evolution but its concrete instantiations at this stage of its development.

possibilities from which some are actualized while others remain in a state of potentiality), but in the sense that their expression in actuality is unsubordinated to any higher necessity pre-determining their form.

Metaphysically, this view can be called *ontological mutabilism*: every worldly process can change as an adaptive response to changes in other corresponding processes in the environment. From this point of view, the notion of a 'permanent adaptation' does not make sense in this context, and the same goes even for laws of nature. There are no eternal, immutable, necessary structures in the universe.

Another philosophically important consequence of this view is that it enables us to accept and justify (in evolutionary terms) the radical view to the effect that even our most basic conceptual resources for understanding the world, such as the causal modalities, have an ultimately expressive, normative and thus not semantically representational function: they are tools that successfully enable us to collectively orient ourselves in the surrounding world by constructing a cognitive map of the future course of events that reduces uncertainty in that it makes us more efficient in controlling the imminent course of events (see also Price, 1991, 2012).[6] Thus, this Darwinian rendering of PSR shows how its *non*-traditional *metaphysical* consequences go hand in hand with the (Sellars-inspired) *non-metaphysical* construal of orthodox PSR as an indispensable normative, practical and critical tool for successfully coordinating empirical inquiry.

[6] Interestingly, for Price, this perspectival expressivist view of causation implies, among other things, that treating the past as fixed and the future (or some part of it) as open is also a perspectival fact that has to do with what enables us to manipulate out environment, to put it under our control (Price, 2005). This does not mean that the future or the past considered aperspectivally are deterministic (it can be argued that determinism is also a perspectival viewpoint). Rather, it means that the indisputable fact that the past is not under our control while the future is need not imply that the fixity of the past and the openness of the future are opposing properties of discrete domains that do not really exist. Indeed, there are processual views that account for this practical asymmetry in another way, namely, as a result of the fact that the present is a contracted past, while the future is itself an enveloped present (Bergson, 1896; Deleuze, 1968). Note that on these processual views, in contrast to the received metaphysical views about time, both the past and the future *do* exist, albeit not actually but virtually (in the Deleuzian sense of the term). We shall see later how Deleuze's views on modality can mesh up with our Peirce-inspired evolutionary processual metaphysics.

6.1.3 *The Modal Status of Being as a Whole: Being as Intensive Multiplicity*

We now move on to the important question of what happens to the modal notions of necessity and contingency as applied to being *as a whole*.

From our 'ontological mutabilist' point of view, and given the orthodox actualist conception of the modal terms 'necessity' and 'contingency' in terms of possible worlds semantics, the modal status of being as a whole is more perspicuously described as neither necessary, nor contingent, and admits of no meta-modal characterizations ('necessarily contingent' (Meillassoux), 'contingently necessary' (Badiou), etc.). Another, Hegel-inspired suggestion here is to say that being as a whole can be best described as exhibiting 'the identity of absolute contingency and absolute necessity' (Hegel, 1816). This is because, on the one hand, processual being, as totally unconditioned, is without ground, and in this sense 'absolutely contingent', without sufficient reason. But precisely because of its unconditioned status, being can also be said to be 'absolutely necessary' in the sense that there is no ground or possibility for it *not* to be (Brown, 2021, pp. 86–89). Yet, I take it that it is ultimately less misleading to describe this unconditional status of being as neither necessary nor contingent, rather than as 'the identity of absolute contingency and absolute necessity', since it is difficult to make sense of what modal distinctions really amount to if they lose their contrastive nature. A third option (inspired by Whitehead (1929) and Deleuze (1968)) is to avoid reifying necessity and contingency not by metaphysically eliminating them, but by transforming them from static logical forms of thought or metaphysical principles of being to dynamic morphogenetic principles of becoming. Below, I shall develop a version of this view.

We can begin by noting that our commonsensical modal notion of possibility, understood as a potentiality with the exact same structure as actuality, which differs from the latter only in that it is (yet) realized in actual state of affairs, is misleading because it is based on a representationalist or actualist model of understanding possibility. This commonsensical understanding of possibility turns the latter into a static and already individuated entity (e.g. a possible horse, the form of horseness) inhering in some abstract realm of potentialities waiting to be realized and conceived on the model of the actual (actually existing horses) by abstracting the actual entity's essential properties from its accidental properties. But on this model, possibility is understood as an exact copy of the form of the actual existing entity to which it applies, the only difference being that possibilities 'subsist' as part of the essence of this entity, but they are not yet realized. Obviously, this model just conceives possibility based on a static, essentialist conception of the being of actual entities. It does not independently account for the special modal status of actuality, neither can it explain the very *becoming* of an actual entity—and the peculiar entanglement of possibility and actuality which this becoming presupposes. At best, it makes explicit the categorial status of the modality of possibility in the context of an object-property categorial framework. But, as was argued in Chap. 3, the object-property framework is a reflection of a metaphysics of static essences whose basic categories are not at the end of the day unconditional (they cannot account for their own necessity as well as for the *haecceity* of particular objects). In a similar vein, it can be argued that it is equally misleading to attribute the modality of *necessity* (in the sense of being 'true in all possible worlds') to the virtual principles of becoming, or *contingency* ('true in some possible worlds') to actualized worldly states of affairs. This is because the understanding of the possible worlds in question is at bottom actualist—a mere structural copy of the way we understand the structure of actual empirical states of affairs, the only difference being the range of worlds this structure is true of.

I think that we can throw light on the precise way in which necessity and contingency are dynamic morphogenetic principles of becoming by explicating the fundamental concept of *intensive multiplicity* (which I use here roughly in a Deleuzian sense). Indeed, the fact that intensive fields (Chap. 2) are structured as a multiplicity makes them an ideal successor 'category' not only of classic modal categories like possibility, necessity and contingency, but also of the traditional categories of essence, unity and quantity (one-many). The structure of intensive fields is that of a *continuous heterogeneous n-dimensional multiplicity*, which harbors spatiotemporal dynamisms (space and time are themselves multiplicities) and undergoes intrinsic ontogenetic variations expressible in differential equations. Multiplicities are fields of variation, self-differentiated systems that are prior to the division between one and many: they are not parts of a greater whole or totality, they do not have ultimate parts (their parts are themselves *continua* that have continuous parts), and they are not defined by reference to prior unities, forms, ready-made objects, properties or supplementary dimensions (e.g. extrinsically defined spacetime coordinates, essences, natural kinds).[7] They are continuous, heterogeneous and

[7] Thus, the notion of continuous heterogeneous multiplicity can be viewed as a successor categorial concept that replaces the traditional Aristotelian and Kantian categories of *quantity*, namely, unity (one), plurality (many) and totality (all).

intensive, as opposed to numerical, homogeneous and extensive ones (Deleuze, 1968). In contrast to the case of metric space which can be divided up into equal parts (numerical extensive multiplicity), intensive multiplicities are topologically articulated and cannot be divided up without changing in nature.[8] Due to their continuous, heterogeneous and intensive nature, the elements of the multiplicity have neither separate identity (as they are not discreet), nor sensible form (intensity is not itself sensible), nor conceptual signification (since the abstract universalizability of concepts presupposes a non-continuous and homogeneous mode of articulation). They are defined by differential relations, physically expressed as differences in potential, and they are distributions of singularities embedded within one another that can be unfolded by a series of critical transitions. Intensive multiplicities express the unfolding of processes through an n-dimensional state space (representing the relevant ways in which the processes can change) and their corresponding degrees of freedom. In turn, singularity points (e.g. boiling points[9]) are physically expressed in the long-term tendencies of the processes in question (e.g. their 'seeking' minimal free energy) as described by their *attractor*, their 'topological form', that can be realized in many different physical forms[10] (DeLanda, 2002, pp. 9–44). As this continuous n-dimensional intensive multiplicity self-differentiates (differentially repeats itself), it immanently produces, through a series of spatiotemporal symmetry breaking cascades (phase transitions) in its singularity points, and under broadly evolutionary principles, the infinite heterogeneity of actual forms.[11] Note that the mode of intelligibility here is more akin to a morphogenetic immanent

[8] Thus understood, multiplicities, as particular intensive movements of change inhering in the actual, are the successor concepts of the classical metaphysical concepts of possibility and essence and are situated at a metaphysically deeper explanatory level than the latter.

[9] A boiling point is a singularity because what it is to be such a point is to be the difference between the liquid and gaseous state of a chemical substance. Indeed, from this point of view, to be a certain substance is to change in various ways relative to temperature, pressure, etc. It is these differences within a chemical substance's structure that make it what it is to be that substance.

[10] Note here that the direction and stability that attractors confer to the unfolding of the processes under their influence is itself an outcome of broadly evolutionary processes. Which attractor a system happens to be in at any one time is determined, in part, by its evolutionary-selectional history (DeLanda, 2002, p. 38).

[11] Importantly, this process of intensive individuation which accounts for the production of actual objects also brings about their powers and capacities. The latter do not pre-exist as such in virtual multiplicities.

explication (unfolding), rather than to a model of a logical, transcendental or conceptual *explanation*.

Given the above, and having reconceived the modal notions of necessity and contingency in such a way as to avoid their commonsensical actualist interpretations, we are now in a position to sketch our conception of the modal status of being as a whole. I suggest that the folding-unfolding of intensive multiplicities under evolutionary principles is *virtually necessary* while its particular modes of being (beings) are *actually contingent* but only on condition that the terms (virtual) 'metaphysical necessity' and (actual) 'nomological contingency' are *counter*-commonsensically reconceptualized in such a way that virtual necessity *implies* actual contingency and vice versa. This metaphysically novel way of understanding modality might seem to resemble Meillassoux's view that everything is necessarily contingent. Yet, the abovementioned reciprocal implication between virtual necessity and actual contingency is not correctly described as a *vertical meta-level* modal relation between being and beings (e.g. 'necessary contingency' (Meillassoux), 'contingent necessity' (Badiou) of everything that exists), but rather as a *horizontal first-level* immanent relation in which virtual necessity and actual contingency immanently point to one another as distinguishable yet inseparable poles of the productivity of becoming. Thus, from this point of view, what turns out to be metaphysically necessary (the ultimate ground) is the virtual necessity of the existence and self-differentiation of being into beings along with the 'infinitesimal chance tendency toward habit-taking', as the latter is 'explicated' by the unfolding of intensive multiplicities, and self-reinforced, diversified and constrained by broadly evolutionary principles of becoming. And this, of course, is the exact *opposite* of what an ultimate ground is traditionally conceived to be (again, one is here reminded of Peirce's 'Tychism'[12] (Peirce, 1892)). Conversely, radical contingency, the lack of any kind of

[12] Peirce called his doctrine that chance has an objective status in the universe 'tychism', a word taken from the Greek word for 'chance' or 'luck'. Tychism is a fundamental part of Peirce's mature view. As we saw, for Peirce nature is not a static world captured by universal natural laws but rather a dynamic and dicey world of evolved and continually evolving habits that exhibit considerable creativity. Note here the radical difference of this conception of chance from the respective Aristotelian conception in which chance is accidental and derivative, understood in terms of inexplicable deviation from (failure of actualization of) the ordinary causal-dispositional powers of objects. From this radical Peircean (and Deleuzian) standpoint, chance is essentially unsubordinated to law or prior probability distributions and lies at the heart of being.

real necessity and thus the universal ungrounding at the level of the actual world, proves to be the ultimate *essence* of every particular actual entity. Recall that here we are working with a *successor* concept of contingency according to which the latter is not understood in terms of the possibility that things 'could have been otherwise' (as if there is a ready-made stock or totality of possibilities from which some are actualized while others remain in a state of potentiality), but just denotes the fact that actual entities are not subordinated to any higher necessity pre-determining their form.[13]

Again, it must be stressed that this reciprocal implication between metaphysical chance and necessity must take place immanently, horizontally and not at a modal meta-level (*a la* Meillassoux), as otherwise we risk the danger of collapsing the logical truth that 'everything could *be* otherwise' into the Meillasouxian controversial metaphysical hypostatization of this logical truth as expressed in his 'principle of unreason' according to which 'everything could *become* otherwise', and where miraculous events can occur at any and every moment without any reason, cause or other kinds of (e.g. evolutionary) worldly constraints, save the fact that no self-contradictory entities can exist. By contrast, in the above Deleuze-inspired horizontal model of reciprocal implication between virtual necessity and actual contingency, the causal history of the actual world, as virtually retained in its actual present state, constrains the expression of virtual tendencies. Thus, from this point of view, the 'rules' (possibility space) of the 'cosmic game' change with each roll (iteration) of the cosmic dice, at every passing moment. In this way, novelty is introduced as an essential feature of the world which is yet subject to constraints imposed by the retention of the entire past in the present. This retention constrains—without uniquely determining—the (constantly changing) field of possibilities that open up for the future (see Chap. 4).[14]

[13] Note that in this (counterpart) sense, contingency and necessity are not at bottom conceived in terms of the counterfactual ability of things to be—or not to be—otherwise than they actually are. Becoming can well be unsubordinated to pre-formed essences (and in this sense contingent) while at the same time its unfolding, being essentially occurrent, is not perspicuously described as something that could or could not be otherwise.

[14] Of course, since processes, even if monadic and perspectival in their intrinsic structure, affect and are affected by a host of other processes at all spatiotemporal levels and scales, this constraint of the actual on the expression of the virtual leaves enough degrees of freedom to the latter. Yet, it does not amount—and in fact it is completely opposed to—Meillassoux's

Lastly, note that on this conception of metaphysical grounding being is unconditional and cannot be itself created or 'come to existence' by something other than it or emerge out of 'complete nothingness'. Creation *ex nihilo* cannot account for the existence of being as such. This is because, in a framework that operates with this principle, either the existence of the world is already presupposed—and thus not really accounted for—in the form of a potentiality (in the mind of God or in a transcendent realm of possibilities) waiting to be realized, or the world pops up into existence out of complete nothingness. But the notion of complete nothingness, while not presupposing being as a ready-made potentiality, is actually derivative from the notion of being: it is constructed by first positing being and then negating it, and thus again cannot independently account for it (Bergson, 1907). Importantly, once we realize that the notion of creation *ex nihilo* does not make sense we dispel the mystery and the supposed deep profundity and wonder at the very existence of the world ('why is it something rather than nothing?'). This stance of wonder at the very existence of the world would be justified only if it was possible for the world as a whole *not* to exist at all, that is, if absolute nothingness was a coherent possibility. But this possibility is only illusory.

I think that, overall, the above Deleuze-inspired morphogenetic rendering of contingency and necessity is compatible with—and can indeed enrich—the Peircean understanding of PSR as entailing an evolutionary monist and ontologically mutabilist account of being. An important difference, which I take to be in favor of my model, is that Deleuze does not seem to have room for a positive *non*-metaphysical role for modal and causal necessity, as he does not even consider the possibility for understanding the necessity involved (as expressed in counterfactually robust laws of nature) in non-representational, normative, pragmatic and critical terms, as expressing regulative ideals of inquiry indispensable for its self-correcting function and development. At best, Deleuze would presumably consider PSR in its traditional form as a 'transcendental illusion' arising from the 'representationalist' metaphysically inflationist demands and

identification of metaphysical with logical possibility, where the latter, after being stripped of any reference to actuality and its becoming (material, spatiotemporal, evolutionary, historical, psychological, social constraints), is then viewed as a model for understanding the very essence of actuality.

assumptions of philosophical thought.[15] Yet, *pace* Deleuze, even if it is correct that more often than not PSR metaphysically reifies essences (forms, capacities, laws, powers) as the latter are conceived in the context of a substance-property categorial framework, it is also true that the function of PSR, even in its traditional form, need not be understood as simply an illusion. PSR is indeed objectionable and metaphysically inflationist if the necessity of its grounding principles is interpreted in a directly representationalist fashion, as purporting to cut nature at its joints (see Chap. 3). Yet, as we saw, PSR is also, and primarily, a principle with normative, critical and prescriptive content, which helps us to reliably navigate the space of causal modalities, by making this space explicit, and to critically revise our core commitments about the structure of this possibility space, thus, indirectly, about the world and ourselves. In this sense, PSR is a useful, indeed indispensable (yet non-representational) meta-tool of our cognitive practices, without which we could not formulate and bring to semantic self-consciousness the basic principles (Deleuzian intensive multiplicities, Peirceian combination of continuity, chance variation and natural selection) that can account for the morphogenetic production of actual entities. And, as we saw, this non-metaphysical reinterpretation of the necessity involved in the causal modalities in the context of classical PSR is not only compatible with this non-classical Darwinian ontological mutabilist reading of PSR but is actually one of its most important *consequences*.

6.2 A Novel Account of Grounding: Against Neo-Humeanism, Neo-Aristotelianism and Neo-Hegelian Absolute Idealism

Importantly, the above Darwinian-Tychistic reading of PSR, with its (Deleuze-inspired) reconceptualization of the relation between metaphysical possibility and necessity (where metaphysical necessity implies

[15] In Deleuze's view, representational thinking pervades vast regions of classical metaphysical philosophy (*philosophia perennis*) and has even been internalized into our commonsensical view of ourselves-in-the-world. According to this conception, which finds philosophical expression in such otherwise different philosophical systems as those of Aristotle, Kant, ordinary language philosophers and even Hegel, the world and ourselves can only be made intelligible in the context of the fundamental categories of identity, essence, object (substance) and its essential and accidental properties, as the latter as represented by a unified subject of experiences in a world governed by laws (natural or moral). Representational thinking understands difference as always mediated by—and hence as essentially subordinate to—a prior identity, analogy, contrariety, resemblance and negation.

nomological contingency and vice versa), can be utilized to solve core problems in contemporary analytic metaphysics of powers. In particular, it will be suggested that our above-sketched novel conception of PSR can provide a viable alternative to neo-Aristotelianism, neo-Humean and neo-Hegelian absolute idealist readings of PSR.

6.2.1 Humeanism, Realism and Idealism in Power Metaphysics

As is well known, contemporary analytic metaphysics of powers is divided into two main competing positions about the metaphysical status of powers: neo-Humean anti-realists about powers according to which the causal modalities (powers, dispositions) of objects must be reductively understood in terms of regularities or constant conjunctions (David Lewis (1983) is the main proponent of this view) and neo-Aristotelian realists who treat powers as metaphysically ultimate explainers, and thus as grounds or ultimate truth-makers of the behavior of actual objects and processes (with Lowe (2006) and Heil (2012) being well-known supporters of this view). Recently, it has been suggested (Mudler, 2022) that a significant novel contribution to this debate over the metaphysical status of powers comes from contemporary analytic neo-Hegelians, such as Rödl (2007) and Kimhi (2018), who also take powers as metaphysical ultimate explainers but construe them in an absolute idealist manner (as opposed to the metaphysical realist stance of neo-Aristotelianism about powers), which internally connects the power of thinking to powers as expressed in objects *in re*. The absolute idealist slogan which succinctly expresses the above view is: 'to be is to be intelligible'.

We can begin to see how our virtual-evolutionary PSR framework can illuminate this discussion about the metaphysical status of powers by recalling that the shift toward powers in analytic metaphysics was a result of attempting to understand the mode of *generality* that an adequate explanation must have in order to function as a metaphysical *ground*. As we shall see, based on this framework, we can understand neo-Aristotelian realism as a dialectical response to neo-Humeanism (correcting the explanatory anomalies of this latter view) and analytic neo-Hegelianism as essentially the same kind of response to (what it conceives to be) the shortcomings of neo-Aristotelian realism. Our own virtual-evolutionary PSR version will, accordingly, dialectically emerge as a view that attempts to 'sublate' analytic absolute idealism about powers.

To see how this whole dialectic gets off the ground, note first that to serve as a metaphysical ground, explanation of particular matters of fact must rest on something general even on the neo-Humean picture. Of course, the only way for a neo-Humean, for example, of a Lewisian stripe, to arrive at something general as a ground is by quantification over the elements of what Lewis calls the 'mosaic' of particular matters of fact, involving nothing but categorical (non-dispositional) properties, whose instantiation at one specific time and location has no implications as to which properties are instantiated at other times and locations[16] (Lewis, 1983, 1986). As is well known, the neo-Humean view attempts to account for causation and the counterfactual robustness of the laws of nature in terms of regularities—that is, the actual spatiotemporal arrangement of the atomic intrinsic qualities of the universe (whatever the latter might turn out to be). The neo-Humean deems quantification over regularities unproblematic because it is simply a convenient way of listing (potentially infinitely) many of the matters of fact he endorses as the foundation of his worldview. Of course, these supposedly unproblematic matters of fact form nothing more than constant conjunctions and relations of temporal or spatial contiguity—adding thus more matter of factual relations into the equation, without any non-brute-fact-like generality to account for them. And this is precisely the basis of the familiar objection against neo-Humeanism: this sort of generality by quantifying over these Humean-like particular matters of fact seems to merely *summarize* them, rather than explaining them in the sense of grounding (Mudler, 2018; Stroud, 2011). The generality of the constant conjunction form seems to be parasitic on its instances instead of providing their explanatory ground. The Humean might argue that she can explain a given event by *subsuming* it under general patterns of the mosaic of particular matters of fact, but this process of subsumption necessarily ends somewhere, presumably at the level of the

[16] Lewis calls this view 'Humean supervenience' and defines it as follows: '[It] is the doctrine that all there is to the world is a vast mosaic of local matters of particular fact, just one little thing and then another. … We have geometry: a system of external relations of spatiotemporal distance between points. Maybe points of space-time itself, maybe point-sized bits of matter or ether or fields, maybe both. And at those points we have local qualities: perfectly natural intrinsic properties which need nothing bigger than a point at which to be instantiated. For short: we have an arrangement of qualities. And that is all. There is no difference without difference in the arrangement of qualities. All else supervenes on that' (Lewis, 1986, pp. ix–x). On this view, ordinary modally thick material objects are mereological sums of these basic intrinsic qualitative non-modal properties.

most general worldly regularities, and there is no explanation of them (Bhogal, 2021). This does not seem to satisfy the explanatory demands of a metaphysical ground, since for something to function as an explanatory ground in this metaphysical sense it must provide non-causal explanations of certain target phenomena by appeal to ontologically more fundamental phenomena or principles on which the existence or truth of the former depend.[17]

Powers are introduced precisely as a way of addressing this problem. Proponents of this view argue that powers can account for the behavior of particular events by providing an (ultimate) explanatory ground for the intelligibility (e.g. lawlike character, casual relations) of this behavior. The popular examples of powers or dispositions on offer in philosophical discussions—fragility, solubility, elasticity, inflammability—are all designed to make explicit this explanatory connection between powers and the actual behavior of objects that exhibit them. Powers are potentialities that inhere in actual objects and point toward their own manifestations in actual episodes of breaking, dissolving, stretching or burning. They thus function as ontological posits that explain their own manifestations, making intelligible how effects (particular events expressing the behavior of objects) can arise out of or derived from their ontological ground (powers of objects).[18] For powers are the ontological sort of thing that, under certain conditions, brings about their own manifestation (Molnar, 2003). In this sense, powers can be viewed as 'ontologized explanations'.

Interestingly, we can discern an internal division within the camp of power metaphysics, depending on whether the emphasis on the above dictum that 'powers are ontologized explanations' goes to the fact that they are *ontologized* explanations (powers bringing about their manifestations independently of minds) or to the fact that they are ontologized *explanations* (where explanation is a rational practice that seeks to make worldly events and their relations intelligible to the mind). While most power metaphysicians are metaphysical realists about powers in that they believe that powers do not metaphysically depend on the existence of logical categories, recently, certain analytic neo-Hegelians, such as Rödl and

[17] For a comprehensive introduction on the topic of metaphysical grounding in contemporary analytic metaphysics see Correia & Schnieder, 2012.

[18] Indeed, from this point of view, causal relations themselves are the mutual manifestings of the objects' reciprocal powers. Causes are understood in terms of interactions between empowered objects, with effects being the manifestations of these reciprocal powers (Heil, 2021, pp. 120–123).

Kimhi, have taken issue with this position, arguing that an 'absolute idealist' construal of powers is a more coherent metaphysical position than the realist one (Mudler, 2022). Both absolute idealists and realists about powers hold that powers are ontologically basic unexplained explainers of particular matters of fact. Yet they disagree over the purport and the nature of the distinction between metaphysical and logical or rational categories. What is the relation between the powers of things or substances and the powers of thinking? Are they metaphysically distinct? Are they asymmetric (the latter being dependent on the former)? If, as it surely seems, our categories of powers of things *in re* mirrors our categories of the powers of thinking (in Brandom's rendering, our 'alethic modal' categories mirror the 'deontic modal' ones[19]) what is the nature of this categorial mirroring? Is it ontologically asymmetric or not, and if yes, what is the direction of dependence? Realists about powers think of metaphysical power categories as mind-independent. Powers of nature are elements of the natural order, and not of the rational order. Thinking is not identical to being but is instead an activity in which some of the beings engage, and others not. And the beings that engage in thinking are distinguished by the others which do not by having a distinctive set of powers, the powers characteristic of thinking (reasoning, free will, intentional action). By contrast, according to absolute idealists, thought is not some power among others. Nor does it fall under any other ontological category, because those categories (e.g. of beings and their distinctive powers) are *themselves* categories of thought. From this point of view, the *nature* of a thing is identified with its *concept*, and powers are understood as being such concepts in their *explanatory* capacity (Rödl, 2018).

In this way, building on a different account of the way in which generality is implicated in explanations if the latter are to function as metaphysical grounds, absolute idealists remind us of the fact—registered in the tradition of *philosophia perennis*—that generality is the mark of the *intellect*, of the understanding, and that explanation is one of its fundamental acts. From this point of view, the realist goes astray in positing mind-independent powers or mind-independent relations holding between particular matters of fact and 'necessitating universals' existing 'out there' (Armstrong, 1978). Nor should we be tempted into positing such mind-independent

[19] We will explore this issue further in Chap. 7 in terms of a categorial mirroring between the space of reasons that articulates objective knowledge and the space of causes that expresses objective facts.

universals in the first place. For this seems to make the relation between powers *in re*, as manifested in particular matters of fact and the power of thinking essentially external, whereas this 'categorial mirroring' relation (to which neo-Aristotelian realists are also committed) is most coherently viewed as an internal one. For, to be a 'matter of fact' in the first place is for something to be an element in a system of other such facts which are internally connected through laws of nature (Rödl, 2012). Indeed, as Sellars saw—recapitulating Kant's argument against Hume—'to grant that there is knowledge of the here and now is to grant that there are general truths of the sort captured by lawlike statements' (Sellars, 1967, §54). The same point is also stressed by Brandom, who, following Sellars, holds that the ability to use ordinary empirical descriptive terms such as 'mouse', 'green', 'rigid', and 'mass'—and understanding the matter-of-factual relations in which they participate—already presupposes grasp of the kinds of properties and relations made explicit by modal vocabulary, including most importantly, lawlike counterfactually robust properties and relations expressing the dispositions, capacities and powers of objects *in re* (Brandom, 2015). From this point of view, shared by otherwise very differently minded philosophers such as Sellars, Brandom and Rödl, *contra* Hume (and neo-Humeanism about powers), there is no antecedently intelligible stratum of empirical discourse that is purely descriptive and involves no modal commitments. And if this way of viewing things is correct, it also presents a challenge to neo-Aristotelian realism about powers as the latter seems to understand the relation between metaphysical and logical (rational) categories as merely *external*, rather than an internal one. In this way, realist defenders of powers fail to see that the very idea of a power comes as part of a package in which metaphysical categories are internally connected to rational ones. What is meant by our claims about what objective facts, laws or powers there are in the world is dependent for its very sense on aspects of our norm-governed rational practices of asserting claims and making inferences, and thus, on the power of thinking.[20]

[20] Note that understanding metaphysical and logical categories—or the powers of things and the power of thought—as internally connected does not *entail* absolute idealism *a la* Rödl. For example, it is consistent even with a radically anti-idealism and materialist position according to which this internal relation holds only for the 'phenomenal world', but not for the 'in-itself' which lacks logical form. Indeed, Sellars is committed to a variant of this view, where the phenomenal world is the world of the manifest image while the world as it is in itself, *contra* Kant, is knowable, lacks logical form and revealed by the ideal scientific image (Sellars, 1967).

Thus, it seems that we should take seriously the fact that neo-Humeanism fails to deliver the proper sense of generality required for metaphysical grounding explanations, and seek to understand the grounding relation of dependence of one matter of fact on others, as expressed by our understanding of causation and the counterfactual robustness of the laws of nature, as an explanatory requirement derived from the essence of thinking itself; a requirement that can be thought of as itself *embodied* in the very *idea* of such particular matters of fact related in lawful—counterfactually robust—ways. Empirical matters of fact themselves are to be seen as including something general on which explanation may rest. And these are precisely their powers as expressed in the counterfactual robust laws of nature that are manifested in their behavior. Moreover, even though, as was mentioned above, understanding metaphysical and logical powers as internally connected does not by itself entail Rödlian absolute idealism, it naturally tends to lead toward this cluster of views: for Rödlian absolute idealism neatly accounts for this internal connection by viewing metaphysical and logical powers as ways of making explicit the explanatory and normative dimensions of thinking, and, ultimately, as ways of articulating the (formal) identity of being and thinking. And this view, unlike most neo-Aristotelian realist ones, is not justified merely on the basis of pre-theoretical intuitions or speculative metaphysical postulation (or a 'reflective equilibrium' between them), but through a more robust transcendental reflection on the very possibility of what it is for thought to have an object (to be about worldly objects). In this way, the idea of a power—say, the ability of lions to roar—is shown to be at work in the very idea of something being an object of thought, namely, the idea of being something objective or real. Grasping a matter of fact (a lion roaring) is always already, grasping its explanatory ground: the powers that explain why, how and under what circumstances the substance of predication does what it does. It thus seems that from this point of view powers, by internally connecting the necessity and generality of the explanans (powers) with the contingency and particularity of the explanandum (manifestation of powers in matters of fact), are ideally suited to function as metaphysical grounds. And by conceiving these metaphysical grounds as equally pertaining to thinking and to being, this view thereby makes grounding at once absolute and reflexive. In a word, it construes grounding in terms of a *self-grounding* principle.

Yet, this neat absolute idealist rendering of power as a self-grounding principle that internally connects thinking and being comes at a high cost:

according to this line of thought, thinking is not just another power among others (just as thinkers are not just objects among other objects). Thought is essentially self-conscious: just as we can think about lions, cats or aluminum, we can also think about thought. But we do not thereby focus on just another topic among many possible topics for thought. For, I can neither think about lions or cats without thereby already knowing myself thinking these thoughts—whereas whether I think about lions or cats or whatever else I want is not necessitated by the form of thought. It is precisely *in* thinking about lions that I comprehend myself *as* thinking. Importantly, these are not two different thoughts: I am not, on the one hand, thinking about lions and their powers, and, in an independent higher-order thought, thinking myself to think just that. When I turn my attention away from lions and their powers toward thought, I do not turn my attention from one topic to another (from worldly powers to the power of thinking), but rather to the very idea of power (Rödl, 2007, 2018). This is why according to Rödl the power of thought is not a power among others.

As is obvious, absolute idealist power metaphysics differs in a rather peculiar way from other brands of (realist, neo-Aristotelianist) power metaphysics. Although it presents us with a neat understanding of the internal relation between metaphysical and logical powers and offers a novel way of considering powers as metaphysical grounds, it is not without problems: for example, it seems strange to say the least to hold that the real world out there to which our thought aim to describe is *literally* articulated in a form *identical* to that of thought—namely, conceptual-judgmental form. Are the benefits of this absolute idealist view over its realist opponents such that they counterbalance the costs of a) ontologically and explanatorily *reducing* objectivity, reality and truth to self-conscious judgmental-conceptual form and b) understanding the power of thought and the objective world disclosed by it as something that can bootstrap itself into existence, incapable of being existentially understood in terms that do not already presuppose it? Does it not seem here that Rödlian judgmentally articulated self-consciousness (which simultaneously discloses nature and everything that can be 'talked about' objectively) comes alarmingly close to functioning as 'the last stand of Special Creation'? (Sellars, 1963, p. 6). A second, related—and to my mind even more serious—problem for Rödlian absolute idealism is that it does not live up to its own standards of rationality, namely, to the task of providing metaphysical grounding that satisfies the principle of sufficient reason.

Absolute idealists are not radical enough in their rationalism in that they do not really reach a *completely presuppositionless* viewpoint regarding the conditions of intelligibility of being, since they give an account for the 'whatness' of beings (the 'essence' of things) that remains externally related to their 'thatness' (their 'factical' existence). It will be suggested that our Darwinian-cum-pragmatic account of PSR, modally interpreted along broadly Deleuzian lines (where the virtual necessity of the principles is immanently connected to the absolute contingency of every actual entity and vice versa), is in a better position to satisfy the explanatory ideals built in the principle of sufficient reason, and thus, can provide a more robust metaphysical grounding principles than Humeanism, neo-Aristotelian realism and absolute idealism about powers.

6.2.2 Morphogenetic Vs Logical-Subsumptive Grounding

Presumably, absolute idealists, or indeed, philosophers such as Brandom (2019), McDowell (1994) and Pippin (2018), who are not absolute idealists in Rödl's sense but all agree that objective reality is conceptually articulated, will find any attempt to formulate a notion of metaphysical grounding which is essentially non-conceptually articulated incoherent or at best a futile attempt at an ineffable pointing to the world as it exists in itself independently of the conceptual structure we impose on it. But this attempt only seems futile if we already take it for granted that the logical form of judgments is isomorphic to the ontological form or reality (see also Morris, 2005). And, as we saw in Chaps. 3 and 4, we are not obliged to take this view for granted. Instead, properly transcendental considerations about how thinking can at once be about the world and within the world lead us to the postulation of an alternative ontological framework, based on the notions of process (instead of object) virtuality (instead of possibility) and multiplicity (instead of essence), that is resolutely non-conceptually articulated since its basic 'categories' all self-consciously depart from Kant's traditional categories as delineated in his table of judgments. Thus, even if the absolute idealist's dictum 'to be is to be intelligible' does have a grain of truth (as a necessary requirement and concise expression of what a self-grounding principle might be), the intelligibility involved in the isomorphism or homeomorphism between thinking and being need *not* be conceptual. It could well be non-conceptual, topological in form, in which case self-grounding would be itself conceived in process-topological rather than objectual-judgmental terms, and where the latter would be understood as a necessary moment (normatively guided representational stabilization for facilitating our collective epistemic coordination) in the unfolding of the latter. In this case, ironically, the resulting view would amount to an *aufhebung* of absolute idealism itself.

Of course, even if the above are in the right direction, they do not by themselves provide a positive account of how the non-conceptual articulation of reality serves as a *constraint* in virtue of which our conceptually articulated judgments (propositions) about reality are true (or false). Note, moreover, that for the constraint in question to be a metaphysical ground it must not just constrain true proposition as a mere *cause*. Neither can its necessitating power as a ground come in the form of a necessitating *reason* for their being true, since reasons as traditionally understood clearly have conceptual structure.[21] So, in what sense exactly can non-conceptual reality ground its own appearance in conceptual form, that is, in the form of states of affairs or facts represented by true propositions? We have seen how this problem might be dealt with by conceiving non-conceptual becoming as essentially self-differentiating and self-fracturing (Chaps. 4 and 5). To that it must be added that from our point of view this problem cannot be solved merely by positing 'internal necessitating relations' between propositions and non-propositional truth-makers such as facts or states of affairs (Armstrong, 2004), since, as we shall see in Chap. 7, facts and states of affairs, to the extent to which they, and their constituent parts, are understood as the worldly structural correlates of true propositions (the non-propositional 'portions of reality' to which propositions correspond when they are true), are *also* conceptually articulated, being essentially a 'transposed language of norms' projected, in the form of alethic modal norms, into the world (Brandom, 2019). In fact, from this standpoint, the whole 'skeletal' semantic framework in terms of which the objective world is disclosed to us in the context of our everyday lifeworld (senses conceived as expressing contents referring to objects, properties and relations as intentionally related constituents of facts or states of affairs) is, at bottom, conceptually articulated (see Chap. 7), and hence,

[21] Interestingly, the requirement that the necessitating power of a metaphysical ground must be neither causal nor strictly logical (relations of logical or conceptual entailment) is built into many versions of contemporary truth-making theory: according to it, to function as a metaphysical ground, the truth-maker must express a non-propositional reality which has the 'ontological' power to necessitate the truth of a proposition (Armstrong, 2004).

cannot function, on pain of circularity, as a metaphysical ground of conceptually articulated true propositions. Also, this problem cannot be solved by postulating 'one totality fact' (which, akin to substance, internally relates all particular determinate facts) as the ultimate non-propositional truth-maker of the truth of particular propositions (for variants of this view see Armstrong, 2004, Schaffer, 2010, Cameron, 2010). For, even if this avoids incorporating extensional relations between distinct terms into facts (thus rescuing necessary connections from the Humean challenge), it reinstates the logical isomorphy between the non-propositional truth-makers and true propositions, thereby undermining the very logic of introducing truth-makers as explanatory grounds of propositional truth in the first place (Bell, 2016).

I take it that the key to a deeper understanding and resolution of the above debate between neo-Humeanists, metaphysical realists and absolute idealists about powers lies in questioning the intelligibility of an unthematized background assumption that, despite their significant differences, all parties take for granted (even if only to reject it): the assumption in question concerns the kind of *generality* that a metaphysical ground must have in order to be explanatory. It is taken for granted by all parties of the debate that the only model for understanding how the generality implicated in a metaphysical ground can be explanatory is by construing it as a *universal* that is *logically* related to its instances (particular matters of fact) in a *subsumptive* manner. Particular matters of fact are explanatorily grounded by being subsumed under the grounding universal. Lacking any alternative model for understanding the kind of generality implicated in metaphysical grounding, we can easily understand:

(a) why neo-Humeans flatly reject this account of grounding: it does not provide a compelling reason for considering the necessity of the ground as a real connection *in re*, but just verbally solves the grounding problem by mere postulation;
(b) why neo-Aristotelian realists about powers, not being content with a mere logical relation between (universal) ground and (particular) grounded facts, attempt to 'ontologize' the former (so as it be really and independently explanatory) by conceiving grounding powers as real and radically mind-independent;
(c) why absolute idealists reject a mind-independent conception of powers in which grounding powers are merely externally related to the power of thinking, holding instead that a properly comprehen-

sive and reflexive grounding metaphysics requires that thinking and being be internally related ('to be is to be intelligible').

That is, neo-Aristotelian realism and absolute idealism about powers can be understood as primarily disagreeing over the nature of that which provides metaphysical 'flesh' to the following skeletal explanatory requirement about the grounding relation between general and particular to which both parties subscribe: the requirement that the generality of grounds is explanatorily connected to the particularity of what is grounded only if the former is viewed as a universal that is logically related to its instances (particular matters of fact) in a subsumptive manner. But, as we saw earlier, both realism and absolute idealism are problematic positions regarding their conception of power as ground: while neo-Aristotelian realists correctly see the need for the ground being independent from what is grounded (thus their conception of powers *in re* as independent of power of thought and its judgmental-conceptual form), they end up with an account of powers in the order of being that is *structurally isomorphic* to—and thus not sufficiently independent of—the power of thinking, while at the same time conceiving of this relation between non-conceptual and conceptual powers as *external*, thus making the explanatory relation in question unintelligible. And while absolute idealists (correctly in my view) are sensitive to the seriousness of the above problems for realists about powers, they themselves face equally serious problems since their conception of grounding, notwithstanding its greater internal coherence and self-reflexive character, ends up sacrificing the explanatory independence of the ground in relation to the grounded, and indeed collapses the non-conceptual dimension of grounds *in re* to the judgmental articulation of the power of thinking in a way which makes the structure of self-consciousness a kind of metaphysical *causa sui*. In this way, it loses from view an essentially *asymmetric* dimension of the relation between thinking and being: being's existence (power) is unconditional, while thinking's existence (and the disclosure of a truth-evaluable objective world it makes possible) is conditional upon being's existence (power).[22] While the world can always be construed as a structural correlate of thinking, this

[22] Indeed, as we saw in Chaps. 2 and 3, the ontological relation between thinking and being is more intimate: thinking is an expression of a self-fracturing of being. But this does not abolish the conditionality of thinking and the unconditionality of being. It just clarifies the sense of the conditionality and unconditionality in question.

correlational structure as a whole itself depends for its very disclosure on the existence of a non-correlational, pre-manifestational world.

I take it that the key to making progress in the direction of a more adequate account of grounding is to insist on the explanatory independence of the ground from the grounded, and thus on the resolutely non-conceptual articulation of grounding principles, while at the same time understanding the explanatory relation of the ground (as universal) to what is grounded (as particular) not as a logically mediated subsumptive relation but as a relation of *immanent morphogenetic production*. Our model for this novel grounding relation is the Deleuzian virtual, which, in its expression as continuous intensive multiplicity, is a real generative power of the actual and explains its infinite heterogeneity of actual forms and structures of things by precisely not structurally resembling with the latter. The grounding explanation here is a transcendentally informed morphogenetic one (e.g. it can come in the form of topological variations of the rates of change of a developmental process or systems of processes), as opposed to a merely 'logical' or 'conceptual' understanding of grounding in which the ground structurally resembles the grounded and only accounts for its condition of possibility. And the morphogenetic unfolding of these immanently actualized grounding principles can be understood along the Darwinian-Tychistic model developed earlier: as a becoming that lacks causal and logical structure and immanently accounts for the actual natural spatiotemporal order (in evolutionary-selectional-dynamic systems theoretical terms) and the emergence and function of causal and modal necessity (in pragmatic-anthropological terms[23]) without itself structurally resembling the latter.

In this way, we can avoid ontological commitment to powers as ultimate grounding principles, since the latter can be understood as necessary grounding principles only *within a certain categorial framework* (the substance-property framework), which however is not absolute, unrevisable or ultimate, and which can in fact be *itself* explanatorily grounded by appeal to a successor process framework where Aristotelian potentialities of objects appear as reified expressions of their occurrent virtual powers of 'affecting and be affected' by other such objects, and where objects

[23] For more details about the pragmatic-anthropological underpinnings of modal necessity as the latter is encapsulated in our conception of powers, see the discussion below.

themselves are understood as reifications of the relatively stable co-occurring processual force-field that constitutes them (Chap. 4). From this vantage explanatory point of view, powers can also be more positively characterized as having a normative coordinating function in our epistemic practices, along the lines of our neo-Sellarsian analysis of role of the causal modalities in the PSR framework developed in Sect. 6.1: powers, as expressed in our understanding of the dispositional properties of objects and the capacities and abilities of living beings (including humans), are indispensable skeletal semantic tools of the object-property categorial framework that enable us to understand and represent to ourselves the behavior of non-living things and living beings as characterized by the contrast between enabling and inhibiting factors, under which an object or living being manifests or fails to manifest its distinctive powers, and the related contrast between a range of normal conditions, under which things behave in a predictable way, and abnormal conditions, under which the manifestation of the thing's powers is hindered by known external objects or becomes unpredictable due to unknown factors. The ordinary modal framework of powers is thus in essence a semantic framework condition—at once normative and explanatory—of understanding and predicting the behavior of things and living beings in a way that simultaneously normatively distinguishes the 'good' (normal, enabling) cases from the 'bad' (abnormal, inhibiting) ones regarding the manifestation of what things are able to do. In this way, by internalizing the modal framework of powers, and by thus coming to represent to ourselves individually and collectively the conditions of their manifestation and the changing contexts and circumstances in which these conditions hold or do not hold, we become capable of registering, systematically controlling and redirecting our behavior toward the paths that we take to be best for the successful actualization of our own conative and cognitive powers. And while the power framework, especially in its semantic dimension in which the world is understood as a modally robust space of actual and possible facts which our judgments may succeed or fail to represent truly, comes first 'in the order of knowledge', in the sense that it is epistemologically necessary for arriving at a processual picture of ourselves-in-the-world, it does not thereby enjoy any priority 'in the order of being'. On the contrary, as we saw in Chap. 4, it points toward an ontology that is not structured along representationally functioning power categories.[24]

Now, returning to our immanent morphogenetic grounding model, it must be stressed that, *contra* neo-Humeanism, it does not base its fundamental ontology on first-order quantificational logic according to which the only way of explaining the relation between ground and grounded is by appeal to mereology and mereological parts of ordinary lifeworld wholes, where the parts in question are described in the extensionalist language of fundamental physics (Lewis, 1991, 1994). My problem here is not with extensionalism. Neither do I hold that a proper grounding explanation of a lifeworld thing-kind (lion, knife) must reproduce its ordinary meaning along intentional essentialist hylomorphic lines (as Wiggins (1967) and Baker (1997) have suggested). I myself have argued that a process framework which provides successor categorial concepts replacing those of the object-property framework is more explanatorily fruitful in ontologically grounding ordinary things-kinds and life-forms, and I have insisted that this necessitates a *change* in their very meaning and their reconceptualization along a different, though not unrelated, successor categorial axis—eschewing traditional modally thick categories (powers, dispositions), and indeed, conceptually articulated content altogether. My problem is specifically with the *mereological* understanding of this

[24] Aristotle's whole philosophy can be understood as a systematic delineation of the conceptual framework of potentialities as the latter are expressed at all levels of being and in all kinds of beings. Of course, this framework—which is no exaggeration to say that a variant of it is still internalized by all of us in our ordinary upbringing and initiation in language—has its shortcomings and explanatory dead ends: abnormal conditions and inhibiting factors are only negatively characterized; abnormal conditions express the absence of normal conditions and inhibiting factors are conceived as obstacles to the proper manifestation of dispositions. But this introduces an explanatory asymmetry in being and within beings that is taken as a brute fact and blocks the road to inquiry. Moreover, relatedly, thing-kinds do not provide an independent theoretical explanation of the causal force of dispositional properties but merely a categorial means of systematizing and unpacking our already existing common-sense conceptual knowledge about how dispositions are connected to objects as their qualitative powers, which under certain conditions are, ceteris paribus, manifested (Sellars, 1957). It is precisely due to these limitations that we need successor categories for thing-kinds (objects) and their powers-dispositions. Moreover, the framework of powers has controversial practical and political implications. For example, an individual or a community that does not exhibit the essential powers or capacities of its distinctive way of being is deemed 'unnatural' and 'deviant' not only in a descriptive-statistical sense but in a normative and pejorative sense, which supposedly justifies their being deprived of the full range of rights attributed to persons (think, e.g. of the place of homosexuals within a heterosexual society).

non-modal, non-conceptual articulation of the Humean world. Mereology does not seem to provide the conceptual resources for identification of at least counterpart successor concepts of ordinary lifeworld thing-kinds that can serve as explanatory grounds for the latter. There is something fishy about mereological wholes, as they do not invoke any sort of *unity* of the wholes they discuss: there is a mereological sum of this book, Jupiter, Julius Caesar's left little finger, and some random, distant electron. No sort of constraint is provided for the kinds of combinations or arrangements of particulars that can be meaningfully said to individuate one (objectual, processual) whole, distinguishing it from another in a way that satisfies the requirement that grounding be genuinely explanatory. For example, mereology allows for arbitrary formation of new particulars from old ones, independently of spatiotemporal or causal connections, it gives no account of alteration except adding or subtracting parts (e.g. moving any of the parts does not change the mereological whole) and needs for formal purposes a 'null individual' which is stipulated to be a part of every whole (Fine, 2012), which however does not seem to do any genuine explanatory work for ontological grounding purposes. In Chap. 5, I suggested that a non-mereological process-topological account of the relation between being and beings and of levels-emergence can be a better model for understanding the relation between ground and grounded.

Lastly, it must be noted that our Intensive Tychistic Darwinism as a grounding model provides in resolutely non-conceptual (and *a fortiori* non-judgmental) terms the sufficient reason of the actual, both in its non-propositional and in its propositionally structured dimension, but it does not (and it is not its job to) provide a sufficient reason for the truth of particular propositions, whatever their subject matter might be. To expect such an account is to misconstrue the nature of the grounding relation between (intensive mutable and multiple) being and (actual) beings. This relation is one of *immanent reciprocal—yet asymmetric—determination* in which the former morphogenetically produces actual processes and the latter, as an ever-growing actual history of worldly processes, act as a filtering constraint on the configuration of virtual multiplicities that are at each moment to be selected as 'live options' for morphogenetic actualization. What is entailed by such a conception of grounding is that there is always room for the emergence of new truth-makers and truths at the level of actuality. There is no such thing as a ready-made actual world with eternally fixed truths which our propositions attempt to unearth or correspond to. The only 'essential' thing about the world is that it is an ever-growing plane of self-differentiating and self-fracturing becoming, an ontologically mutable world, in which the only thing that is eternally 'repeated' is its evolutionarily induced topological variations of the spatio-temporal unfolding of processes, leading to changing adaptational strategies, novel worldly and human habits, emergent worldly, psychological and societal structures and, in general, to novel possibilities, concepts, interests and purposes that cannot be anticipated in advance, but can be best dealt with in what pragmatists called 'an experimental spirit', as part of our never-ending striving to strike a dynamic allostatic balance between our internal, external and social environment.[25]

References

Armstrong, D. M. (1978). *Universals and scientific realism, volume II: A theory of universals*. Cambridge University Press.

Armstrong, D. M. (2004). Through the open door again. In J. Collins, N. Hall, & L. A. Paul (Eds.), *Causation and counterfactuals* (pp. 445–457). MIT Press.

Baker, R. L. (1997). Why constitution is not identity. *The Journal of Philosophy, 94*(12), 599–621.

Beatty, J. (2016). The creativity of natural selection? Part I: Darwin, darwinism, and the mutationists. *Journal of the History of Biology, 49*, 659–684.

Bell, J. (2016). From difference-maker to truthmaker (and back). In J. Bell, A. Cutrofello, & P. Livingston (Eds.), *Beyond the analytic-continental divide: Pluralist philosophy in the twenty-first century* (pp. 129–150). Routledge.

Bergson, H. (1896). *Matter and memory*. Zone Books.

Bergson, H. (1907). *Creative evolution*. Dover Publications.

Bhogal, H. (2021). Induction and the glue of the world. *Australasian Journal of Philosophy, 99*(2), 319–333.

Brandom, R. (1994). *Making it explicit: Reasoning, representing and discursive commitment*. Harvard University Press.

Brandom, R. (2011). *Perspectives on pragmatism: Classic, recent and contemporary*. Harvard University Press.

Brandom, R. (2015). *From empiricism to expressivism: Brandom reads Sellars*. Harvard University Press.

Brandom, R. (2019). *A spirit of trust: A reading of Hegel's phenomenology*. Harvard University Press.

[25] I submit that in this ontologically mutable world the only invariantly meaningful end, individually or collectively, would be what Dewey (1930) called 'growth': the reorganization of experience that increases our ability to direct its future course, enables its enrichment and differentiation and extends the possibilities of experience.

Brown, N. (2021). *Rationalist empiricism: A theory of speculative critique.* Fordham University Press.

Cameron, R. P. (2010). From humean truthmaking theory to priority monism. *Nous, 44*(1), 178–198.

Carroll, L. (1895). What the tortoise said to achilles. *Mind, 4*(14), 278–280.

Christias, D. (2023). *Normativity, lifeworld and science in Sellars' synoptic vision.* Palgrave Macmillan.

Correia, F., & Schnieder, B. (2012). *Metaphysical grounding: Understanding the structure of reality.* Cambridge University Press.

Dawkins, R. (1976). *The selfish gene.* Oxford University Press.

Deacon, T. (2013). *Incomplete nature: How mind emerged from matter.* Norton.

DeLanda, M. (2002). *Intensive science and virtual philosophy.* Continuum.

Deleuze, G. (1968). *Difference and repetition.* Columbia University Press.

Dewey, J. (1930). From absolutism to experimentalism. In J. A. Boydston (Ed.), *John Dewey: The later works, 1929–1930* (Vol. 5). Southern Illinois University Press.

Fine, K. (2012). The pure logic of ground. *Journal of Symbolic Logic, 5*(1), 1–25.

Gould, S. J. (1977). *Ever since Darwin: Reflections on natural history.* Norton.

Hegel, G. (1816). *The science of logic.* Cambridge University Press.

Heil, J. (2003). *From an ontological point of view.* Clarendon Press.

Heil, J. (2012). *The universe as we find it.* Clarendon Press.

Heil, J. (2021). *Appearance in reality.* Oxford University Press.

Kimhi, I. (2018). *Thinking and being.* Harvard University Press.

Lewis, D. (1983). New work for a theory of universals. *Australasian Journal of Philosophy, 61*(4), 343–377.

Lewis, D. (1986). *Philosophical papers: Volume II.* Oxford University Press.

Lewis, D. (1991). Parts of classes: .

Lewis, D. (1994). Humean supervenience debugged. *Mind, 103*(412), 473–490.

Lowe, E. J. (2006). *The four-category ontology: A metaphysical foundation for natural science.* Clarendon Press.

McDowell, J. (1994). *Mind and world.* Harvard University Press.

Meillassoux, Q. (2008). *After finitude: An essay on the necessity of contingency.* Continuum.

Molnar, G. (2003). *Powers: A study in metaphysics.* Oxford University Press.

Morris, M. (2005). Realism beyond correspondence. In H. Beebee & J. Dodd (Eds.), *Truthmakers: The contemporary debate* (pp. 49–66). Clarendon Press.

Mudler, J. M. (2018). The limits of humeanism. *European Journal for Philosophy of Science, 8*(3), 671–687.

Mudler, J. M. (2022). Absolute idealist powers. *Australasian Journal of Philosophy, 100*(3), 471–484.

Peirce, C. S. (1887). A guess at the riddle. In C. Hartshorne & P. Weiss (Eds.), *Collected papers of charles sanders peirce.* Harvard University Press.

Peirce, C. S. (1892). The doctrine of necessity examined. *The Monist, 2*(3), 321–337.
Peirce, C. S. (1989). Design and chance. In *Writings of Charles S. Peirce: A chronological edition, volume 4: 1879–1884* (pp. 544–554). University of Indiana Press.
Pippin, R. (2018). *Hegel's realm of shadows: Logic as metaphysics in the science of logic*. The University of Chicago Press.
Price, H. (1991). Agency and probabilistic causality. *British Journal for the Philosophy of Science, 42*, 157–176.
Price, H. (2005). Causal perspectivalism. In H. Price & R. Corry (Eds.), *Causation, physics, and the constitution of reality: Russell's Republic Revisited* (pp. 250–292). Oxford University Press.
Price, H. (2012). Causation, chance and the rational significance of supernatural evidence. *Philosophical Review, 121*, 483–538.
Rödl, S. (2007). Eliminating externality. In J. Stolzenberg & K. Ameriks (Eds.), *International yearbook of german idealism 5* (pp. 176–188). De Gruyter.
Rödl, S. (2012). *Categories of the temporal: An inquiry into the forms of the finite intellect*. Harvard University Press.
Rödl, S. (2018). *Self-consciousness and objectivity: An introduction to absolute idealism*. Harvard University Press.
Schaffer, J. (2010). The least discerning and most promiscuous truthmaker. *The Philosophical Quarterly, 60*(239), 307–324.
Sellars, W. (1957). Counterfactuals, dispositions and the causal modalities. In H. Feigl, M. Scriven, & G. Maxwell (Eds.), *Minnesota studies in the philosophy of science* (Vol. II, pp. 225–308). University of Minnesota Press.
Sellars, W. (1961). The language of theories. In H. Feigl & G. Maxwell (Eds.), *Current issues in the philosophy of science* (pp. 57–77). Holt, Rinehart & Winston.
Sellars, W. (1963). Philosophy and the scientific image of man. In W. Sellars (Ed.), *Science, perception and reality* (pp. 1–40). Routledge.
Sellars, W. (1967). *Science and metaphysics: Variations on kantian themes*. Ridgeview Publishing Company.
Sicha, J. (Ed.). (1980). *Pure pragmatics and possible worlds: The early essays of wilfrid Sellars*. Ridgeview Publishing Company.
Stroud, B. (2011). *Engagement and metaphysical dissatisfaction: Modality and value*. Oxford University Press.
Whitehead, A. N. (1929). *Process and reality: An essay in cosmology*. Free Press.
Wiggins, D. (1967). *Identity and spatiotemporal continuity*. Blackwell.

PART III

Normative Spaces, Hyperstition and Process

CHAPTER 7

Norms and Facts as Fractured Projections of Processual Reality

In this chapter, I turn my attention to the issue of the seeming incommensurability between two equally indispensable ways of understanding ourselves-in-the-world: our normativist self-image (space-of-reasons intelligibility) and our causal-naturalistic self-image (space-of-causes intelligibility). My suggestion is that both these frameworks are best understood as fractured projections of processual reality. To this end, I first stress the irreducibility, self-contained character yet also 'parallelism' of the space of reasons (norms) and the space of causes (facts) and suggest that these relations as a package deal are an essential part of a metaphysical view (itself a sophistication of our common-sense view of the world after modernity) that Brandom calls 'bimodal hylomorphic conceptual realism' (Brandom, 2019). I then propose that Brandom's conceptual realism and modal realism that articulate our ordinary descriptive metaphysics about reasons, causes and their relations can be understood as indispensable 'ladder languages' that can yet be eventually transcended by a naturalistic process metaphysics—where both causes and reasons are understood as different ways of being processes.

© The Author(s), under exclusive license to Springer Nature Switzerland AG 2025

D. Christias, *Towards a Hyperstitional Process Nominalism*, Palgrave Perspectives on Process Philosophy,
https://doi.org/10.1007/978-3-032-06350-2_7

7.1 The Normative and the Causal as Irreducible, Self-Contained Yet Parallel Orders

The expressions 'space of reasons' and 'space of causes' are meant to draw our attention to the fact that our understanding of the world and ourselves is not univocal but is bifurcated in essentially two distinct, self-contained and incommensurable modes of intelligibility. To place an item (worldly fact, human behaviors) in the space of reasons serves the purpose of displaying phenomena as having a special kind of intelligibility, the kind of intelligibility a phenomenon is revealed as having when we enable ourselves to see it as manifesting responsiveness to reasons as such (McDowell, 1994). This special kind of intelligibility is revealed if we realize that an important part of our behavior does not seem to be intelligible merely as an instance of a causal or law-governed process. Human thought and action seem to be more than another way of causally affecting or being affected by the world or other human beings. Our thinking is answerable to criteria of assessment that are essentially social, holistic and normative. As Sellars puts it: 'there is no thinking apart from common standards of correctness and relevance, which relate what *I do* think to what *anyone ought* to think' (Sellars, 1963, pp. 16–17). And McDowell, echoing Davidson (1980), emphasizes the essentially holistic character of this social normative process of assessment: 'We place beliefs and desires in the space of reasons by putting them in a context that includes other beliefs, other desires, and valuations, in the light of which the beliefs and desires we are aiming to understand are revealed as manifestations of rationality on the part of their possessors' (McDowell, 2019, p. 4). Thus, there are social, holistically structured intersubjective standards of correctness and relevance, the fulfillment of which is necessary for our perceptions to count as veridical, our beliefs as true and our intentions as successfully issuing in actions. The contents of our beliefs can be ascertained as correct or incorrect (true or false) depending on how the world is (which is itself decided on the basis of intersubjective criteria of correctness); they are justified or unjustified, depending on whether they are supported by adequate grounds or reasons; and they can be judged as consistent or inconsistent, depending on whether the concepts used in articulating this content logically exclude one another. Correspondingly, our intentions can be fulfilled or unfulfilled depending on whether they issue in the action expressed in their content. Thus, what we do is answerable to what we intend and what we believe is answerable to how the world is, where both beliefs and intentions are not merely causal states of human organisms but contents articulated and individuated by normative relations between concepts (consistency, coherence relations among the concepts involved in it) as well as between these contents and the way the world is (correctness-incorrectness of the belief, fulfillment or non-fulfillment of an intention) (Maher, 2012).

By contrast, to place an item in the space of causes (or natural law) is to see it as an instance of a law of nature or a causal process. This mode of intelligibility articulates our modern, scientific conception of the world and ourselves as complex physical systems characterized by all kinds of inclinations and habits related to perception, inference and action, yet devoid of normativity and the characteristic distanced orientation it makes possible toward circumstances that otherwise would have simply induced inclinations (this is why normativity is internally related to rationality, freedom and deliberation). For example, to say that planets in the solar system revolve around the sun, plants lean toward sunlight, humans fight or flight in response to life-threatening circumstances or that supposedly higher ethical values are ideological rationalizations for serving our vested interests is to place these processes in the space of causes (or the realm of law). In all these cases, what happens becomes intelligible without reference to reasons, norms, values, or, in general, to how things *ought* to be, but is understood as an instance of ways in which things unfold in conformity to laws of nature, in an extended sense of the term that includes (statistical) laws of evolutionary history, of human nature, history and society.[1]

Can we reduce norms to facts? Reasons to causes? Space-of-reasons intelligibility to space of causes intelligibility? Many philosophers have convincingly argued that this cannot be done, at least if it is construed as

[1] Some phenomenologically oriented thinkers (Dreyfus, 2013; Merleau-Ponty, 1962) take it that the space of causes and the space of reasons are both intellectualized abstractions of a more primary lifeworld reality, namely, 'the space of motivations', in which agents are directly motivated to act in a certain way in virtue of perceiving context-bound, situation-specific environmental affordances and solicitations. While I agree that there is an important sense in which the phenomenological concept of motivation can illuminate discussions on the relation between reasons and causes, I think it cannot altogether abolish the epistemological need for the distinction between reasons and causes. This distinction reemerges *within* the space of motivations in the guise of a distinction between motivationally salient affordances and solicitations to act that can be registered and consciously recognized by the agent, and those that convey an unconscious, latent, meaning that can be made explicit only by the construction of novel (socio-psychological) concepts that are not already part of our self-understanding in the context of our ordinary lifeworld.

an explanatory reduction of normatively laden facts (space-of-reasons contents) to their non-normative causal-naturalistic correlates (Sellars, 1997, Davidson, 1980, McDowell, 1994, Brandom, 1994). If we could indeed successfully reduce the socio-normative character of our perception, thought and action to their causal (historical-psychological-neurophysiological) correlates in nature this would mean that the non-normative content of these latter physical, psychological and natural-historical structures does not differ in kind from the normative content in terms of which we make our behavior and our relation to the world rationally intelligible. And this would mean that we only differ in degree of complexity from non-human systems which lack normative capacities and the respective kinds of intentional directedness and answerability to the world. Our supposedly normative responsiveness to the world would be revealed as having an essentially affective character that may informationally covary with evolutionary-neurophysiological-psychological facts of our causal history but is not rationally answerable to the world (correct-incorrect, true-false, justified-unjustified, capable of a distanced orientation toward the world). But attempts to offer an explanatory reduction of the above kind seem to change the subject, since the non-normative (neurophysiological, evolutionary, psychological, social) correlates of normative fact, by not preserving in their conditions of identity and individuation the normative-functional conditions of individuation of normative facts, far from offering a deeper understanding of them in terms of natural facts, essentially lose them from view, eliminate them altogether from our worldview (Turner, 2010, pp. 186–192, McDowell, 1994, Brandom, 2015). Thus, what is explained in those cases does not seem to be the *same* as what was originally thought to be in need of explanation. At most, what is explained in this case in terms of non-normative phenomena are the enabling conditions of normative phenomena (e.g. the neural firings and brain activity in cases of veridical perception), rather than that which *constitutes* them as the normative phenomena they are (McDowell, 1999). The properties of the purported *explanans* are essentially such (e.g. non-normatively described neurophysiological states, biological adaptations, habits of action) that they do not even make room for distinctions essential to the very being of the (normative) phenomenon to be explained. For example, in case the explanandum is that of intentional action, the very distinction, essential to the being of this normative phenomenon, between intentional action performed on the basis of reasons potentially in conflict with one's subjective desires and impulses, and unintentional, impulsive or habitual bodily movements is construed, at best, as a distinction which is subjectively imposed on reality and is lost from view as an objective distinction, if the content of the explanans is non-normative. Thus, the very phenomenon (intentional action) to which this distinction is essential for its being as such is lost from view if its content is reduced to its non-normative correlates. The point holds of the space of reasons in general, considered as an interconnected, holistically articulated framework built around the concept of a person and its normatively individuated capacities: conceptualized perception, thought, action, and the related freedom from inclinations that they entail. As Sellars characteristically puts it, the irreducibility of space-of-reasons intelligibility consists in that 'anything which can properly be called conceptual thinking can occur only within a framework of conceptual thinking in terms of which it can be criticized, supported, refuted, in short, evaluated. To be able to think is to be able to measure one's thoughts by standards of correctness, of relevance, of evidence. In this sense a diversified conceptual framework is a whole which, however sketchy, is prior to its parts, and cannot be construed as a coming together of parts which are not already conceptual in character' (Sellars, 1963, p. 6). This means, among other things, that thoughts cannot be reduced to complex patterns of images, and that the genesis and rationality of conceptual thinking cannot be accounted for in terms of associative connections (a combination of sensory and habitual elements) which are not themselves conceptual, or by a direct non-conceptual action of the physical environment on the individual. Moreover, the irreducibility of space-of-reasons intelligibility implies that the way in which individual minds are related to the world (both in the sense of being connected to it and in being *about* it) cannot be understood in terms of a direct causal action of perceptible nature, as perceptible, on individual minds. Direct associative connections between mind and world through sensation and habit can account for associative connections and the non-rational accommodation of ourselves and other animals in the environment, but not for the rational connections of conceptual thinking. This reflects that fact that from the point of view of space-of-reasons intelligibility we are rationally open to the world in the sense that our perception and thinking is *about* the world and rationally answerable to it (judged as true or false depending on how the world is), and not merely something that causally or informationally covaries with worldly processes.

Now, this irreducibility of space-of-reasons intelligibility to space-of-causes intelligibility and vice versa also finds expression to the fact that those intelligibility spaces are self-contained, in the sense that the fundamental concepts and categories of each such mode of intelligibility are a) mutually supportive and interdependent and b) sufficient to describe the world, ourselves and our relation to the world. These forms of intelligibility are *whole* ways of viewing ourselves-in-the-world and their categorial elements are holistically related and mutually supported. Even if each side takes the other (correctly in my view) to leave something essential out of the picture (the causal-naturalistic worldview leaves out persons and their rational relation to other persons and the world, while the normativist worldview cannot account for our fundamentally a-rational ways of being attuned with others and the environment), it is still the case that the worldviews in question are internally coherent wholes and account in their own way for everything they take to be worth of preservation in the scheme of things. In this sense, these different ways of making the world and ourselves intelligible are self-contained. They are also self-contained in the sense that elements of one mode of intelligibility cannot be mixed with those of the other, on pain of creating unnecessary philosophical problems and confusion.[2] In Davidson's slogan 'nothing can count as a reason for a

[2] Of course, there are many neo-Aristotelians who believe that not only can we mix elements from these different ways of intelligibility without engendering confusion, but that this is the key for dissolving the philosophical problems that ensued from modernity onward due to the dualistic hypostatization of the causal-naturalistic and the normativist worldview. Categories such as 'natural norms' that pertain to specific animal species and related notions of 'teleological' or 'rational' causation that can be differently actualized in different forms of life depending on their general way of life (Foot, 2001; Thompson, 2008) have been frequently taken as both more natural, closer to the way in which we understand ourselves-in-the-world in the context of our everyday lifeworld, and crucial for dissolving contemporary philosophical problems about the place of norms within nature. I do not have the space here for a proper examination of neo-Aristotelianism about norms and nature. I tend to think that while neo-Aristotelians offer interesting and for the most part penetrating and accurate philosophical systematizations of our everyday lifeworld, they are committed to certain controversial forms of Categorial Givenness (they adhere to a picture according to which we have 'inside' a priori knowledge of the basic categorial structure of our form of life and of the basic categorial structure of being), and they do not succeed in delineating a proper metaphysics all things considered (see, e.g. Chap. 6 in which I argue that neo-Aristotelian power metaphysics does not provide a proper metaphysical grounding for understanding nature and our place in it). However, as we shall see in Sect. 7.2 Brandom's neo-Hegelianism revives a different kind of neo-Aristotelianism that respects the *sui generis* character of reasons and causes, while at the same interpreting this distinction in dynamic terms, as part of a histori-

belief except another belief' (Davidson, 1986, p. 126). Anything whose way of being is not properly normative (worldly event, bodily sensations/ feelings/affects, brain states which informationally covary with worldly states, a-rational habits, expectations, images) cannot stand in normative relations (and to be a reason for believing something is a species of such a relation) to conceptually articulated perceptual beliefs, thoughts, intentions or actions. Conversely, items that are normatively laden cannot stand in causal-naturalistic relations to other such items as described and explained by empirical-scientific inquiry and thus cannot directly affect and be affected by the world as described by science. Instead of a vertical relation between reasons and causes in which, for example, reasons function as causes effecting changes directly in the natural world of bodies, we have two series of horizontal relations which connect reasons to other reasons and causes to other causes that form a nexus akin to Spinoza's psychophysical parallelism: the inferential relations of thought contents run parallel to—yet without directly communicating with—the causal relations of bodies (worldly objects, including our brains and bodies).

7.2 Spinozist Vs Neo-Hegelian Readings of the Parallelism between Reasons and Causes

There are two ways of understanding this parallelism between mind and body or between normative and non-normative facts (which is what the mind-body problem becomes after the normative turn taken by analytic philosophers influenced by Sellars).

The first is an (updated) Spinozist one according to which the parallelism in question between these self-contained realms (the normative realm of thoughts and their inferential connections, the non-normative realm of bodies and their affective states) consists in the fact that every inferentially articulated content of thought, though irreducible to accompanying bodily non-normatively articulated states, images, memory traces, events or relations to the environment, is in some sense identical to a complex of such bodily states (and vice versa). A similar view is expressed by Sellars in his 'norm/nature meta-principle' (O'Shea, 2007) which states that 'espousal of [normative] principles is reflected in [non-normative]

cally evolving process of conceptual development and determination in which what we believe and the explanatory reasons we have for believing it mirrors the lawful connections and properties of the objects that are the contents of these beliefs (see also Chap. 8).

uniformities of performance' (Sellars, 1962b, p. 48). According to this view, semantical rules (language-entry, language-language, language-exit moves) generate by their very nature (ceteris paribus) semantical *uniformities* which belong in the 'order of nature', thereby establishing an internal (but not direct, 'vertical') connection between the space of reasons or the order of signification and the non-normative causal order (Sellars, 1962b, p. 48). Importantly, this kind of *parallelism without isomorphism* between the normative realm of thought and the non-normative realm of bodies does not consist merely in denying any real causality between the mind and the body. It also thereby disallows any primacy of the one over the other. If any superiority of the mind over the body is rejected, this is not in order to establish a superiority of the body over the mind. As Deleuze reminds us in the case of Spinozistic parallelism (and the same goes for the parallelism between the normative and the causal discussed above): 'The practical significance of parallelism is manifested in the reversal of the traditional principle on which Morality was founded as an enterprise of domination of the passions by consciousness. It was said [e.g. by Descartes] that when the body acted, the mind was acted upon, and the mind did not act without the body being acted upon in turn. On the contrary, for Spinoza, what is an action in the mind is necessarily an action in the body as well, and what is a passion in the body is necessarily a passion in the mind. There is no primacy of one series over the other' (Deleuze, 1988, p. 18). Moreover, the parallelism between the normative and the causal order does not imply any devaluation of the normativity of thought in relation to the causal-affective field of bodies. Actually, it implies a devaluation of consciousness, of our conception of what normative thinking amounts to in relation to thought proper, to what thinking really is. In this way, just as there are unknown causal-affective relations in the body and between bodies of which we are unaware, there are also corresponding normative relations in thinking of which we are also unaware, that parallel the former. Deleuze brilliantly suggests that this amounts to a discovery of the unconscious of thought just as profound as—and parallel with—the unknown of the body (Deleuze, 1988, pp. 18–19). Transposed to our normative-causal parallelism, it can be argued that the unconscious of thought amounts to the unacknowledged and unrecognized inferential commitments of the contents of our thoughts. Although certain inferential consequences or incompatibilities in the content of our beliefs, thoughts or intentions are expressed or contextually implied in our normative practices, they remain implicit and unacknowledged as such in these practices, with the result that when made explicit (acknowledged as such), they demand a change of these inferential relations, and thus a change in the content of our beliefs, thoughts and intentions in order to retain their compatibility and restore the lost balance between what we are *really* committed to by holding certain beliefs, thoughts, intentions and what we *think* (i.e. what we consciously acknowledge) we are committed to by doing so.

Now, there is another, different way of understanding the parallelism between the normative and the causal order, which this time is more Hegelian than Spinozistic. It is distinguished by the latter in that it is a parallelism *with* isomorphism, as it were (both orders, while otherwise heterogeneous, are conceptually articulated), while Spinozistic parallelism is one without isomorphism: the causal realm is a non-conceptual order while the rational realm is a conceptual order. Now, our model for this second way of construing the normative-causal parallelism is Brandom's 'bimodal hylomorphic conceptual realism' (BHCR), a view he attributes to Hegel, and in fact amounts to an updated analytically domesticated Brandomian version of Hegelianism. For Brandom's Hegel the parallelism between the causal and the normative come as a package deal and is a sophisticated attempt to reunite of our common-sense view of the world after the split between the causal and the normative realms in the wake of the scientific revolution and Kant's Copernican revolution, without abolishing the *sui generis* character of these orders of intelligibility. To my mind, this Brandomian neo-Hegelian construal of the normative-causal parallelism systematizes in an illuminating manner and better captures our contemporary lifeworld's conception of the relations between the causal and the normative orders, although I take it that an updated Spinozistic rendering of the normative-causal parallelism is closer to the truth in terms of ultimate metaphysics.[3]

From the standpoint of Brandom's Hegel BHCR, the causal order is a realm of (non-normative) *facts* (about the properties of objects) subject to natural *laws* expressible in *alethic modal terms* while the normative order is a *discursive* realm subject to *rules* expressible in *deontic normative* terms.

[3] As we shall see later, the category of process plays a similar role to Spinoza's substance in being the unconditional metaphysical principle which makes meaningful the view that the realms of the normative and the causal are different dimensions and expressions of *one and the same* metaphysical reality. By contrast, Brandom's neo-Hegelian framework is based on a substance-property metaphysics, which, despite its dynamic and historically informed character, is strictly incompatible with a process metaphysical framework.

The metaphysical relations between these realms, captured by BHCR, is a way of retaining their irreducible and self-contained dimension, without turning the latter into a metaphysical dualism—that is, without making essential relations between the distinguished realms unintelligible, as was arguably the case with Descartes (in his understanding of the realms of body and mind) and with Kant (in his rendering of the realms of facts and norms). The deontic normative senses of 'incompatible' and 'consequence' articulate the attitudes of knowing (thinking, intending) subjects, and they concern what commitments one can be jointly entitled to and when commitment to one content entails commitment to another. The corresponding alethic modal sense of relations of incompatibility and consequence articulates objective facts of the form 'the coin is pure copper', which express objective modal facts, for example, the fact that an object that is pure copper cannot be pure aluminum, and objective facts about the causal powers of objects, for example, that the object in question, by being copper is thereby an electrical conductor (and not an electrical insulator), that it melts at 1084 degrees Celsius, etc. The normative realm is the one licensing the knowing and practical attitudes of thinking and intending subjects and expresses the 'form' in which we think about the world and ourselves, while its corollary, the causal realm, articulates the causal powers of objects as constituents of facts, through lawful relations of physical necessity, impossibility and causal consequence between objects *in re*. In this way, it expresses the modal structure of the objective world (the 'enmattered form'). Moreover, these realms are reciprocally sense-dependent, though not reference-dependent: one cannot *understand* the objective pole without understanding the subjective pole and vice versa. Yet the *existence* of the objective pole does not itself depend on the existence of the subjective pole since they are true counterfactuals in the former that remain true independently of the existence of human semantic practices. And as is obvious, the alethic modal categories of causal powers of things *in re* (causal order) mirror the deontic normative categories of the powers of thinking (normative order). For example, the concept of a law of nature is sense-dependent (but not reference-dependent) on the concept of explanation. In order to understand what a natural law is, one must understand how statements of natural laws function in counterfactually robust inferences in explanations.[4] The same goes for the concepts of

[4] It is instructive here to note that statements of laws—which as we shall see amount to statements about the causal powers (dispositions) of the objects involved—are implicit in our

property and universal, on the one hand, and the concept of what one is doing in classifying by applying predicates on the other, for the concepts of object or particular, and the concepts of referring and singular term, and for the concepts of fact or state of affairs and the concept of what one is doing in asserting by uttering declarative sentences. They come as a package in that one cannot understand either one without understanding the other. It is in this precise sense that alethic modal properties that articulate the objective causal order and deontic normative ones that express the normative order are two sides of the same coin or parallel one another. In sum, the fundamental metaphysical categories of the object-property framework reflect (i.e. they are reciprocally sense-dependent though not reference-dependent with) the deontic normative categories of thought as follows: these normative categories as expressed in our semantic practices of using singular terms and predicates are the semantic and epistemic condition of our ability to refer to objects and to classify their properties. The practice of using declarative sentences to assert, make claims or judgments is the semantic and epistemic condition for organizing worldly processes into facts that might obtain or not obtain, and using universally quantified

understanding even of ordinary concepts such as dog, cat, book, etc. To understand what a book is I must understand that if I release my grip on the book, then it will fall to the floor. And for this to count as a counterfactually robust inference, as it must if I am to understand the nature of ordinary objects and its causal powers that distinguish it from other objects, I must make some distinction between collateral circumstances that would defeat the inference (a table is moved under it, someone else catches it, it is snatched up by a passing bird) and those that would not (it is Tuesday, it is slightly cooler today than it was yesterday, my car has been moved slightly further away). Of course, we might be wrong about whether any of these particular auxiliary hypotheses actually would or would not defeat the inference to the conclusion. Counterfactually robust inferences are non-monotonic. But if we make no distinction of this sort at all we would not even understand the concepts 'book', 'falling', etc. These distinctions are precisely what is made explicit by the vocabulary of subjunctive and counterfactual conditionals (Brandom, 2015, p. 193) and show the internal connection between our conception of ordinary objects, their causal powers (capacities, dispositions) which express their nature, and their law-governed behavior. The scientific revolution, by conceiving objects as bereft of causal powers of their own and as obeying exceptionless laws of nature, challenged this picture, but it is not clear what are we to make of this as regards ultimate metaphysics. Should we become Humeans or can we save the appearances in the context of a neo-Aristotelian power metaphysics? In Chap. 6 we saw the metaphysical pros and cons of both options. What concerns me here is the metaphysical view about the relation between causes and reasons that better captures our understanding of the world and ourselves in the context of the ordinary lifeworld, and I think Brandom's neo-Hegelian view, by incorporating while at the same time overcoming neo-Aristotelianism without relapsing to Humeanism, does a pretty good job on that front.

subjunctively robust conditionals of the form 'if copper were heated in 1084 degrees Celsius, it would melt' is a semantic and epistemic condition for explaining the unfolding, interference and co-occurrence of worldly processes in terms of the obtaining of certain lawful natural connections exhibiting the causal powers of objects. In this way, the usually contrasting notions of *representation* and *expression*, of representing how things are through alethic modal categories and expressing what we are doing in using concepts in the context of our linguistic practices through deontic normative categories, are also revealed as two sides of the same coin (Brandom, 2019, pp. 207–210).

Lastly, it must be stressed that this parallelism between the causal and normative realms is not static but engenders a dynamic process of critically controlled conceptual change in response to experiential anomalies: the normative order expresses how things are for consciousness (what we believe them to be) while the causal order expresses how things are in themselves. Any discrepancy between these two orders expressed in the form of incompatible commitments triggers an effort to revise our commitments and thus the contents of the respective concepts in order to restore normalcy. This happens, for example, if one believes that sticks are necessarily (by their own nature) rigid and thus remain straight and perceptually notices that if we submerge sticks in the water they bend. In this situation one finds himself with commitments incompatible by his own lights. So, something has to give for restoring coherence in our commitments about sticks. And the same goes for every concept. These 'experiences of error' institute a historical process of attempting to overcome incompatible commitments by changing our commitments, and in the process forming new commitments that better track the reality presupposed by and implicitly referred to by our previous commitments. In this process, one sense (concept) is found to be implicitly defective by its own standards (as it leads in holding incompatible commitments) and is replaced by another, whose application leads sooner or later again to a new set of incompatible commitments, and so on indefinitely. And while this process is inevitably one of an experience of error and failure, it is also the process and the practice whereby conceptual contents are progressively determined and truth progressively discovered.

From this point of view, there is no unbridgeable gulf between facts and norms, causes and reasons. At the same time, their distinctive *sui generis* mode of intelligibility is retained. The two realms do not mix up, do not vertically communicate (reasons are not *sui generis* kinds of causes), and thus they retain their self-contained dimension. Yet, the built-in parallelism or 'categorial mirroring' between the alethic modal categories of the space of causes and the deontic normative categories of the space of reasons shows that there is no ontological gap to be bridged in the first place between them. The distinction between deontic normative categories and alethic modal ones, just as that between how we take the world to be and how the world really is (appearance-reality distinction), is *functional* and not ontological. Our deontic normative categories always already present to us a world or lawfully connected objects with causal-dispositional properties. The only legitimate question here is epistemic: whether what we *take* the world to be as reflected in our deontic normative commitments articulating the contents of our empirical beliefs coincides with the real modal connections between worldly objects. This is an everlasting task with no discernible end, since we can never erase all contingency from experiential immediacy, nor can we be certain that we do not hold incompatible commitments as this is too difficult to discern synchronically and impossible to access diachronically.[5] But this does not turn the distinction and *sui generis* character of reasons and causes into an untenable dualism. The philosophical heavy lifting here is done by 'conceptual realism', a thesis amounting to a semantically informed metaphysics, according to which conceptual articulation is not an exclusive feature of thoughts, but also of reality itself. The *modal* relations of physical possibility, impossibility and necessity to which determinate states of affairs stand in to other states of affairs are a species of *conceptual* articulation. From this standpoint, to say that reality is non-conceptually articulated does not make sense as it makes all content indeterminate. Moreover, since Brandom's Hegel works here with a 'non-psychological conception of the conceptual', the view that the world *itself* is literally conceptually articulated, through the modally robust relations of exclusion and consequence exhibited by the modal properties of objects, with the latter understood as constituents of facts, does not presuppose for its existence the existence of

[5] Importantly, as we shall see in what follows and in Chap. 8, this appearance-reality distinction by operating *within* the conceptual realm (where both terms of the distinction are conceptually articulated) is revealed as *itself* a kind of appearance from the standpoint of a (non-conceptually articulated) process metaphysics. This view is structurally analogous to Kant's conception of the phenomenal world, which also operates on a distinction between reality (objective facts) and appearances (what is merely taken as objective fact, without being so). This distinction is only valid for the phenomenal world, and is shown itself to be a 'conceptual appearance' of the in-itself.

language or concept-users. The latter are only presupposed for making sense, for ***understanding*** the conceptual articulation of what really exists, not for its existence as such (recall: the world is sense-dependent, but not reference-dependent, on our normative practices).[6]

7.3 Norms and Facts as 'Ladder Languages' Expressing Ways of Being Processes

As was mentioned earlier, I take Brandom's Hegel model of parallelism between facts and norms, causes and reasons, to systematize in an illuminating manner our lifeworld conception of the relations between the causal and the normative order. And the conceptual realist connotations of this bimodal conception of the relation between reasons and causes, though departing from our lifeworld understanding of the issue, provide a way for accepting the *sui generis* character of causes and reasons while at the same time construing them as different forms which a *single* (conceptually articulated) reality can take. This is why this reality needs to be thought of as *conceptually* articulated. According to this view, if we construed reality as non-conceptual in character, or as bifurcated into a conceptual and a non-conceptual realm, this would reinstate an untenable ontological dualism between reasons and causes. But this form of solution to the problem of understanding the metaphysical relations between causes and reasons runs the risk of creating other perhaps more serious metaphysical problems about the existential status and independence of

[6] The conceptual realist dimension of Brandom's Hegel understanding of the relation of reasons and causes shows that this view does not just represent our lifeworld but takes a stand on certain issues that are left undecided from the point of view of the lifeworld. I take it that from the standpoint of the ordinary lifeworld, reality as it is in itself is not viewed as conceptually articulated, but rather as non-conceptual in character. Yet, at the same time, our lifeworld conception of thought is close to Hegel's view in that we do not take thinking to have an external boundary, and we do not think that formulating our understanding of how the world is in terms of judgmental or conceptual content leaves something essentially out regarding how the world really is. Our lifeworld understanding of this issue is not fully coherent (since this is not needed for practical purposes) and can be developed along different and even diverging directions. One of these is Brandom's Hegel conceptual realism, but only at the cost of rendering the world itself as having conceptual structure. This is not automatically problematic. A philosophical position is judged by the illumination it provides to our understanding of the world and our place in it, not by its degree of accuracy in representing the structure of the lifeworld—since the latter in philosophically crucial points does not take a clear stand and can be developed along different directions philosophically.

the world from our ways of thinking about it, at least in the context of this investigation, which is resolutely naturalistic and nominalistic in its metaphysical orientation. And as we saw in Chaps. 4 and 6, we have good reasons to believe that an evolutionary processual metaphysical framework is a better candidate for providing metaphysical grounding principles compared to a traditional substance-property one based on neo-Aristotelian, Kantian or Hegelian categories, since it can accommodate the latter as a special case of itself (it can account for its explanatory successes and can explain its explanatory failures), while the opposite is not true. From the viewpoint of the substance-property Aristotelian, Kantian or Hegelian categorial framework, evolutionary processual metaphysics does not even make sense as an option to be rejected.

Does this mean that the alternative updated Spinozist model of the relation between reasons and causes, in which these parallel orders are not univocally conceptually articulated, can more successfully accommodate reasons and causes in a coherent and comprehensive worldview? I do not think so, at least not without substantial reworking. In this case, the problem is that without an account of the way in which the conceptual order of reasons is ontologically connected with the non-conceptual causal order, the mere fact of their epistemological parallelism and compatibility does not throw much light on the issue of how it is so much as possible for their supposed ontological *identity* to acquire a *determinate* sense. This becomes apparent especially if we take into account the fact that the rational and the causal order, however interdependent and parallel, constitute *mutually exclusive* ways of understanding ourselves-in-the-world. And if this is true, the whole Spinoza inspired parallelist rendering of the relation between reasons and causes threatens to break down. How could the contents of two mutually exclusive ways of understanding ourselves-in-the-world be determinately conceived as ontologically identical bereft of any conceptual tools for univocally characterizing such a reality?

Is there a way out of this impasse? If both the updated Brandomian-Hegelian and the Spinoza-Sellars renderings of the parallelism between the rational and the causal order are at bottom problematic, where are we to look for a more coherent and comprehensive understanding of the ontological and epistemological relations between reasons and causes? I think we can make progress in this direction by combining two key lines of thought: the first is to see both norms and facts, reasons and causes, construed in Brandom's neo-Hegelian terms, as 'ladder languages' that are not at bottom representational but make explicit the modal structure of the substance-property categorial framework. In this way they express the normative infrastructure of thought that introduces a common measure between mind and world (through conceptual articulation), makes possible the *self-correcting* dimension of our descriptive-explanatory practices and epistemically orient them toward the identification and occasional revision of counterfactual conditionals (what is causally necessary, possible or impossible) regarding the behavior of worldly objects and human subjects. The counterfactual conditionals themselves, that delineate our conception of the causal powers of objects, reflect our inferential commitments about the nature of things (the kinds to which objects belong based on their special causal powers) and express our pragmatic attitude of 'sticking-our-neck-out'—of willingly placing ourselves at risk—about the future behavior of objects[7] (Sellars, 1957, 1964). By making these inferences explicit and by being capable of revising itself with respect to its inferential commitments about counterfactual conditionals, the substance-property categorial framework is an indispensable *semantic* and *epistemic* means for improving the descriptive and explanatory resources necessary for *arriving* at an adequate understanding of ourselves-in-the-world. But, as we saw, it turns out that the most adequate picture of ourselves-in-the-world in terms of comprehensiveness and explanatory scope is one in which being is processual (becoming) and beings are ways of being processes (Chaps. 3, 4, and 6).

From this point of view—and this is the second key point for addressing the problem of the place of reasons and causes in the scheme of things—we can accommodate both reasons and causes in a comprehensive worldview by transposing them in a processual key. Thus understood, reasons and causes would be both ontologically efficacious by being recategorized as *modes of becoming*. Although the difference between reasons and causes would not be altogether eliminated, as the counterparts of the rational and causal realms would still express *intensively different ways* of processual existence (notice the adverbial characterization of their being), this processual rendering of reasons and causes would show that viewing them as incommensurable ontological kinds of things is highly misleading. Moreover, while to say that the rational and causal realms of intelligibility run parallel with one another is epistemologically true as a feature of the substance-property 'ladder' categorial framework, it makes no sense at the

[7] This attitude forms what we might call our 'anticipatory awareness' of the world, which practically pays off by hugely augmenting our ability to predict and control our environment.

level of an ontology of becoming since the latter is a paradigm case of a resolutely *univocal* understanding of the world. Interestingly, our thinking here is in its spirit (though not in letter) Spinozistic: the category of process plays a similar role to Spinoza's substance in being the unconditional metaphysical principle which makes meaningful the view that the realms of the normative and the causal are different dimensions and expressions of *one and the same* metaphysical reality. Yet, by being recategorized as ways of being processes, reasons and causes ultimately turn out to be expressions of processual being, thereby ceasing to be *distinct orders or realms* at all as far as their being is concerned. For example, it can be suggested that the categorial distinction between freedom as self-determination based on reasons for belief or action and causal affection from and between worldly objects would be reconceived in terms of intensive differences in degrees of freedom between complex sign-using organic systems (humans) and inorganic dissipative process (matter) or organic self-regulating process (animals) with regard to their power to act and be acted upon (their *Conatus*) and in terms of increasing decoupling of certain organic (e.g. brain) processes from immediate survival needs. And this means that with regard to their ontology, reasons and causes are neither parallel orders without categorial isomorphism (Sellars-Spinoza) nor parallel orders with categorial isomorphism (Brandom-Hegel).

But how exactly would the deontic and causal modalities be reconceptualized in a process framework? Ordinarily, causal relations of the form 'the oven's heat causes the potato to become worm' are presumed to be asymmetrical (causes precede effects), non-reflexive (no event causes itself) and transitive (if *A* causes *B*, and *B* causes *C*, then *A* causes *C*) relations between distinct events—where events are themselves modifications of propertied objects (substances). This view articulates our deeply entrenched intuition that causes *bring about* their effects, with some kind of *necessitation*, expressed in counterfactual conditionals ('if the oven were heated, it would become warm'). Causation thus acts as some kind of metaphysical glue holding together two externally related events. I think that the 'power framework', discussed in Chap. 6, provides a more illuminating way of understanding causal relations: from this standpoint, causal relations are symmetrical, continuous and wholly reciprocal expressions of a *single* unfolding process, itself expressing mutual interactions between empowered objects. This is obvious, for example, in the case of salt and water interacting, combining forces, to produce a particular kind of outcome, a saline solution, through the process of dissolving (Heil, 2021, pp. 120–123). Note that this recategorization of causal relations in terms of interactions among empowered objects still operates *within* the object-property framework and has not yet been cashed out in *pure* processual terms. We have already seen (Chap. 6) what happens to powers when they are transposed from the substance to the process framework: roughly, our ordinary quasi-Aristotelian conception of potentiality is replaced by Deleuzian virtuality. And while the understanding of powers in terms of the traditional conception of possibility only allows for novelty within a pre-determined possibility space that expresses the unchanged essence of each substance (object), to construe powers in terms of virtuality is to dynamize the space of possibilities itself. Yet, a question that remains to be addressed is how could we transpose talk of causal interaction (efficient causation) between substances to a process framework?

A (very) short answer to this question would be that instead of construing efficient causation in terms of a local interaction between self-standing objects, in a process framework such interactions would be understood as wavelike ripples in the texture of spacetime (with ordinary objects reconceptualized as processual modes of the unfolding of spacetime) while efficient causation would be reconceived in terms of patterns of interference in the unfoldings of these processual modes. Moreover, the relation between the processes that would be the counterparts of the occurrence of 'causes' and of the occurrence of 'effects' within a substance framework should not be conceived on the model of a prior and separately identifiable cause externally related to its subsequent effect. This, among other things, presupposes the existence of a homogeneous metrical spacetime continuum as a neutral stage-setting within which objects are individuated and identified by their essential properties (causal powers). In the process framework, objects are four-dimensional spatiotemporal 'worms' (Sellars, 1962a), and their causal powers are themselves ways of qualitative differentiation of these unfolding processual worms. And this means that the actualization of an object's 'causal powers' through its 'interaction' with the causal powers of other objects is to be reconceived in a process framework in terms of qualitative 'disturbances' and 'alterations' of the very *spatiotemporal unfolding* of the process that is the counterpart of this object. If we want to retain the notion of causation to describe these patterns of qualitative spatiotemporal alterations between interfering processual fields, we could perhaps call the relation between these interfering processes one of *expressive* causality (Chap. 2): Here what is expressed (the 'cause') is not an independently existing and prior condition of the expression (the 'effect'), for example, in the form of an antecedently existing possibility or power in the cause waiting to be expressed in the effect. Rather, the relation between the expressed and its expression is one of progressively unfolding differentiation which constantly changes (even if minutely) the very spatiotemporal qualitative space which occurrently expresses what we ordinarily call 'the object's power to affect and to be affected'.

On the other hand, the normative deontic modalities, as expressed in the normative statuses of authority (the status of making a move in the space of reasons) and responsibility (the status of accepting that a move has been made), within a process framework are reconceptualized as *processes of recognitive normative attitudes* of treating or holding normative statuses as authoritative for justifying our counterfactual reasoning, our practices of asserting and referring, etc. Thus understood, normative statuses function as regulative ideal markers, which presuppose the existence of sign-making practices, and are held in place by the aforementioned normative attitudes. There is a crucial ontological asymmetry between normative statuses and attitudes. Only the latter do exist (i.e. have real causal efficacy) in the world of becoming. Or, better, as we shall see in Chap. 8, normative statuses exist only as manners in which normative attitudes appear to themselves. Normative statuses themselves depend on their causal efficacy on the attitudes of acknowledging and attributing them to ourselves and others. And normative attitudes, at bottom, are ontologically emergent regimes of *affective* qualitative experience, expressed in the recognitive attitudes of *holding* or *treating* oneself and others as responsible or authoritative for believing or doing something. In a nominalistic world of pure processes, normativity is sustained by orectic normative we-attitudes, as expressive of emergent affective fields in human organisms. These are the real causal forces that sustain social normative structures (practices, institutions). Thus, at bottom, the causal power of normativity lies exclusively in the *enactment* of those statuses in the corresponding attitudes.[8] And, as we shall see in more detail in Chap. 8, normativity is a distinctive kind of *hyperstitional* process, namely, a process the unfolding of which expresses predictions and expectations of the

[8]While the attitudes in question are non-normative in their ontology (they are non-conceptual affective acts), they can legitimately be called 'normative' in the sense that the attitudes of, for example, *undertaking* a certain commitment is something that goes beyond a mere desire and can even express a willingness to risk or sacrifice one's other desires.

future that are themselves a causal force for bringing about and entrenching the expected behavioral regularities in the present.

Finally, it must be noted that the orectic we-attitudes that collectively sustain normative statuses are themselves situated at a resolutely non-conceptual level, in which there are no such things as already formed subjects standing over against an objectively existing and intersubjectively available external (or social) world. These orectic we-attitudes can be described as situated in what Simondon calls the realm of the *'transindividual'* (Simondon, 2020). This is a realm antecedent to the divisions between the subjective-intersubjective-objective, or the division between the (sovereign) individual and the (intersubjectively recognized) community. In fact, it is this transindividual realm that makes *both* individual persons and intersubjectively accessible social institutions (language, money, ownership) and relations (rational answerability between language and an external public objective world) possible. This transindividual realm exists in the form of pre-individual processes of affective resonance or attunement between organisms, that are progressively differentiated through conceptually articulated processes of abstraction, stabilization and resolution, into individual subjects, intersubjective relations, social objects (values, institutions), and the external world as conceptualized by the community. This transindividual realm is based on a minimally non-conceptually shared mimetic ability of iconically representing others. This ability of forming collective non-discursive we-attitudes can be naturalistically grounded in our ability to picture the world from the standpoint of others in the community, antecedently to our conceiving others as constituting an intersubjectively recognized community or conceiving ourselves or others as already formed subjects. The formation of subjectivity and intersubjective relations are conceptual achievements themselves based on this more primordial 'affective feel' of ourselves as acting in a we-mode, as direct embodiments of the group.[9] This primordial feature does not go away ontologically when we develop the conceptual achievements in ques-

[9] Indeed, as was mentioned in Chap. 5, this shared mimetic attunement which non-conceptually traverses human beings, co-constituting individual and group mentality, is not transparent to itself, and is akin to Castoriadis' (1975) anonymous social imaginary. This transindividual dimension of human interaction and meaning-making practices, due to its in-built non-transparency, generates social forms that it does not intend, that is, forms that are opaque to the practices that generate them. It is the task of philosophy and (social) science to de-reify these 'objectively' existing social forms and bring them to conceptually articulated intersubjective self-consciousness (to the extent to which this is practically and informationally possible), where they may be subjected to critical scrutiny.

tion. On the contrary, since the latter cannot but be enacted in this primordial affective group level if they are to have any causal efficacy, they must acquire a significant affective-cum-group dimension if they are to be counted as real causal worldly forces and be culturally reproduced from generation to generation.

Of course, we must always bear in mind that the above reconceptualization of the deontic and causal modalities in processual terms does not imply that the semantic framework that makes essential use of them is a metaphysical relic of no need to a nominalistic process account of ourselves-in-the-world. The central skeletal semantic framework in terms of which we conceptually articulate our contact with the world (senses conceived as expressing contents referring to objects, properties and relations as constituents of facts) and where the deontic modalities run parallel and are categorially isomorphic to the causal modalities does indeed, as Brandom suggests, have an expressive and critical function for enabling the language to respond to anomalies in a critically controlled manner that is at once categorially stable, self-critical and dynamically evolving in creative unpredictable ways. By conceiving senses as ***internally and normatively*** related to referents,[10] and as progressively converging through a dynamic historical process of constantly updating the former in an effort to restore the incompatible commitments to which they lead, this Brandomian-Hegelian rendering of the function of our skeletal semantic framework is indeed in a position to account for this dynamic rational stability-through-change of our epistemic practices. Yet, *contra* Brandom, this semantic framework need not necessarily be interpreted as ontologically committal, and thus as metaphysically necessary. And, as we shall see in Chap. 8, the dynamic dimension of critically controlled conceptual change (something akin to Hegel's determinate negation) has a counterpart dynamic dimension at the level of process ontology proper, where the engine of dynamic change is expressed in terms of Deleuze's *non*-Hegelian dialectic of (virtual) problems and (actual) solutions, fueled by intensive differences whose restoration comes in the form of morphogenetic actualization of novel structures,

[10] According to the Brandomian-Hegelian model, sense is normatively and internally related to reference in that the notion of what things are *in* themselves is understood in terms of the notion of how what things are *for* us *ought* to be. Importantly, *contra* Frege, senses and referents are not different kinds of things. In effect, Brandomian-Hegelian referents are expressively ideal senses (Brandom, 2019, pp. 435–436).

which in turn create new problems that call for new kinds of solutions ad infinitum (Chap. 4).

7.4 Reasons and Causes as Adaptive Critical Discursive Tools

Lastly, it must be stressed that from our viewpoint the semantic incommensurability between the realm of causes and the realm of reasons is an incommensurability in *use*, and thus, ultimately a *practical* incommensurability, grounded in different but equally fundamental pragmatic needs and interests. The incommensurability in question can be ultimately explained in terms of the long-term collective benefits of the division of intellectual labor in human societies in the wake of the scientific revolution.[11] It is not understood in representationalist terms, for example, as involving the representation of incommensurable domains of reality, but in terms of difference in the *semantic functional roles* of propositions about reasons and causes, that is, of their placement in the context of practices with different criteria of individuation, subject-matters and procedures of explanation and assessment, which is itself ultimately justified by the collective benefits incurred by the intellectual division of labor.

The pragmatic division between reasons and causes forms a central part of the self-correcting function of inquiry in its diachronic unfolding. Reasons are sometimes revealed as mere rationalizing causes of our behavior, while the ensuing knowledge from this reduction of reasons to causes is reincorporated in the space of reasons as a factor to be taken into consideration in judging if our reasons for believing, thinking or acting are sufficiently justified. The explanatory reduction of reasons (for believing and acting) to causes (to a-rational desires, power relations) brings to light

[11] The pre-modern world did not have a sharp categorial distinction between reasons and causes. In the context of the pre-modern world, natural happenings were conceived teleologically, as striving to conform to their internal life-plan or purpose, and the categorial difference of the powers of natural objects from those of animals or humans was conceived in terms of a hierarchy of values and not in terms of a difference between factual causes and value-laden reasons. This latter categorial distinction was introduced after the huge success of the scientific revolution in predicting and explaining the course of natural phenomena. The world was now to be understood as populated with value-free material unobservable objects or bodies governed by mathematically described natural laws and related through mechanistic causation, eliminating the need for invoking teleological, value-laden normative categories for explaining their behavior (Koyre, 1957).

the ideological elements in the function of space-of-reasons concepts, by showing the divergence between how certain space-of-reasons concepts *actually* operate (how reasons actually contribute to our 'here and now' behavioral patterns) as opposed to how they *ought* to operate—in accordance with the normative standards built into them. Thus understood, this kind of *indirect* interaction between reasons and causes *presupposes* their pragmatic incommensurability (the impossibility of *direct* vertical connection between them), and functions as a dynamically self-correcting tool for putting into question not only particular beliefs within the space of reasons but the adequacy of the (descriptive-explanatory function of the) very *concepts* used in the framework for expressing these beliefs.[12]

This self-correcting function of the reasons-causes distinction is intimately related, as species to genus, to the more obviously socially significant value-fact distinction. It can be argued that this latter distinction is also justified on pragmatic grounds, by a kind of self-correcting 'experimental' attitude it makes possible regarding how to best live as a community given what we know about how the world and ourselves regularly behave and interact. Indeed, in the wake of the scientific revolution and the Weberian disenchantment of nature, and given that we do not yet have a widespread consensus about the norms and values that are actually conducive to collective welfare, as well as about the causal means (policies) that might work better in instilling these values to the majority of humanity, there emerges another important pragmatic utility in insisting on the irreducibility of values to facts and vice versa. For, in the absence of the existence and widespread acceptance of a scientific social theory about what maximizes the collective welfare of humanity, and about the most efficient means for reaching such a goal, the next best thing is to institute and inculcate the value-fact distinction in the following experimental spirit characteristic of democracy at its best: by not being exclusively guided by substantial ethical values in its methodology and modes of explanation,

[12] This analysis is compatible with—and indeed presupposes—the Brandomian parallelist account of the relation between reasons and causes put forward in this chapter and puts some critical and pragmatic flesh on its formal semantic bare bones. As we shall see, if we cash out this pragmatic dimension in terms of problem-solving adaptive processes, we can internally connect it with process ontology. From this standpoint, the Brandomian level of the semantic categorial mirroring between reasons and causes is ultimately shown to be the manner in which processes that differentiate norm—from fact—discursive habits appear to themselves so that they constitute a unified practice (a 'unity-through-diversity') that conceptually articulates the world in a way that makes the practice answerable to the world.

empirical inquiry and its sophisticated extension, science, is thereby free to propose alternative ways of understanding the world and how to best collectively thrive in it, which can then be communally judged and recalibrated or revised on the basis of their pragmatic consequences in actual practice.[13] By contrast, a society that collapses the fact-value distinction, bereft of built-in mechanisms for challenging entrenched values and norms, runs the risk of becoming ossified and totalitarian, gradually losing contact with reality (which, as Philip Dick used to say, has the annoying feature of being something which 'when you stop believing in it, does not go away'[14]).

Now, what are these different yet equally fundamental needs and interests served by placing worldly and behavioral patterned regularities in the realm of causes as opposed to placing them in the realm of reasons? Causal discourse, in a sense that is to be distinguished from discourse about reasons, is intimately linked—at least from modernity onward—to our practices of describing and explaining (Sellars, 1957). It can be suggested that the needs served by our sign-making practice of describing and explaining reality are needs of survival and well-being within a hostile and unknown external and internal environment and are best satisfied by providing causal explanations (which, from the scientific revolution onward, proved to be more fruitful in this regard than, e.g. normative or teleological explanations[15]). On the other hand, the needs served by our sign-making practices in which signs function as evaluative appraisals of our behavior

[13] Note that the justification of the fact-value distinction is itself value-laden, which is not to say that there is no such distinction, but that it must not be turned into a dualistic dichotomy, for example, supposedly justified by formal logic alone.

[14] In anticipation of our discussion of normativity as a hyperstitional regime in Chap. 8, we must note here that Dick's slogan about reality shows the radical difference between conceptually articulated hyperstitional structures (normatively individuated things) and the non-conceptual world of sense and nature. As we shall see in Chap. 8, cycles of negative and positive feedback are an essential part of non-conceptually articulated structures too. It is just that the latter exert an asymmetrically powerful and direct influence in the (normative-cultural) structures of the former that often, due to our ignorance of the processes involved and the ways they affect us, compromises the allostatic balance of human organisms individually and collectively. There are many ways in which this allostatic balance can be rehabilitated which do not involve the construction of reliable conceptual maps for navigating the environment (see appendix). Yet, in the long run, establishing increasingly reliable maps for navigating non-conceptual nature is the best means for achieving long-term survival and flourishing.

[15] Impersonal causal explanations of our behavior—including neuroscientific, sociological or depth-psychological ones—explain experientially anomalous (and thus conceptually

are practical needs of survival and well-being in a narrower sense, namely, needs for successful mutual interaction and coordination among fellow human beings in a trustworthy and ever-expanding manner and are satisfied by placing these behavioral appraisals into the realm of reasons. It can be argued that the normative appraisal of our behavior makes possible, due to contingent but deeply ingrained facts of our biological and cultural evolutionary history (e.g. the fact that we do not have direct access to the factors influencing our own and others' thoughts and actions), speeds up effective communication among human beings and collective interaction in an orderly and group-like manner. It does that by cultivating mutually and ever-increasingly predictable expectations about the linguistic and non-linguistic behavior of other human beings, which otherwise would remain significantly more opaque and less effectively interpretable for practical purposes.[16] Reasoned discourse thereby provides us with a means of coordinating our verbal dispositions in cases of conflict or disagreement, thus shaping them in mutually beneficial directions: the efficient coordination of cooperative ventures, learning from each other different perspectives, making collective plans or decisions, pooling knowledge by indicating credible individuals worth relying on or deferring to and so on (Koren, 2021; Mercier & Sperber, 2017). This is why these practices of normative appraisal are socially selected and reproduced as a set of behavioral patterns from generation to generation, eventually becoming second nature to us.

Indeed, from the point of view of the evolutionary process naturalism discussed in Chap. 4, both reasons and causes can be understood as patterns of discursive habits selected and retained through a *self-differentiating process*—which becomes progressively self-conscious after the scientific revolution, the ensuing Weberian disenchantment of the world and ourselves, and the attempt by early modern and contemporary philosophy to find a place for these world-transforming events into a coherent story of ourselves-in-the-world. By progressively differentiating the expressive

incomprehensible) variations or lifeworld generalizations couched in person-level normative terms. For a justification of this view see Sellars (1963) and Christias (2016, 2023).

[16] This line of thought is similar to Dennett's account of rationality in terms of our adopting the intentional stance toward ourselves and others (Dennett, 1989). Yet, in our view, normativity is not just a stance justified for exclusively instrumental purposes. Normativity is a real collective world-making power (at the affective level of we-attitudes) and cannot be reduced to a mere instrument for facilitating independently fixed goals by already formed individuals, as the latter themselves have an ineliminable normative dimension.

tools needed for better orienting ourselves in the world (descriptive-explanatory tools of causal discourse) in an ever-expanding collaborative and trustworthy manner (prescriptive-expressive tools of reasons discourse), this self-differentiating process makes it possible for *both* sets of tools to better serve these purposes, thereby creating a flexible, dynamic, mutually beneficial and self-reinforcing symbiotic discursive relationship. We already briefly mentioned some important features of this symbiotic discursive relationship between reasons and causes earlier, and we shall further examine this issue further in Chap. 8. For the time being, it will suffice to say that the practical incommensurability between reasons and causes is fully compatible with—and indeed presupposes—their ontological *univocity*, as different ways of being processes. It is precisely by means of a self-division effected by this univocal processual plane at the level of sign-making practices that the realm of reasons and the realm of causes obtain their very identity as distinct yet parallel realms of making sense of one and the same reality and their credentials as *phenomena bene fundata*—as 'well founded' projections of fractured immanent sign-making practices.

References

Brandom, R. (1994). *Making it explicit: Reasoning, representing and discursive commitment*. Harvard University Press.

Brandom, R. (2015). *From empiricism to expressivism: Brandom reads Sellars*. Harvard University Press.

Brandom, R. (2019). *A spirit of trust: A reading of Hegel's phenomenology*. Harvard University Press.

Castoriadis, C. (1975). *The imaginary institution of society*. Polity Press.

Christias, D. (2016). Can Sellars' argument for scientific realism be used against his own scientia mensura principle? *Synthese, 193*(9), 2837–2863.

Christias, D. (2023). *Normativity, lifeworld and science in Sellars' synoptic vision*. Palgrave Macmillan.

Davidson, D. (1980). *Essays on actions and events*. Clarendon Press.

Davidson, D. (1986). A coherence theory of truth and knowledge. In E. Lepore (Ed.), *Truth and interpretation: Perspectives on the philosophy of Donald Davidson* (pp. 307–319). Blackwell.

Deleuze, G. (1988). *Spinoza: Practical philosophy*. City Lights Books.

Dennett, D. (1989). *The intentional stance*. MIT Press.

Dreyfus, H. (2013). The myth of the pervasiveness of the mental. In J. Shear (Ed.), *Mind, reason and being-in-the-world: The McDowell-Dreyfus debate* (pp. 15–40). Routledge.
Foot, P. (2001). *Natural goodness.* Oxford University Press.
Heil, J. (2021). *Appearance in reality.* Oxford University Press.
Koren, L. (2021). *Practices of reason: Fusing the inferentialist and scientific image.* Routledge.
Koyre, A. (1957). *From the closed world to the infinite universe.* Johns Hopkins Press.
Maher, C. (2012). *The Pittsburgh school of philosophy: Sellars, McDowell.* Routledge.
McDowell, J. (1994). *Mind and world.* Harvard University Press.
McDowell, J. (1999). Naturalism in the philosophy of mind. *Neue Rundschau, 100,* 48–69.
McDowell, J. (2019). Sellars and the space of reasons. *Analysis. Claves de Pensamiento Contemporáneo, 21*(8), 1–22.
Mercier, H., & Sperber, D. (2017). *The enigma of reason.* Harvard University Press.
Merleau-Ponty, M. (1962). *Phenomenology of perception.* Routledge.
O'Shea, J. (2007). *Wilfrid Sellars: Naturalism with a normative turn.* Polity Press.
Sellars, W. (1957). Counterfactuals, dispositions and the causal modalities. In H. Feigl, M. Scriven, & G. Maxwell (Eds.), *Minesota studies in the philosophy of science* (Vol. II, pp. 225–308). University of Minesota Press.
Sellars, W. (1962a). Time and the world order. In H. Feigl & G. Maxwell (Eds.), *Minnesota studies in the philosophy of science, vol. III: Scientific explanation, space, and time* (pp. 527–616). University of Minnesota Press.
Sellars, W. (1962b). Truth and correspondence. *Journal of Philosophy, 59*(2), 29–56.
Sellars, W. (1963). Philosophy and the scientific image of man. In W. Sellars (Ed.), *Science, perception and reality* (pp. 1–40). Routledge.
Sellars, W. (1964). Induction as vindication. *Philosophy of Science, 31*(3), 197–231.
Sellars, W. (1997). *Empiricism and the philosophy of mind.* Harvard University Press
Simondon, G. (2020). *Individuation in light of notions of form and information.* University of Minnesota Press.
Thompson, M. (2008). *Life and action: Elementary structures of practice and practical thought.* Harvard University Press.
Turner, S. (2010). *Explaining the normative.* Polity Press.

CHAPTER 8

Normativity as a Hyperstitional Dynamic Regime

In this chapter, I take up the challenge of articulating a successor concept for normativity that construes the causal efficacy of the latter in a resolutely immanent naturalistic manner—yet without reducing it to mere physicalism, for example, of a Lewisian variety. To this end, I mobilize Land's (2011) notion of 'hyperstition' as such a successor concept and draw a picture of normativity as an emergent intensive regime which, by way of we-attitudes as its manner of causally affecting the world, institutes a hyperstitional dynamics which can turn 'lies into truths' (i.e. false empirical generalizations into true ones). This is a way of illustrating and developing my basic immanent materialist view of normativity according to which its causal power lies in the enactment of normative statuses in the corresponding (we)attitudes (Chap. 7). In this connection, I note an unexpected convergence of viewpoints with a conception of normativity articulated by Sellars (1949) according to which norms are embodied generalizations which tend to make themselves true, by inhibiting the occurrence of the events that would falsify them (Sellars, 1949, pp. 123–124). From this point of view, normativity is understood in terms of what we might call '*living commitments in the making*', and its function can be compared to that of a self-fulfilling prophecy.

Further, it will be suggested that the above hyperstitional account of normativity can illuminate the phenomenon of ideology. Specifically, building on Kukla's (2000) suggestion that at the core of the function of

© The Author(s), under exclusive license to Springer Nature Switzerland AG 2025

D. Christias, *Towards a Hyperstitional Process Nominalism*, Palgrave Perspectives on Process Philosophy,
https://doi.org/10.1007/978-3-032-06350-2_8

ideology lies the misrecognition of a norm for correctly perceiving, thinking and acting as always already binding, it is argued that this way of construing normativity can throw light on the phenomenon of ideology.

Moreover, it will be argued that, in line with my general fractured immanent model of ourselves-in-the-world, hyperstitional dynamics is best seen as bifurcated in a non-conceptual and conceptual register: non-conceptual-level hyperstitional dynamics can be understood in terms of predictive processing (Clark, 2016, 2019), whereas conceptual-level hyperstitional dynamics can be construed in terms of Brandom's neo-Hegelian model of content determination and conceptual change (Brandom, 2019; Christias, 2025).

8.1 Normativity as a Hyperstitional Phenomenon

The term 'hyperstition', introduced by Nick Land (2011), is a neologism that combines the words 'hype' and 'superstition' to describe the action of successful ideas in the cultural field. It is akin to Dawkins' concept of a meme—a bodiless self-replicating unit of culture, propagated by imitation.[1] Hyperstitions are narratives able to effectuate their own reality through the workings of feedback loops, generating new socio-political (but also linguistic, economical, psychological) attractors. When enacted in the socio-political arena these engendered positive feedback cycles gradually make hyperstitions 'real'. Examples of hyperstitional feedback circles are financial speculation and Judeo-Christianity. The very fact that a sufficient amount of people *expect* that a state might go bankrupt in the imminent future or *treat* Jerusalem as a holy city with a special world-historic destiny might itself function as an affective catalyst that spreads the attitudes in question to even larger amounts of people, thereby gradually turning a mere 'fiction' into a 'truth'. This is so in the sense that the belief to the truth of this story has causal consequences in the real world, irrespectively of—and indeed precisely due to—the fact that the origin of the binding force of the belief might be mislocated by those who hold it.

[1] Imitation here is understood in a broad sense that includes language learning, persuasion and certain means of enforcing social obligations. Interestingly, Dawkins compares memes to parasites. As he puts it: 'When you plant a fertile meme in my mind you literally parasitize my brain, turning it into a vehicle for the meme's propagation in just the way that a virus may parasitize the genetic mechanism of a host cell. And this isn't just a way of talking -the meme for, say, "belief in life after death" is actually realized physically, millions of times over, as a structure in the nervous systems of people all over the world' (Dawkins, 1976, p. 207).

The cybernetic mechanism that drives hyperstitions can be described as follows. It all starts with an image of the future: a rumor, a prophecy, a prediction or expectation, that is, a 'fiction' (of course, to predict or to expect something is not literally a fiction, as is a rumor or a prophecy, but for the moment we shall stick to Land's terminology). In some cases, the hype will cause more and more people to believe that 'fiction' to be potentially real. If the belief is sufficiently strong at the affective level, people will act in anticipation. In doing so, they increase the reality of the fiction thus feeding the hype and so on. The mechanism is cybernetic in that it is driven by feedback loops in the form of cycles of hype and anticipation. In this manner, hyperstitional processes form orienting paths with which to navigate forward in a collective manner. Of course, as is obvious, this process is not headed toward a singular destination nor is it the expression of a pre-established law of nature or history that guarantees its success.

Land, as well as left accelerationists inspired by his work (notably Srnicek & Williams, 2015), understand hyperstitional processes as distinctive of the political domain or more generally of cultural phenomena such as capitalism, religion, myth, conspiracy theories, etc. I want to suggest that the way hyperstition works can illuminate a far more pervasive sociocultural phenomenon, distinctive of our very humanity as such, namely, *normativity*. Normative phenomena as such, including fundamental ones such as the normative structure of our linguistic practices, the categorial articulation of our lifeworld and our very being as persons (to the extent to which persons are normatively defined bodily and mental processes), are hyperstitional in their ontology. This view reinscribes normativity in nature without eliminating its distinctive character. Rather, it illuminates the distinctive immanent causality of normativity, which to my mind, consists in the fact that the causal power of normativity lies exclusively in the *enactment* of normative statuses in the corresponding (we)attitudes through the generation of iterated positive or negative feedback loops[2]and feedforward control loops.

[2] Here it must be noted that though hyperstition is primarily a positive feedback process, it is not exclusively bound to it and can interact with negative feedback mechanisms in several ways. Negative feedback can emerge as a secondary effect, as a reaction or even as a mechanism that reshapes or redirects the trajectory of a hyperstitional process. This suggests that hyperstitional processes can sometimes encounter or even necessitate negative feedback loops to sustain themselves in the long run. However, as we shall see, typically, in hyperstitional normative processes, negative feedback loops, by relatively stabilizing a system of norms, ultimately serve as a means for the system's expansion through self-correction of its

8.1.1 *The Being and Causal Power of Normativity*

Before attempting to show that normativity is essentially a distinctive kind of hyperstitional process, let me start from briefly explicating what I take normativity to be. Normative phenomena form the central core of our sense of personhood and are not limited to culturally enforced rules of conduct, conventions or policies. As might be expected, normativity is involved in action performed on the basis of reasons or in perceiving and judging a behavior in moral terms. But it is also inextricably implicated in a decidedly non-moral and non-reflective manner at the far more basic level of perceptual, conceptual and intending acts. At first approximation, at an abstract level, to say that all the above distinctively human acts have an essentially normative dimension is to say that in conceiving something as a perceptual, thinking or intending act, we do not merely describe or explain it, but we see it as incorporating standards that serve to assess its empirical realization—as correct or incorrect, normal or abnormal, true or false, justified or unjustified, fulfilled or unfulfilled. These assessments are registered, undertaken, attributed and criticized (thereby conferring normative statuses of responsibility and authority to holders of the respective conceptual contents) by being placed in a social network of reciprocally recognized rights and duties ('cans' and 'oughts') pertaining to meaning, knowledge and action. As we saw in Chap. 7, this is what Sellars calls the 'space of reasons' (Sellars, 1956, see also McDowell, 1994, Brandom, 1994). This space is inhabited by persons, which are themselves constituted through being mutually recognized as sufficiently skillful navigators of this space, that is, as capable of being bound by norms concerning perceiving, thinking, intending and acting—in-the-world. Importantly, this recognition *itself* has a normative structure: when I recognize myself or others as persons I do not describe or explain an antecedently existing fact or essence ('natural kind') of personhood (Sellars, 1963, p. 40). I make authoritative demands, acknowledge and attribute commitments and entitlements to myself and others that are conducive in making my behavior comply to the standards of being a person, thereby constituting myself and others as persons proper.

Moreover, although the normative dimension of the content of a perceptual, thinking or intending act is clearly distinct from and irreducible to

basic parameters (norms) that *changes* its set points. Thus, as will be shown, the long-term tendencies of normative hyperstitional processes are best represented by positive away-from-equilibrium feedback cycles.

describing and explaining these acts in empirical-scientific (e.g. evolutionary-historical-psychological) terms, it is also the case that the *causal force* of the normative statuses of responsibility and authority with respect to belief-thought-intention contents comes exclusively from their actual enactment in corresponding recognitive *we-attitudes* of *acknowledging* and *attributing* normative statuses of authority (to the world or others as arbiters for the justification of our beliefs) and responsibility (for offering reasons for our beliefs) by *undertaking* commitments to do so (Brandom, 1994). Thus, to my mind, there is a crucial ontological asymmetry between normative statuses and attitudes. Only the latter do exist—that is, have real causal efficacy—in the 'really real' world of becoming. Normative statuses themselves depend on their causal efficacy on the attitudes of acknowledging and attributing them to ourselves and others.[3]

Are these recognitive we-attitudes naturalistically respectable properties immanently arising from a nominalistically described processual reality? Given our model of nominalistically respectable emergence (Chap. 5), I take it that the answer here is positive, even if, as I think we must, we understand normative we-attitudes as emergent intensive properties, not ontologically reducible to their physicalistically described 'components'. Indeed, we can speculate that, just as there is reason to believe that sensible qualitative features be causally efficacious as such only if they exist as emergent (yet still nominalistically respectable) properties of sentient organisms of a certain complexity (Chap. 5), the same goes for *affective* or *orectic* qualitative experience, in its guise as 'desire for recognition' of desiring organisms as expressed in recognitive attitudes of *holding* or *treating* oneself and others as responsible or authoritative for believing or doing something (see also Chap. 7). If this is correct, it follows that orectic normative we-attitudes, as expressive of emergent affective fields in human organisms, are the real causal forces that sustain social normative statuses (being bound by norms so that one's performances can count as correct or incorrect) and the corresponding social practices and institutions. As was mentioned earlier, it is the enactment of those statuses in the corresponding attitudes that constitutes the causal force of normativity. And the enactment in question can be understood in cybernetic terms,

[3] This ontological asymmetry, among other things, enables us to explain how normative statuses could come to exist in the first place, whereas their mutual ontological interdependence (Brandom, 2019) makes this a complete mystery as, together with the corresponding normative attitudes, they must always already be presupposed as having been in place.

that is, in terms of the generation of iterated positive-cum-negative feedback cycles and feedforward control loops in *projective picturings*[4] of the physical and social world.[5]

Crucially, this does not imply that normative statuses are just epiphenomenal, illusory states. On the contrary, in the form of internalized impersonal semantic, epistemic, practical and ethical standards of correct performance, they are indispensable tools for making explicit, regimenting, coordinating, criticizing and thus for collectively restructuring normative we-attitudes toward sustaining practices and institutions that are more conducive to our epistemic and general welfare (Christias, 2025). This 'relative autonomy' of the level of normative statuses over normative attitudes (which tends to be reified by many philosophers that emphasize the irreducibility of the normative to the causal-descriptive-explanatory level) is indispensable as a tool for regulating the latter at a collective level for at least two reasons: (a) Normative we-attitudes are not ontologically transparent to themselves; (b) they are context-bound, incapable of being abstracted from their ineradicably monadic and perspectival qualitative form and content. By contrast, normative statuses, in the form of thinkable conceptual contents, can—as abstract functional roles within a publicly accessible social space, namely, the space of reasons—represent any possible fact therein objectively, as the same for everyone, irrespectively of the subjective perspective of the corresponding normative attitudes that ontologically sustain the conceptual contents in question. Normative statuses thereby constitute an objective possibility space, the space of reasons, which, by greatly enhancing and facilitating the individual and collective stabilization, individuation and coordination of 'monadic' normative attitudes, better satisfy our fundamental (survival-related) needs, interests and purposes at the collective level, and, by enabling the formation of *novel* such needs, interests and purposes, prove also indispensable for satisfying our need for individual and collective flourishing *beyond* mere survival.

[4] For an analysis of the concept of 'projective picturing', which is a blend of Sellars' naturalistic account of picturing with that of a generative model for anticipating reality as this notion is used in predictive processing theories in cognitive neuroscience see Sect. 8.3.2.

[5] See also Sachs (2022). According to Sachs, the relations between normative content and naturalistic picturing be illuminated if we construe linguistic behavior as triangulated cybernetic behavior. We ascribe semantic content to any material utterance or inscription (including our own) that can be coordinated with other utterances or inscriptions that are functionally integrated into the sensorimotor feedback loops of other cybernetic systems, where the criteria of coordination lie in successful cooperation.

Relatedly, it can be argued that by assigning authority or responsibility to behavioral patterns or worldly states of affairs, normative statuses function as *summarizing*-cum-*regulating markers* of the complex motley of interrelations between normative attitudes (Christias, 2025). By functioning in this way, this relatively autonomous manner of appearing of the recognitive relations between normative attitudes not only makes possible new ways of regulating our collective behavior and attention (salience points) but also canalizes our individual and collective behavior toward what the community deems to be the most efficient paths for achieving our purposes. Normative statuses, for example, do this by explicitly flagging the circumstances, the behaviors and the individuals that the community deems as reliable or unreliable sources of knowledge and action. And all this increase in degrees of freedom and expressive power can be achieved without the need for ontological commitment to anything beyond the complex interrelations between normative attitudes and the modes in which they appear to themselves. This is because, as we shall see in Sect. 8.3, normative statuses, while relatively autonomous from normative attitudes, are ontologically on a par with the latter, in the sense that they are at bottom a *subtractive manner of appearance* of the recognitive interrelations between the latter. And since manners of appearance do exist in the really real world, the world of normative statuses and contents is fully real, an outcome of the world-making powers of orectic normative attitudes.

8.1.2 Normativity and Hyperstition

Now, the intimate connection of normativity to hyperstition becomes clearer if we combine the fact that normative statuses exist only in their enactment in normative we-attitudes, with the *pragmatic* insight, which this time comes from early Sellars (1949), to the effect that the normativity of rules is at bottom something existentially concrete, that is, something that *lives* in behavior, or, in Sellars' words, 'an embodied generalization written in flesh and blood, or nerve and sinew, rather than in pen and ink', which 'tends to make itself true' (Sellars, 1949, p. 123). This conception of rules and normativity as enacted in behavior and as having the status of what we might call *living commitments in the making* can be usefully understood in terms of hyperstitional dynamics.

To see this, we must examine more closely this pragmatic account of rules as outlined in Sellars' relatively neglected article 'Language, Rules and Behaviour' (Sellars, 1949). In this article (123–124), Sellars holds that although rules are to be distinguished from empirical generalizations in that rule-governed behavior is behavior that occurs because of the rule as opposed to behavior described by empirical generalizations which, at best, merely conform to rules, there is also a sense in which a rule is not something completely other than a generalization. As Sellars characteristically puts it: 'The mode of existence of a rule is as a generalization written in flesh and blood, or nerve and sinew, rather than in pen and ink' (123). What distinguishes rules from empirical generalizations is that the propositions that are expressions of rules contain certain special terms such as 'correct', 'proper' and 'right'. And these special features in the formulation of rules give expression to the fact that the rule is an *embodied generalization which tends to make itself true* (123). More specifically, by containing these terms, the 'embodied generalizations' in question tend to *inhibit* the occurrence of events that would *falsify* those generalizations. In this connection, Sellars makes the extremely interesting suggestion that at the core of the formulation of rules lie factually *false* empirical generalizations which express in factual language how the world should ideally be in certain respects. For instance, the core-generalization on which the moral rule 'one ought to tell the truth' is built is 'people always say what they believe' which is, of course, false, while the role of the rule is precisely to inhibit the occurrence of events that would falsify this generalization.[6]

Another—perhaps more important—case of norms at the core of which lie factually false empirical facts which gradually tend to make themselves

[6] Sellars' account of rules as factually false empirical generalizations that 'tend to make themselves true' primarily focuses on the world-guiding dimension of rules, that is, their function as conduct regulators. It might be objected that rules such as purely conceptual or regulative principles (e.g. the theological claim that God is omni-benevolent or certain mathematical axioms) do not directly function by tending toward empirical realization but by structuring our conceptual framework. In these cases, deviation does not directly trigger corrective feedback in the same way; instead, apparent counterevidence is reinterpreted within the conceptual framework. In response, it can be argued that even these latter purely conceptual or regulative principles do have an *indirect* world-guiding function. They are not mere contemplative propositions. They structure our conceptual frameworks, regulate our interpretations of experience and reconfigure our social and interpretive comportment toward reality. The difference lies in their mode of operation: whereas moral rules aim to directly inhibit falsifying events, these more abstract rules work by transforming the interpretive space and social behavior through which the world is lived and understood. An example of this kind of regulatory rules (about the proper description of our perceptual experience) is provided below in the main text.

true if the norm is enacted in embodied behavior is provided by the following example, again taken from Sellars: John is a necktie salesman who, not yet understanding the distinction between what the color of neckties actually is and how it merely looks (seems to be) under non-standard conditions of lighting, is initially at a loss when his colleagues point out to him that a certain necktie merely looks green but is not actually green under certain non-standard conditions such as bad lighting. John has the tendency to say that in these circumstances the necktie just changed in color, say from red to green ('it was red before, yet now it is green') (Sellars, 1956, §16). Yet, by accepting this norm as making legitimate demand of him and by incorporating in his behavior what it says about how he *should* respond to these situations, the norm performatively tends to make itself true (turn itself into a 'fact'). This is so in the sense that while before endorsing this norm John's experience, lacking the is-seems distinction, is *not* correctly described by saying that 'it merely looked to him that the tie was green, but the tie was not actually green', *after* endorsing the norm John's experience *is* now correctly described in this way. Thus, again an initially false empirical generalization at the core of a (perceptual) norm ('what changes under x,y,z conditions of observation is the relation of our perceptual system with the object, not the object itself') tends to make itself true by being enacted (endorsed) in living behavior.[7]

This pragmatic account of normativity shows that the ultimate *raison d'être* of rules lies in transforming the world and in shaping our habits of action in such a way that an initially false empirical generalization, which lies at the core of the formulation of the rule, be gradually turned into the respectively true empirical generalization, or, in other words, so as to bring it about, by shaping the right habits of perception, inference and action, that what the rule prescribes be concretely realized in the world in

[7] It may seem puzzling to treat the is-seems distinction as involving an initially false empirical generalization, since the external objects did not, in fact, change color. Yet note that in this example the empirical generalizations in question are not descriptions of external states of affairs but descriptions of the world *as experienced by perceivers*. Before adopting the norm, John's 'world' was one in which color-stuffs themselves changed with lighting conditions. After adopting it, his world is reorganized into one where objects have constant colors and variations are attributed to appearances. This is a genuine change in the world-as-experienced, and the rule inhibits the old judgment patterns ('the tie has changed color') that would falsify the new generalization. In this sense, the norm does function self-fulfillingly: by enacting it in his perceptual judgments, John makes it the case that the generalization becomes true within his conceptualized experience.

the form of embodied individual and collective pattern-governed behavior. And if this analysis is extended to normative phenomena in general, it amounts to conceiving normativity in all its dimensions (semantic, epistemic, practical, moral, aesthetic) in hyperstitional terms: as phenomena whose build-in function, through the 'law of effect' (the role of rewards and punishments in shaping behavior), is to *gradually bring about their own reality* in a manner akin to a self-fulfilling prophecy by being performatively enacted in pattern-governed behavior. For example, from this point of view, values of all kinds (epistemic, moral, aesthetic, practical) can be understood as embodied desires that, by being understood as norms for action, tend to make themselves true. This view points toward an interesting internal connection between facts and values: viewed hyperstitionally, values are the (collectively desired) facts of the future (Feenberg, 2017, p. 8).[8]

Moreover, this proposal has the following double advantage over the Landian conception of hyperstition: a) Hyperstitional phenomena are no longer conceived as applied only to specific domains of culture (religion, capitalism, mob psychology, viral posts/videos/stories, etc.) but encompass all social practices, including most importantly their very condition of possibility, that is, *linguistic* practices as expressed in *meaningful discourse* instituted by semantic rules for the correct application of signs in concepts and propositions. b) In this sense, by conceiving even semantic normativity as hyperstitional, we bring the ideal dimension of normativity down to earth (we construe it in immanent, materialistic, non-givenist terms) while at the same time accounting for its irreducibility to non-normative facts, its inexhaustibility to the current factual *status quo*, by highlighting the

[8] Note also the interesting connection of this hyperstitional conception of normativity with Deleuze's (Deleuze, 1968) or Badiou's (Badiou, 1988) 'commitment to the truth of the event' which, if faithfully sustained by a community of believers, *makes* the content of the commitment true, that is, gradually realizes it in the actual world. Perhaps, Deleuze, and most certainly Badiou, would want to distinguish events proper, which have a potentially emancipatory and unobjectifiable dimension, from those which do not, but at this level of analysis, I do not see how this dimension can be disentangled from its opposite (reterritorializing, objectifiable) dimension. From my point of view, such a disentanglement can be only relative and depends on whether the unfolding of the process of the 'commitment to the event' which gradually makes it true unleashes joyful or sad passions at the individual and collective level (see Chap. 9).

essentially creative, performative and enactive dimension of its mode of existence. And, *pace* Land, we do this by incorporating the radicality entailed by the performative and enactive mode of existence of the normative, without having to *literally* hold that hyperstition turn 'fictions', 'rumors' or 'lies' into truths. For, obviously, the way in which, for example, our words acquire their meaning through collective hyperstitional practices of attributing and acknowledging commitments and entitlements involving the proper way of using written or spoken signs need not be literally construed as a process that turns fictions or lies into truths—if for no other reason than that it is precisely the hyperstitional institution of semantical discourse that makes possible the very existence of fictional stories, narratives, speculations and lies. Yet the crucial insight of hyperstitional dynamics, namely, the fact that predictions and expectations of the future are themselves a causal force for bringing about and entrenching the expected behavioral regularities in the present, is preserved and indeed radicalized, as it applies to normativity as a *whole*, namely, to that which makes us *distinctively human*: persons with sensitivity and commitment to (semantic, epistemic, ethical) norms, whose minds are understood normatively, as *loci* of normative authority and responsibility.

Thus, the above account of normativity in terms of hyperstitional dynamics is not only a resolutely non-metaphysical way or understanding the distinctiveness of humans as persons endowed with minds (since it does not present humanity as a separate ontological realm within nature), but, more radically, it conceives this distinctive and irreducible normative character of ourselves as persons—and indeed, of any complex material process that can count as playing the normative-functional role of personhood—as an always precarious yet potentially ever-entrenching diachronic *achievement*. From this point of view, the very *reality* of persons as a normative status has an essentially performative and processual element: it depends on the each time achieved thresholds of entrenchment of recognitive we-attitudes of *treating* ourselves *as* being bound by rules (for using words in the same way, for keeping our promises, etc.), that is, of *trusting* that future behavior will conform to present predictions and expectations as incorporated in present normative standards, criteria and demands. This means that the very *being* of ourselves as persons—as well as the being of all normatively laden concepts and categories that are derived from the category of personhood—depends on the activation, sustenance, collective transmission and enhancement of the above recognitive orectic we-attitudes from generation to generation.

This is at once a liberating and potentially frightening and paralyzing thought. On the one hand, it shows that we are the authors of our destiny: we are collectively self-determining beings.[9] Our flourishing as persons-in-the-world depends on our *collective commitment* to *make* the natural and social world a place worth living and striving for. Yet, the fact that we are authors of our own destiny, and the abyssal freedom and responsibility that comes with the realization that there are no supernatural, natural or social forces to the authority of which we can defer to successfully address our deepest concerns and problems, can well go hand in hand with individual and collective existential anxiety: there is no happy-ending cosmic narrative to give us hope and consolation, there are no laws of nature, biology of history that can guarantee the sustainability and meaningfulness of our practices (both in the semantic and in the ethically loaded sense). We have only ourselves to blame if humanity does not progress toward an increase of individual and collective freedom and autonomy and instead regresses into a state of increasing inequalities in accessing vital epistemic and non-epistemic goods. And while this situation demands a violent maturity of humanity and calls for cultivating attitudes of a high level of sensitivity to the interconnectedness and interdependence between human beings and a corresponding solid commitment to the promoting of collective welfare and mutual recognition in the face of ideological and material obstacles to their realization, it is also the case that the realization that we are essentially normatively hyperstitional beings, by liberating us from the view that we have a pre-determined essence that fixes our place in the cosmic order, opens the way for unprecedented enhancement of our individual and collective freedom, autonomy and creativity. Our existential predicament lies precisely in whether we shall prove worthy of our 'best versions' of ourselves, in the full knowledge that there is no fixed essence or goal (except the most general one of maintaining a dynamic allostatic balance with the environment) which determines once and for all our fundamental needs, desires,

[9] It does not follow from this that the self-determining character of normativity is an end in itself. Rather, it can be best conceived as an evolutionarily selected multipurpose tool for speeding up collective coordination (at all levels, be they semantic, practical, moral), whose *raison d'être* is ultimately the diminishment of 'sad passions' and the enhancement of 'joyful passions' (Spinoza). Or, in quasi-Marxist terminology, the ultimate 'purposeless purpose' best achieved by the semantic-practical tool of 'artificial normative enhancement' is the obtaining of ever-increasing levels of collective resonance (skillful coping) with the natural and social environment so as we be able to freely exercise our sensory and intellectual powers as joyful ends in themselves (Christias, 2023). See also Chap. 9.

hopes and interests. And this is perhaps the most important hyperstitional wager we placed on ourselves by becoming what we currently are: increasingly self-conscious post-enlightenment beings.

8.2 Ideology as the Reification of Hyperstitional Normative Force

As was mentioned earlier, I take it that, as a bonus, this hyperstitional account of normativity can illuminate the phenomenon of ideology.

But what exactly do we mean by 'ideology'? A common model for understanding the phenomenon of ideology is the doxastic one according to which cognition is construed in terms of particular epistemic attitudes that subjects hold toward determinate propositions. In this context, ideology is construed as a propositional system of beliefs or representations about the world and ourselves that distorts reality by falsely representing in the subject certain ideas and ideals as independent of our social relations and our commerce with nature. Along with other theorists (Haslanger, 2017), I believe that this model misconstrues the nature and function of ideology. Ideology is not a propositionally structured system of ideas and ideals that form our self-conception within a social world already populated by subjects (Railton, 2000; Shelby, 2003), but an unreflective yet motivationally effective background system of settled attitudes which *conditions* and *orients* our conception of our being-in-the-world by immediately informing our actions within certain social contexts (Stanley, 2015), and by thus indirectly shaping our self-conception again at an unreflective level. Our explicit web of beliefs, values and ideals about the world and our place in it are the surface expressions of this background orientational system and cannot directly account for it. This background system of settled attitudes at once discloses an (ideologically laden) natural and social world in which we have our everyday being *and* by doing so hides from view the relation of this world to the asymmetrically recognitive we-attitudes that make it possible, as the latter are expressed in modes and relations of material and 'immaterial' (semantic, cognitive) production actualized in the form of social institutions such as economy, politics, law, state, education, science, etc. Thus understood, ideology does not merely express a false consciousness of an already formed subject which distorts its representation of reality. Instead, ideology has a recognition-as-misrecognition structure (Althusser, 1971), which discloses ('recognizes') both an everyday lifeworld *and* us as sovereign subjects standing over against it, precisely *by* concealing ('misrecognizing') the very causal-historical-social conditions of possibility and functioning of this double subject-lifeworld disclosure.

Now, building on Kukla's (2000) suggestion that at the core of the function of ideology lies the misrecognition of a norm for correctly perceiving, thinking and acting as always already binding, I shall argue that our hyperstitional account of normativity can throw light on the phenomenon of ideology:

On Kukla's Althusserian understanding of ideology, at the core of this phenomenon lies a misrecognized attempt to *make* a certain fact true by treating (recognizing) it as *always already binding*. More specifically, according to this view, the basic function of ideology is the *mis*recognition (covering up) of the active effort to make something true under the guise of its recognition as 'always already' true in a way that is normatively binding (e.g. in the way metaphysical *essences* are thought to be both necessarily existing and normatively binding independently of being recognized as such).[10] This conceals the fact that the normative validity of the empirical generalization which 'ought' to be established in each case (e.g. correct use of words, appropriate behavior of individuals and collectives in appropriate circumstances) would not have any normative force, in the sense of motivational power, independently of its enactment in corresponding normative attitudes. Interestingly, an ideological misrecognition of the above kind is an ideal way for recognizing and consolidating specifically normative patterns in our behavior as it is true in the case of normative facts that if we treat ourselves as *bound* by them, that is, we hold our behavior and ourselves *accountable* to them, we thereby, and through their constant performative enactment and reproduction as behavioral standards of correctness, bring them gradually into existence (Kukla, 2000, pp. 168–170; Althusser, 1971).

[10] The paradigmatic case of how this misrecognition process works in constituting normative authority is provided by Althusser (1971): according to him, subjects get inducted into normative space through a process of misrecognition, where other authoritative subjects (mis)recognize a (potential) subject as already bound by norms and capable of navigating normative space. The subject, in turn, (mis)recognizes itself as properly identified in that recognition, and as already having been the subject it is recognized as being. In this way, this process brings subjects into existence by placing them within normative space and binding them in the appropriate ways, thereby serving to bring about its own correctness (Kukla, 2000, p. 167).

This Althusserian conception of ideology is foreshadowed in our above conception of normativity in terms of lived commitments that tend to make themselves true (though, of course, this understanding of normativity also contains the resources to criticize and liberate us from the mythical force and lure of ideology). For, to treat something as always already making legitimate demands on us thereby concealing (even from ourselves) the fact that it is its treatment and enactment in collective embodied behavior *as* binding that constitutes its normative force turns out to be an exceedingly efficient way to gradually bring it about in the actual world in collective pattern-governed behavior. How can recognizing something through acts of constitutive misrecognition in which its material conditions of existence and enunciation are concealed is nonetheless an efficient means of bringing about conformity with the recognized content? A very sketch explanation of this fact might be the following: normative force as always already binding is understood as situated out there in the independent world (this is what being always already binding independent of our attitudes *means*). In this way, the world is effectively understood by analogy to a person that can impose its binding normative demands on us, the only difference being that this person-like world is independent of any and every actual person or human community synchronically or diachronically. Again, this is precisely what it means for a demand to be always already binding: it means that there is a supposed origin of this normative fact (it is 'already' true) that we are unable to locate however deep we delve into the past (it was 'always' true). Thus, normative force is seen as normatively imposed or enforced by a higher and less parochial worldly authority than individuals and human communities.[11] And since the necessity involved in this normativity is understood as a feature of the world (or the subject) itself, it is construed as a normatively binding *metaphysical essence*, the criteria of conformity to which carry metaphysical significance in that the deviation from them amounts to a deviation from what the 'cosmic order', 'human nature' or the 'authentic self' prescribes. Moreover, by having this presumed metaphysically authoritative status the 'always already binding' norm thereby *both mandates* our taking active measures for its universal imposition at the level of actual embodied behavior and *justifies* us in doing so, while at the same time alleviating the burden of personal responsibility for any seeming 'collateral damage' (pain, suffering) that our

[11] This, importantly, includes the 'authentic' self as the normative sovereign core of the individual that we only have to unearth to become 'who we really are'.

actions to this end might entail. This latter attitude can be readily explained precisely because from this point of view the damage inflicted appears legitimized by a higher objective authority, as opposed to being a result of mere coercive force for satisfying our vested interests.

Of course, as we have seen earlier, this is not to say that this performative dimension of normativity appears *only* in the form of ideological recognition-as-misrecognition. Indeed, it can be argued that the ultimate goal of inquiry or a regulative ideal of our practices in general is the minimization of this ideological understanding of normative force and its reconceptualization as a set of patterns of behavior that are not understood as obligations imposed on us by a transcendent force (metaphysical essence), but as cases of immanent communal *activity* that aims to *change* existing physical and social reality so as to promote our common good as sapient and as sentient agents. On this conception of normativity, *we* are *responsible*, individually and collectively, for the manner and justification of their realization in the physical and social world. The norms to which we are committed are prescribed neither by God nor by Nature or History. We are only answerable to one another for the commitments and the actions we undertake.[12] What is more, the recognition of the *hyperstitional* dimension and character of norms makes the adequate identification of the *content* of what *in fact* maximizes our collective epistemic and general welfare all the more urgent, as systematic, ideologically elicited errors in our conception of it will institute self-reinforcing feedback loops with ever-increasing negative effects to the lives of ever-increasing number of humans, animals and the ecosystem as a whole.

We can conclude this section by making more salient the intimate connection between a central theme of the present book, namely, the myth of the Given, and ideology. The Given, understood as the view that our thinking, acting and categorizing the world is legitimized only if it is guided by metaphysical necessities having normative force independently

[12] Note that this is consistent with holding that initiation into the space of reasons often involves treating norms as *given*, as if they are just *how things are*. And indeed, this 'as-if Given' dimension may well be unavoidable at the early stages of normative socialization. But it does not undercut the view that normative authority can and should ultimately be understood as immanent and socially instituted, not metaphysically given. Rather, it underscores the developmental and pragmatic necessity of a transitional or 'pedagogical' ideology—what we might call *provisional Givenness*. From this point of view, ideology might be understood as a ladder we climb to enter the space of reasons, but one we can—and must—learn to throw away.

of their being recognized as such by our social practices, can be thought of as a *universal core feature* of ideology. Moreover, in view of our argument in Chap. 3 to the effect that a resolute rejection of Givenness (needed for a strict immanent metaphysics and epistemology) entails a resolutely anti-representationalism in semantics and ontology, a more positive suggestion can be made about the content of ideology in its most general form as mythical Givenness: it consists in construing our fundamental life-world categories as normatively binding in a big-R *representational* sense, that is, as natural kinds or 'natural norms' that cut nature (including human nature) and society at their categorial joints, and in conceiving the deviation from these natural norms as pathologies that alienate us from our very essence.

Note that by having normatively binding force, these natural kinds are supposed to have epistemic power and necessarily so: they are taken to be *epistemic natural kinds*, situated in the mind, offering us an epistemically secure, immediate and transparent cognitive contact with the world. Traditionally, this immediate point of epistemic contact between subject and object has been understood in terms of the existence of a faculty of sensuous or intellectual intuition which supposedly gives us secure and privileged access to the semantic, epistemic and ontological features of the world and the mind. This supposed epistemic natural kind that has authority independently of being part of our space-of-reasons practices is pivotal for the function of ideology as it places normative authority and the bindingness conferred from it inside the *mind*. Thus understood, the mind is itself conceived as the locus of intrinsic normative power (Cartesian mind) or as the privileged medium that through the 'mind's eye' directly perceives the ultimate source of norms (Platonic Forms, Aristotelian essences, Augustine's *adequatio rei et intellectus*, Husserl's phenomenological essences). This view effectively privileges those individuals or groups that recognize themselves as *loci* of this authority over others, conferring to the latter the responsibility to acknowledge the authority of the former over them, but not vice versa. In this way, the social origin and nature of recognitive epistemic relations of authority and responsibility is concealed by being misrecognized as a natural epistemic kind of an extra-social nature.[13]

[13] This is akin to the Hegelian master-slave model of asymmetric recognition as masterfully developed by Brandom (2019). See also Rorty (1979), who was arguably the first to realize that there is something deeply anti-democratic about epistemology due to its reliance on the myth of the given.

And, as was shown earlier, this recognition-misrecognition of normative epistemic power as always already binding independently of our socially shaped space-of-reasons practices is actually a very efficient way of (hyperstitionally) *making* these relations of asymmetric epistemic authority and responsibility *true*: this mythical conception of normative authority can well shape and be embodied in our behavior, in our individual and collective attitudes of acknowledging and undertaking normative commitments and entitlements of all kinds.

One last way of showing the intimate connection between mythical Givenness and ideology is to draw our attention to the fact that the Given goes hand in hand with an objectionable essentialism about meanings, thought and reason: from this standpoint, the contents of meanings, the objects of thought and the contents of the intentions that motivate our actions are seen as timeless forms or essences, complete in their meaning prior to our reflection upon them. The reflection upon them is at best construed as an eternal contemplation, a passive intuition or timeless perception of these forms or essences. Yet, once we reject the Given, we realize that the content of our meanings and intentions and the object of our thought is not antecedently given to our thinking but is discovered only through its own self-positing. The act of positing in question is not to be understood as something static or momentary but as a dynamic temporally extended process of extending and revising our concepts and conceptions of thought and its contents in order to better accommodate incompatible commitments existing in their former versions (Brandom, 2009, pp. 100–101; see also Sect. 8.3).

Indeed, an exceptionally fitting example of such a process is provided by Sellars' myth of Jones (Sellars, 1956), where mythical genius Jones and his peers *discover* their own thoughts, desires and intentions as inner objects to which they have privileged first-person access not by just noticing or paying attention to them (as if they were already categorially in place), but by *positing* them as in a manner akin to theoretical entities to explain anomalies in the verbal behavior of himself and his peers in his 'Rylean' community. Ryleans have a purely behavioristic language without reference to inner mental states. They may well perceive, think, desire or intend to do thus-and-so but they do not thereby classify or categorize these semantically significant states *as* inner mental states to which they have first-personal privileged access. They view these states as attributed to individuals but their contents as essentially public, situated in the social realm. If the myth of Jones expresses a coherent view, if, that is, the intentionality of language is prior to the intentionality of thought in the order of knowledge (Chisholm & Sellars, 1956), it follows that even the contents of our most intimate mental states, namely, those of our own thoughts, sensations, intentions and desires, to which we *do now* have privileged epistemic access and authority, need *not* be considered as always already authoritative from an evolutionary-genealogical point of view. They are themselves our collective conceptual creation to which we learned subsequently to respond reliably in a non-inferential way, thereby creating the (false) impression that they were always already there in our minds waiting for us to notice them.[14] In this way, as Sellars puts it, 'we confuse our own *creative enrichment* of the framework of empirical knowledge, with an analysis of knowledge as it was. We construe as *data* the particulars and arrays of particulars which we have come to be able to observe, and believe them to be antecedent objects of knowledge which have somehow been in the framework from the beginning. It is in the very act of *taking* that we speak of the *given*' (Sellars, 1956, §62). I submit that this shows in a particularly revealing way the core of the phenomenon of ideological misrecognition: our creative enrichment of the ways of experiencing and understanding ourselves-in-the-world by performatively constructing and applying categorial and empirical concepts, and the 'data' these concepts make possible for us to observe, are misrecognized as having been 'out there' or 'in here' all along waiting to be reported, described or noticed by us. They are thereby construed as antecedent objects or recognition that have an intrinsic binding force over us, a self-authorizing character, independently of our normative attitudes toward them.

[14] Of course, in reality, the creation of the concepts of an inner realm of sensations, thoughts, intentions and desires to which each of us has privileged access cannot be literally attributed to an individual genius and did not occur by way of proto-scientific methods of postulating theoretical entities for explaining anomalous phenomena (the occurrence of silent intelligent behavior, phenomena such as illusions, hallucinations, etc.). Sellars' myth of Jones attempts to show through a fictional story how such concepts and our privileged access to them could coherently and plausibly come about in a non-givenist manner while also accounting for the (false) impression that they are Given to the mind. In reality, these concepts presumably emerged in a collective and unconscious manner and culturally survived by a process of unconscious cultural selection due to selective benefits they conferred (in the long run) to the communities that made use of it over those which did not.

8.3 Toward a Fractured Immanent Model of Hyperstitional Dynamics

In Chap. 7 I suggested that the space of reasons and the space of causes can be understood as different, incommensurable, yet equally normatively laden modes of intelligibility whose way of being is in the last analysis processual. The irreducibility of these different modes of intelligibility was explained in terms of irreducibility in their use in the context of our practices, that is, in terms of their suitability in fulfilling different—yet complementary—kinds of equally fundamental collective needs and interests of human organisms. Moreover, as we saw, from the perspective of evolutionary naturalism both of these dimensions of intelligibility can be understood as evolved patterns of discursive habits selected and retained through a self-differentiating process which effectively better serves our vital need for collectively finding our way around the world. The progressive differentiation of the expressive tools needed for better orienting ourselves in the world (descriptive-explanatory tools of empirical-scientific discourse) in an ever-expanding collaborative manner (prescriptive-expressive tools of normative discourse) makes it possible for both sets of tools to better serve their own purposes by creating a mutually beneficial, self-correcting, self-reinforcing and thus feedback-loopy symbiotic discursive relationship. Normative expressive tools create a discrepancy between 'ought' and 'is' which opens a conceptual space in which both the recognition of explanatory anomalies and the means of their explanatory accommodation is made possible. And this amounts to making self-correcting empirical inquiry possible.

Now, in line with what was shown in the present chapter, it can be argued that these different domains of intelligibility (space-of-reasons, space-of-causes intelligibility), being both essentially normative in character, are themselves different ways of being *hyperstitional* processes. Space-of-reasons intelligibility corresponds, roughly, to hyperstitional processes that performatively make 'normative facts' true while space-of-causes intelligibility corresponds to hyperstitional processes that performatively make 'empirical facts' true. Recall that, since the space of causes is a transposed normative space (Chap. 7), empirical facts *also* have a performative normative dimension. Their difference from normative facts is that they are performatively constituted as always already existing in a *non*-normative manner within a *norm-free* nature (and misrecognized as having the unity and structure of propositionally articulated *facts* independently of our instituting practices), whereas normative facts are performatively constituted (and recognized-misrecognized) as always already existing in a *normatively binding* manner.

Yet, my goal in this section is to dig somewhat deeper and highlight what I take to be *another*, even more crucial distinction to which, again, different hyperstitional dynamics correspond. This time, the distinction in question is not that between two kinds of intelligibility (reasons, causes) that both ultimately have conceptual form (Chap. 7), but between modes of intelligibility that have conceptual form (thus including *both* reasons and causal intelligibility) and *non*-conceptually articulated modes of intelligibility. And, in line with my general fractured immanent model of ourselves-in-the-world, I will suggest that hyperstitional dynamics is not an exclusive feature of normative, conceptually articulated facts but is also to be found at the level of our non-conceptual commerce with the world, in which sensing, thinking and acting function in an *iconic*, contentless manner rather than through semantic relations of intentionality, sense and reference. I thus contend that hyperstitional dynamics is also bifurcated into a non-conceptual and a conceptual dimension: non-conceptual-level hyperstitional dynamics can be understood in terms of predictive processing dynamics (Clark, 2016, 2019), whereas conceptual-level hyperstitional dynamics can be construed in terms of Brandom's neo-Hegelian model of content determination and conceptual change (Brandom, 2019; Christias, 2025). We shall see how this bifurcation might be immanently reconceptualized by construing the conceptual dimension of hyperstitional dynamics as itself a 'self-diremptive' feature of its *non*-conceptual dimension, namely, as a subtractive manner in which the latter appears to itself.

8.3.1 Conceptual Hyperstitional Dynamics

I begin with Brandom's neo-Hegelian model of content determination and conceptual change (Brandom, 2019), which, to my mind, constitutes perhaps our best current model of the semantic structure and dynamics of the space of reasons, and has the merit of clearly articulating the dynamic, historically informed and performative conception of content determination, thereby bringing to light its essentially hyperstitional character.

Before delving into the specifics of Brandom's neo-Hegelian model of content determination, a word is in order for its relation to our account of hyperstitional normativity developed in Sect. 8.1.2. There, borrowing some suggestive ideas from Sellars, we identified the causal-projective vector of normativity in terms of its ability to shape practice through anticipatory uptake. Here we shall see how this vector operates in articulating semantic structure, namely, how such uptake gains determinate meaning within inferentially structured practices. Taken together, these two hyperstitional models describe a normativity that is both world-creating and sense-generating.

From Brandom's neo-Hegelian viewpoint, concepts, in all domains of inquiry, are more determinately developed and expressively clarified over time. Thus, determinate conceptual content is both created and discovered through the historical process of making claims and of forming intentions while encountering anomalies (incompatible commitments), making efforts to overcome incompatible commitments by changing our commitments, and in the process forming new commitments that better track the reality presupposed by and implicitly referred to by our previous commitments. In this process, one sense (concept) is found to be implicitly defective by its own standards (as it leads in holding incompatible commitments) and is replaced by another, whose application leads sooner or later again to a new set of incompatible commitments, and so on indefinitely. This is a process of development of senses by the progressive determination and expression of conceptual contents. And this, among other things, means that meanings can only be more determinately established retrospectively, which entails that concepts and their contents (meanings), *contra* Kant and Frege, have a built-in historical, dynamic and processual dimension. The progressive history by which we come to know what we mean and hence come to approximate the truth (where senses converge with referents) is built into the structure of our concepts—through our social practices of using concepts—from the beginning. Truth itself is not a static achievable state or status, but is *itself* a process, which in the form of the 'experience of error' (making explicit and restoring incompatible commitments), gradually converges senses (how things are for us) and referents (how what things are *for* us *ought* to be).

Remarkably, this is a semantics with a built-in *performative* dimension. If we properly understand the conditions of having determinate thoughts and intentions, namely, the conditions for binding ourselves by determinately contentful conceptual norms for judging and acting properly, we thereby come to understand that this presupposes a commitment on our part to *mutually recognitive practical attitudes of forgiveness, confession and trust*. Determinate (objective) content is impossible without reciprocal recognition. And reciprocal recognition, in its fully adequate form, amounts to diachronic reciprocal confession, recollective forgiveness and trust: that is, the determinacy of content presupposes as conditions for its possibility a) the acknowledgment of error and failure, of the distinction that cognition and agency involve, due to causal (evolutionary, socio-historical) factors or which we are not aware, between what things are in themselves and what they are for consciousness (reciprocal confession), b) recasting the previous actual applications of concepts so as to exhibit them as an expressively progressive revelation of the contents of concepts that show up as having been always already all along the ones knowers and agents were binding themselves by (recollecting forgiveness) and c) our capacity of forming the collective commitment to *make* reciprocal confession of and recollective forgiveness for our cognitive and practical failures *true in the future* (trust) (Brandom, 2019, pp. 610–635). I take it that this analysis is in the right direction and makes manifest the essentially *hyperstitional* character of all content determination.[15]

Importantly, Brandom extends this model of content determination to intentional action in a way that highlights the essentially social, dynamic and retrospective dimension of its hyperstitional dynamics. According to this model, fulfilling a complex intention is a cyclical process of intervention according to a plan aimed at a goal, observation of the results of the intervention, adjustment of the plan, further intervention, further observation of its results and so on. It has the dynamic structure of a 'Test-Operate-Test-Exit' loop. Importantly, intentional action has an ineliminable retrospective dimension: even if fully successful (or a failure) in the sense that it accomplished (or failed to accomplice) the intentional purpose for which it was performed, intentional action is always open to future reappraisal with respect to what was actually *done* by the one who performed it. This is because what one actually *does* by performing an

[15] Note that, as we saw in Chap. 7, this performative dimension of content determination is necessarily understood as *non*-arbitrary, as rationally constrained by the (represented) world, since, on this account (which I take to be essentially correct), although the external world is open to us due to our normative attitudes of practically *treating* the world in a certain manner, we cannot so much as *understand* our empirical concepts as determinate unless we conceive of this world as having an objectively determinate form independently of our acts of perceiving, willing and thinking, that is, as consisting of lawfully related determinate facts. Note, however, that this objectivity is a feature of the world *as conceptually represented*. As has been argued throughout the book, to adequately conceive of the world as it is in itself, in its non-conceptually articulated dimension, we must mobilize a process categorial framework.

intentional action is not only dependent on one's intended purpose, but also includes the unintended consequences that inevitably result from performing the action in the public external material and social space.[16] From this point of view, the content of intentional action is not identified and individuated *antecedently* to the process seeking to actualize the intentions involved in it. And in this process of actualization, the content of my action does not depend on me alone but depends as well on its determinate acknowledgment by others who attribute to me responsibility for the performance specified in ways that go *beyond* those in terms of which I understand its content as mine (Brandom, 2019, p. 396). Moreover, since the unfolding of the consequences of an action after its actualization is potentially infinite, the purpose or end acknowledged by the agent cannot be guaranteed to completely coincide with what is actually done or achieved by the action. The content of the latter is always open to renegotiation and reconceptualization from the perspective of others. Thus, what is done in actualizing an intention is not fixed by the agent's conception of it but is essentially *retrospectively reassessable* as to what it really amounts to. The very content of my own actions is not Given but is always at issue.

To my mind, this dynamic, performative account of intentional action, in which the latter is viewed as a temporally extended process, whose content is always retrospectively reassessable is one of the most profound and sophisticated renderings of our conception of intention. It also has the merit of clearly revealing its essentially hyperstitional character. The content of an action is not fixed in advance, for example, in the mind of the person who has the intention or when the realization of the intention in an action has intuitively 'ended'. From a more penetrating point of view, what one has done in actualizing the intention is under constant renegotiation. The unintended consequences of an action can be reconceptualized as to their content and significance by the community that takes the person who performed the action in question as its member (as a member of the 'we', as 'one of us'). Subsequent actions by those who recognize the original agent and attribute the action to him contribute to the content of

[16] What one actually does by performing an action need not coincide with one's intention. By turning on the light one can thereby alert a burglar (of whom one is unaware) although the purpose of this action—the reason why one did it—was only to get something to eat from the refrigerator. The agent may not be held responsible for the unintended consequences of his action, but this does not make the latter mere happenings externally related to what he *did*. It is just that since this deed was not performed purposefully, he is not to blame for it.

the always-evolving deed. And one essential part of recognizing the original agent as *one of us* (a member of our recognitive community), which is itself a hyperstitional fact, is acknowledging one's own *responsibility* to shape the agent's deed by affecting its consequences. Taken to its ultimate consequences, this view shows the power of hyperstitional dynamics, in the form of articulating a 'commitment in the making' expressive of a grand semantic-cum-ethical regulative ideal. Leibniz's metaphysical belief that nothing is for nothing in this best of all possible worlds becomes a regulative ideal which does not express an objective metaphysical fact but the content of a *commitment* which subjects are responsible for *making true*.[17] Brandom eloquently articulates this hyperstitional regulative ideal as follows: 'Recognizing an agent as one of us is practically treating what she *did* as part of what we all are *doing*. Adopting that attitude is acknowledging the responsibility to *make* what was already done come out right, as a constraint on what we have reason to do *now*. What we must have reasons for, what we must justify doing now, is a deed that includes everything *already done* by those we recognize as our fellows. It is part of our task to see to it that those earlier doings make positive contributions to the larger whole that subsumes it -a whole that includes and so is partly constituted by our own *current* and *future* doings. Those earlier deeds are ongoing processes that flow as if streams into the river that is our doing' (Brandom, 2019, pp. 735–736, emphasis mine).

8.3.2 *Non-conceptual Hyperstitional Dynamics: Predictive Processing as Projective Picturing*

I take it that Brandom's neo-Hegelian account of content determination, including the case of intention, nicely captures the hyperstitional character of our space-of-reasons conceptually articulated practices. But we might wonder if this model has an ontological counterpart in the non-conceptually articulated world of processual becoming. Is there a counterpart non-conceptually articulated hyperstitional dynamics at the level of our processual being and experience which is distinct (though perhaps inseparable)

[17] This, importantly, shows that the problem of nihilism is not so much that it is metaphysically erroneous or self-referentially inconsistent, but that it misses the essentially hyperstitional dimension of meaning both in the semantic and in the ethically edifying sense. Or, better, the metaphysically problematic and self-referentially inconsistent semantic status of nihilism are a consequence of its obliterating the hyperstitional dimension of meaning.

from the conceptual-level hyperstitional dynamics expressed in our space-of-reasons practices? I think there is, and that contemporary developments in predictive processing theories of cognition in cognitive neuroscience can throw light on how this might be so much as possible.

This non-conceptual counterpart of conceptually articulated hyperstitional dynamics is needed because, as has been argued throughout the book, we have good reasons to believe that a nominalistic process world bereft of modal, logical and deontic structure is the ontological ground (not cause) of the world as experienced and conceptualized in the content of the lifeworld (Chap. 4). The structure of this latter world is nicely captured by Brandom's bimodal hylomorphic conceptual realism. As we saw in Chap. 7, the metaphysics of the ordinary lifeworld, while distinguishing reasons and causes, draws an essential correspondence between their inferential spaces: both these logical spaces are thought of as different forms that one identical conceptual content can take. What is revealed from the standpoint of the space of causes, the world represented as objectively existing independent of our attitudes toward it, is understood as having literally conceptual form, as being conceptually articulated, through the modally robust relations of exclusion and consequence exhibited by the modal properties of objects, with the latter understood as constituents of facts, independently of language or concept-users. Yet, this resolutely conceptual realist point of view, in which the world itself is conceptually articulated (since causes themselves are seen as modally involved, as essentially implicated in counterfactual inferences and thus as conceptually articulated alethic worldly contents), does not seem to have a principled place for a 'really real' world independently of the form in which it is represented by our linguistic practices. Moreover, relatedly, this picture of the world, being essentially a sophisticated holistic 'mirror of nature' view of the mind, draws a Givenist picture of the relation between mind and world as one of pre-established harmony of form. But, as we have seen (Chaps. 2 and 3), there is no transcendental vantage point or pre-established harmony between mind, language and world to guarantee that the world as it is in itself *must* be ontologically categorized in a way that necessarily corresponds to the way our language is syntactically categorized (Brassier, 2016, Christias, 2023, pp. 126–130, Simonelli, 2022, p. 1048). What is more, as we saw in Chap. 4, it is possible and explanatorily fruitful to sketch an alternative, non-conceptually articulated, process nominalist picture of the world and ourselves as they are in themselves, which rejects a modal realism and does not use the categorial resources and distinctions of the object-property (essence-accident, subject-object, reasons-causes) framework.

In light of the above, it is urgent, in the context of this book, to search for a *non*-conceptual hyperstitional dynamics as an ontological counterpart of hyperstitional processes expressive of our space-of-reasons-cum-space-of-causes practices.[18] I have already stated my proposal: we can search for a non-conceptual hyperstitional counterpart to the conceptually articulated hyperstitional cycles of action-and-cognition culminating in recollective rational reconstruction of intentional action, in the context of predictive processing theories of cognition and action currently thriving in cognitive neuroscience (Clark, 2016; Hohwy, 2013).

According to predictive processing (PP), instead of analyzing the entirety of sensory data from the bottom up, the brain limits information processing in predictions and prediction errors only. It is continually engaged in formulating predictions, or hypotheses (called 'generative models'), about the causes of its sensory inputs, and in testing these predictions against incoming streams of sensory signals, thereby shaping perceptual content, guiding action and driving learning. The brain accomplishes this by implementing some approximate form of Bayesian inference, wherein sensory signals are combined with prior expectations to form posterior representations, that is, representations about what the organism will perceive, think or act. The central idea here is that in order to maintain the system's homeostasis the brain implements approximate Bayesian inference through a process of prediction error minimization, aiming at minimizing the mismatch between how things are and how the brain represents (predicts) them as being. Top-down signals within perceptual hierarchies convey predictions, while bottom-up signals report prediction errors—that is, differences between what the brain expects and what it gets—at each level in the hierarchy. The model itself is installed by some combination of our evolutionary heritage and ongoing lifetime learning. But lifetime learning is itself enabled by the ongoing attempt at prediction, thereby instituting a bootstrap structure in which prediction-based world-knowledge is later used to generate better and better predictions (Clark, 2018). The overall picture here is one of a continual dynamic

[18] Recall that in this work, the term 'non-conceptual' denotes the fact that world does not itself come with a conceptual-logical structure. The term does not positively characterize this non-logical structure. I have been arguing that such a positive ontological characterization can be provided by a process metaphysics.

engagement between the organism and the environment, where the former, like a surfer, is constantly latching on to big and small salient properties (affordances), indicated by precise prediction error,[19] of the 'waves' of sensory stimulation in its environment, hooking its internal generative model up to the changing flow in order to 'ride' the 'wave' most skillfully[20] (Clark, 2016, pp. xiv–xvi).

What is crucial here for our purposes is that these generative models, or methods of *projective picturing*,[21] as we might call them, are not only changing (updating) the mind to accommodate the prediction error signaled by its worldly encounters in perception but also at the same time change the *world*, through action, to fit their projections (predictions) about what the world *ought to be* like so as for the organism to successfully navigate it. This shows how this kind of projective picturing can be dynamic, performative, world-transforming *and* capable of standing in relations of *degrees of adequacy* to pictured worldly processes, *without* direct recourse to notions drawn from representational semantics (propositional truth, meaning, sense, reference), and without the need for

[19] Prediction error signals encode whatever is currently not accounted for by the top-down (and sideways) flow of prediction. To make the best and most flexible use of that flow of prediction error PP architectures simultaneously estimate the *precision* of the prediction error signal itself. Reading a difficult passage in a philosophy book while listening to music, I may need to devote my attention to the visually presented words, temporarily ignoring or downregulating the impact of auditory information. In PP terms, that means increasing the estimated precision—hence the impact—of prediction errors regarding the book-reading task.

[20] I take it that PP can be explanatorily fruitful not only for understanding empirical cognition, but also for highly abstract cognitive tasks such as those involved in logic and mathematics. If we construe the latter as normative practices, as expressive tools for modeling and managing modal relations (what could, must or would be the case), then PP can be seen as the machinery that generates expectations about those modal spaces, tests internal consistency via prediction errors when an inference does not fit the rules or expected outcomes and refines generative models based on high-level errors in abstraction, much like we update beliefs in the empirical world.

[21] Recall that picturing relations, being resolutely non-normative and non-intentional, are not *reference* relations. Reference is a *semantic* concept with *normative* connotations: how things are set standards for the normative assessment of the truth of our judgments, the material correctness of our inferences and the success of our actions. Referential correctness presupposes pictorial adequacy, where the latter is conceived in non-normative terms, for example, as an evolutionarily developed mapping, itself regulated by relations of dynamic homeostatic equilibrium between brain processes and the environment. Here, I supplement picturing with the projective dimension of PP's generative models, as this makes explicit its essentially world-transforming, creative, hyperstitional dimension operative directly at the level of the 'really real' world.

ontologically reifying the conceptual dimension of hyperstitional dynamics. Moreover, crucially, the content of PP's generative models is not only essentially action-guiding, but also *not* language-like. Ordinary speech is tailor-made for communication. By contrast, the probabilistic generative model is designed to dynamically engage the world in rolling, uncertainty-modulated, cycles of perception and action. The representations thus constructed are not actual *re*-presentations or *duplicates* of objects in the world: they are more like *embodied probability density functions.*[22] Even more importantly for our purposes, PP's generative models materially realize the dynamic and performative dimension of conceptually articulated intentions in that they *literally* are active constructors of the worlds they reveal through *active inference* (Friston, 2010). In active inference we *make* a given hypothesis (of the generative model) come true by directly engaging in the respective action that would have such an outcome—for example, by moving around and sampling the world so as to generate or discover the very perceptual or motor patterns that we predict. Thus, the flow of action is *itself* brought about by a select sub-set of predictions, namely, predictions of the trajectory of proprioceptive consequences that would ensue were the desired action to be performed (e.g. I predict that I am getting up from the chair to bring a beer from the refrigerator and move my limbs so as to bring about this state of affairs). Thus, active inference names the combined mechanism by which perceptual and motor systems conspire to reduce prediction error using the twin strategies of altering predictions to fit the world (perception) and altering the world to fit the predictions (action). Action thus emerges as a kind of *self-fulfilling prophecy* in which (a) neural circuitry predicts the sensorimotor consequences of the selected action, (b) as these consequences do not currently obtain, prediction error ensues and (c) the prediction error is then eliminated by moving the body so as to bring about the predicted sequence of sensations. In this way, this continually updated dynamic active inference is world-transforming and institutes a hyperstitional self-organizing dynamics directly at the non-conceptual level, without immediate recourse to semantic-linguistic conceptually articulated scaffoldings. Note, importantly, that in this model, motor commands (volitions) and desired consequences (goals) have been replaced—or better implemented—by

[22] Thus, instead of simply representing, say, 'Cat on mat', the probabilistic Bayesian brain will encode a conditional probability density function, reflecting the relative probability of this state of affairs (and any supported alternatives) given the available information.

proprioceptive expectations. In general, in the PP framework, the central explanatory (causally efficacious) notion in understanding the workings of action is that of *expectation* (embodied in generative models) and not the more traditional commonsensical categories of volition, plan, purpose, goal, etc. These latter categories are different conceptualized ways of 'freezing' and explanatorily simplifying the non-conceptual barrage of different dynamic patterns of expectations so as to provide relatively reliable stable priors to the generative models in their dynamic commerce with a changing external (and internal) world.

Moreover, although the PP framework is silent on this issue, I contend that another important difference between the conceptual and the non-conceptual dimension of hyperstitional dynamics lies in the nature of the motor of dynamic change of the feedback-loopy system: at the conceptual level, the motor of change is akin to Hegelian 'determinate negation', as the latter is expressed in the form of materially incompatible commitments (anomalies) that demand corresponding changes in the contents of these commitments to retain compatibility. This draws a hyperstitional path in which our self-correcting practices gradually turn contingency into necessary (they give contingency the form of necessity, as Hegel used to say) by collectively instituting the aforementioned normative attitudes of confession, forgiveness and trust. On the other hand, the motor of change at the level of active inference is a loss of allostatic balance in the organism-environment complex. And instead of a dialectic of incompatible, revised and updated conceptual commitments fueled by determinate negation, at the non-conceptual level the engine of change, to my mind, is expressed in a more illuminating way in terms of Deleuze's non-Hegelian dialectic of (virtual) problems and (actual) solutions, fueled by intensive differences that effect a loss of allostatic balance whose restoration comes in the form of morphogenetic actualization of novel structures (that reach a new level of allostatic balance), which in turn create new 'problems' that call for new kinds of 'solutions', etc. (see Sect. 4.4). This counterpart hyperstitional path, rather than 'turning contingency into necessity', performatively does the opposite: it gives (what seems to be) necessity the form of contingency. Yet, it also thereby makes contingency an object of affirmation, something with which we can individually and collectively resonate, not because 'it is in our best interest to accept what we cannot control' (Stoics), but because—precisely through *conceptual* hyperstitional dynamics—we become ever-more skillful navigators of the contingencies of the physical and social world.

The foregoing considerations bring me to a concluding methodological note on PP and its relation to non-conceptual hyperstitional dynamics. While I view PP theories as promising candidates for fulfilling the role of the non-conceptual hyperstitional dimension of cognition and action, what ultimately concerns me is the explanatory role, not the specific content, of the theory. If PP theories are replaced in the future by other theories that better explain cognitive phenomena (while respecting their hyperstitional character), so much the better for the general view of cognition that I want to defend here. As things stand now, an important criterion of adequacy for the general processual nominalistic viewpoint adopted here would be, for example, whether current PP theories of our experience of time, space and objects can accommodate the Deleuzian concept of the virtual, and its dynamic non-Hegelian dialectic with the actual, which, in my view, are key dimensions for any adequate nominalistic process ontology (Chap. 4). Another important lacuna of current PP theories is that their conception of information processing cannot account for the fact that, in organisms, information takes an essentially intensive, perspectival and polarized form (Simondon, 2020, pp. 244–256), namely, that of sensory and affective brain processes picturing external and internal environmental affordances whose ultimate *raison d'être* is the satisfaction of the organism's needs. In other words, PP must explain how past experiences are informationally embodied in a qualitative and intensive form (in sensations and affects), and how this kind of information is such as to open a perspectival salience-laden environment which directly affects the way current situations are perceived and organized (Raynolds, 2024). I do not see these difficulties as unsurpassable for PP, but their proper accommodation might well demand its essential supplementation with an embodied and intensive account of information (as a topological 'world-making' property of perspectival brain-body-environmental processes) which for the time being eludes us.[23]

[23] A classic objection to PP is that no matter whether we take the goal of a PP generative system to be the minimization of prediction error or the maximization of prediction success, there will be a scenario, consistent with that imperative, that looks inconsistent with true flourishing, in the sense that it provides an endless but in some intuitive sense boring or trivial supply of the information-theoretical quantity being maximized or minimized (Clark, 2018, p. 530). I take it that it is precisely the lack of this intensive account of information, and its construal in qualitatively neutral Shannon-information theoretic terms that accounts for this problematic consequence of PP. For an interesting recent attempt to construe qualitatively non-neutral (positive and negative) affective states such as feelings and moods of joy,

8.3.3 *Hyperstitional Dynamics as Fractured Yet Immanent*

We mentioned earlier that the causal power of normativity lies exclusively in the enactment of normative statuses (situated at the conceptual-semantic level) in the corresponding normative *attitudes* (situated at the non-conceptual level). The world can have normative authority over our behavior, that is, it can justify or validate our perceptual beliefs about it, only to the extent to which we collectively *treat* it *as* authoritative independently of anyone's attitudes over it. Thus, while normative statuses are conceptually irreducible to any *particular* normative attitudinal spatiotemporal perspective, they are not special kinds of entities with causal powers over and above those of normative attitudes.

It was also suggested that the causal powers of normative attitudes are materially expressed through the generation of iterated positive and negative feedback loops (effecting collective coordination and increasing adequacy) in monadic *projective picturings* of the physical and social world. Normative attitudes effect cycles of non-conceptual hyperstitional dynamics while the corresponding normative statuses at the conceptual (semantic-logical-intentional) level express our way of collectively controlling, stabilizing and regulating these hyperstitional dynamics. We saw earlier (Sect. 8.1.1) that normative statuses are relatively autonomous from normative attitudes and perform indispensable functions in our cognitive economy. They function as stabilizing attractors that open up an objective possibility space as a reference point for 'keeping scope' of (validating, criticizing, aligning, updating) monadic subjective perspectives, and they thereby make possible new ways of regulating and coordinating behavior, memory, attention and anticipation, registering and keeping track of consequence and incompatibility relations between contents that can constitute objective reasons—as opposed to merely associative causal connections, desires and impulses—for belief and action. In this way, they enable us to construct novel and scaled-up collaborative ways of navigating the

sadness or anxiety as expressions of interoceptive error dynamics that track how well we are currently doing at minimizing prediction error versus how well our brains were expecting us to do, see Clark, 2023. From this point of view, which also links affective states directly with action, the feeling, for example, of 'being in the zone' in sports reflects unexpectedly good error dynamics of this kind. This view also aspires to explain why we automatically seek out good learning environments, preferring ones that are neither too predictable nor too unpredictable, and why we value exploration, novelty: we value them to the extent to which they enable and contribute to good error dynamics.

world, understanding others and representing ourselves. We also saw that this does not imply the need for ontological commitment to anything beyond the complex interrelations between normative attitudes and the modes in which they appear to themselves and thus is completely consistent with our nominalist process ontology.

Of course, there must be some kind of internal connection between the non-conceptual level of iconic hyperstitional processes and the conceptual norm-governed level of hyperstitional dynamics, as the latter patently do not belong or refer to a separate ontological realm. The relation between them is not best understood in terms of causal interaction—as this would involve the postulation of distinct top-down causal powers of conceptual hyperstitional dynamics, which would thereby constitute an ontologically distinct natural kind. Instead, it can be argued that public, conceptually articulated, norm-governed hyperstitional dynamics at once opens up and represents the *conceptual space* through which we navigate the (non-conceptually articulated) world (Sellars, 1963). In other words, conceptual hyperstitional dynamics does not directly represent the world but makes salient information about the perceptual-inferential-practical *role* of unfolding behavioral patterns. These roles are expressed in salience (affordance) spatiotemporal patterns in the 'really real' world, and this shows the precise way in which there can be a real connection between public and private-iconic hyperstitional processes without direct mirroring (or vertical relations of causal efficacy) between them. By making salient and explicit certain patterns and possibilities of our perceptual-inferential-volitional behavior while precluding others, and by inciting changes in these patterns and in their overall orientation when found diachronically inadequate, conceptual hyperstitional dynamics thereby serves as a flexible and easy to scale-up tool for increasingly adjusting the flow—including importantly the precision weightings of prediction error signals at different levels of functioning—of non-conceptual hyperstitional dynamics toward allostatically balanced states that serve basic needs of survival and well-being.[24]

[24] Indeed, it can be argued that from a PP point of view, the conceptual dimension of hyperstitional dynamics functions as a *scaled-up surprisal minimization device* both at the individual and the collective level. It is in essence a technological scaffolding inducing a *novel organization of spatiality and temporality* (memory, anticipation, salience patterns) for better regulating non-conceptual hyperstitional dynamics, that is, for collectively coordinating private projective picturings of ourselves-in-the-world (Clark, 2016, pp. 282–284).

We must insist a bit more on this point: conceptual hyperstitional dynamics is indispensable to our lives not only because of its ability to scale up human cooperation through the construction and internalization of a public objective possibility space for assessing our behavior, but also because it marks the advent of a unique and unprecedented manner in which we *represent worldly information to ourselves.* Conceptual articulation marks the transformation of concretely interacting worldly and human behavioral patterns into collectively sustained conceptually articulated modal spaces with the capacity to distinguish themselves as representational spaces from their 'intentional objects' (the world) and from the 'subjective perspectives' upon it (individual subjective histories). This is an expression of a novel self-relating power of cognition to *mirror within itself* its relation to the world, which is itself based on its ability (a) to *open-endedly self-relate*, to take itself and its direct commerce with the environment (first-order language) as an object of investigation (metalanguage), and (b) to incrementally increase the combinatorial possibilities of linking iconic pictures with one another through the discretization of the sound continuum into distinct types and tokens (see below). These novel features of conceptual cognition effectively free us from the in-built communicative constraints of iconic hyperstitional processes, namely, the fact that the latter are context-bound, incapable of being abstracted from their qualitative form and content, and bound up with the adaptive constraints of the 'here and now'.[25] This transformation thereby opens a social space of meanings which facilitates mutual prediction and enables the interaction of iconic perspectives with different informational histories thus making available trajectories of learning and discovery that no individual could alone comprehend. And, most importantly, it enables us to *uncouple* our cognitive abilities from our cultural and biological *milieu*, thus exponentially increasing our expressive powers and our control over adverse environmental contingencies (Wolfendale, 2019, Sacilotto, 2024, pp. 183–184).

[25] Compare here Frege who, commenting on the process of representing perceptual occurrences as signs, connects this ability with freedom of the will. As he puts it: 'We create in this way a firm, new focus about which ideas gather. We then select another idea from these in order to elicit its symbol. Thus, we penetrate step by step into the inner world of our ideas and move about there at will, using the realm of sensibles itself to free ourselves from its constraint. Symbols have the same importance for thought that discovering how to use the wind to sail against the wind had for navigation' (Frege, 1964, p. 156).

Lastly, it must be reminded that the relative autonomy of conceptual hyperstitional dynamics over its non-conceptual counterpart and its power to open novel possibility spaces, express novel self-relating powers and spatiotemporally enhance, inhibit, highlight, occlude, separate and recombine sensory-inferential-volitional patterns in the flow of the latter does not derive from special emergent causal powers of the former but by its being a subtractive manner of appearance of a self-divisional process occurring at the *non*-conceptual level of hyperstitional dynamics. As was mentioned in Chap. 5, this self-divisional process is based on the primitive self-referential ability of our senses to picture part of their own intrinsic activity, namely, the occurrence of particular (visual, oral, tactile) sensory impressions, affects and images, as occurrently instantiated *schemata* that selectively provide novel salience points (enhance, mute, bring to foreground, withdraw to background, highlight as relevant, play down as irrelevant) for *digitally* arranging audiovisual similarities and differences of these sensory and affective patterns in our respective manifolds. This process makes possible the distinctions between types and tokens, contents and acts, statuses and attitudes, language and metalanguage, thus opening up the space of semantics *as such*, enabling action at an open-ended spatial and temporal distance (Blumenberg, 2020, p. 263) and paving the way for a huge increase in both negative and positive freedom (freedom *from* adverse worldly contingencies, and freedom *to* self-express ourselves in novel ways). I have argued that this self-synthesizing, sign-making function amounts to the imaginative ability to see *in* what it is something *other* than it (Castoriadis, 1975), and is effected through evolutionarily explicable processes of compression, stabilization, separation and integration of its manifold sensuous and affective particularity. The digitalized rendering of audiovisual informational continuum through productive imagination (integrative processes) and subtractive imagination (compressive, stabilizing, separating processes) yields the syntactic infrastructure upon which semantic inferential articulation of conceptual content is built when the former is put into use in practice. In this way sentient self-consciousness yields sapient self-consciousness through its own self-fracturing. It is as if non-conceptual hyperstitional dynamics becomes self-dirempted, viewing its own stabilizing, compressing, separating and integrative activity through the lens of the conceptualized mode of hyperstition. Presumably, this is because this occlusion of non-conceptual productive and subtractive activity and the bringing forth only of a sedimented dimension of this activity with a different form of unity, self-relation and combinatorial possibilities (conceptual-inferential articulation) proves informationally far more efficient from an evolutionary point of view for the survival and flourishing of the group. Thus, the causal powers involved in this self-relating sign-making process are not located at the level of general structures such as *types* of things, *contents* or normative *statuses*. Types, contents and normative statuses do not exert a direct causal influence on the world *as such*. Yet they *do* influence it in the 'productive-through-subtraction' manner mentioned above, which *is* an emergent process located at the nominalistic level of concrete processual individuals.

Appendix: Hyperstition, Non-representationalism and Politics

Hyperstition and Social Informational Networks

In this appendix, I want to suggest that the notion of hyperstition, as part of the metaphysical framework of this book, besides offering a philosophically illuminating account of normativity and its place in nature, can also be a part of compelling empirical *explanations* of currently much-discussed yet not well-understood phenomena of great political interest. Land (2011) had already identified hyperstitional processes as distinctive of the political domain, capable of providing important insight on the workings of cultural phenomena such as financial markets, religion, myth, rituals and conspiracy theories. Yet, to my knowledge, there has been no attempt to show how hyperstition and its essential non-representational dimension can throw light on the workings of currently much-discussed yet little understood phenomena such as (a) the proliferation of sealed-off echo chamber communities despite increasing individual access and collective integration of information at a global level, and (b) the easy dissemination and increasing causal impact of fake news.

These phenomena cannot be easily accounted for if information is conceived in representational terms, as having the form of neutral data that represent a pre-existing reality truly or falsely. On this understanding of information, it seems that the way to make information networks more factually credible is by increasing their ability in gathering and processing large quantities of information. On this view, while delusional or deceitful networks might occasionally arise in moments of historical crisis, it is guaranteed—by a process akin to natural selection—that in the long term they are bound to lose to more clear-sighted and honest rivals. Thus, the antidote to most problems we encounter in gathering and processing information seems to be the gathering and processing of even more information (Kurzweil, 2024; Pinker, 2021). Even seemingly rock-bottom ideologically infused disagreements about ultimate values are said to be partially converging toward the truth by being placed in a free market of information dissemination. It seems that given enough time and information, we are bound to get closer to the truth about things ranging from viral infections to racist biases, thereby at once increasing our power and control over nature and ourselves and gaining the theoretical and practical wisdom necessary to make prudential use of this power.

Now, while this view does have some purchase (a doctor wishing to identify the cause of an epidemic by examining a single patient is less likely to succeed than thousands of doctors gathering data on millions of patients), it cannot at all explain the proliferation of sealed-off echo chamber communities and the easy dissemination and increasing causal impact of fake news. For increasing individual access and collective integration of information at a global level, far from undermining the force of these phenomena, actually amplifies their effects (Harari, 2024; Voshughi et al., 2018). But suppose that the informational dynamic networks (and their supporting technological frameworks) implicated in these phenomena are understood as *hyperstitional*, namely, as performatively instituted, hype fueled and feedback-loopy ways of *world-making* by connecting and synchronizing disparate processes (emotions, political views, social needs of belonging-together), thereby facilitating social coordination yet in a manner that is *not* directly representational. From this point of view, the we-attitudes that enact informational feedback cycles in informational networks are not in the business of representing the world truly or falsely, but are ways of (social) world-making that only *indirectly* 'face' reality, through the long-term consequences of confronting behavior that conforms with information patterns licensed by the information network to worldly and social processes as the latter (ontologically directly yet epistemically opaquely) impinge on the behavior in question. At bottom, the role of information networks (language, markets, social media) in our lives is primarily social: entrenching social order by synchronizing previously disparate worldly, behavioral and social patterns of activity. In other words, informational networks perform this function by creating new intersubjectively sustained realities through forging novel interconnected salience patterns, and by endowing these patterns with the abovementioned normative and practical import. The 'goal' of informational networks is achieved if the synchronizations they effect (the 'worlds' they make) have enough efficiency in collectively coordinating the community in finding its way around the world such as to make these networks culturally reproduced from generation to generation. And they can well achieve this goal without directly representing the world. As is obvious, if the phenomena of the proliferation of echo chamber communities and the increasing dissemination of fake news are seen under the light of this resolutely non-representational, normative and hyperstitional account of information networks, they become readily intelligible and only to be expected.[26]

The fact that informational networks that are crucial for the well-functioning of society can be adequate enough for its self-reproduction without adequately representing the world is an expression of the absence of an epistemically Given relation between mind and world: bereft of a transparently given epistemic access to the world or to the non-conceptual dimension of our sensory processes, we cannot directly compare the generative models of the world produced by our informational networks from a neutral external epistemic standpoint (the non-conceptual world, neutral sensory data) to uniquely and independently determine if our generative models on the world are correct or incorrect. Since our only epistemic access to the structure of the non-conceptual world is indirect, through the disruptions that the application of these generative models to the world effects on our overall allostatic balance, expressed in the form of material incompatibilities between elements and levels of the 'anticipatory world-stories' of our conceptual informational frameworks, we can only progress in our knowledge of the world and ourselves through self-correcting inquiry, by critically controlled conceptual change and

[26] Relatedly, active inference can also readily account for ideological biases. In social and political domains, one can minimize prediction error by selecting environments where people believe the same things and act in predictable ways, as well as by selecting information sources that do not challenge one's preconceptions. For example, people are likely to seek out news sources that echo their existing ideological beliefs and frame facts in a way that does not threaten their worldview. This is also exemplified in the formation of echo chambers in which little ideological disagreement is allowed, since selectively associating with others who share one's worldview ensures that one is better able to predict their beliefs and behaviors, minimizing the influence of prediction errors on global beliefs. Of course, as a byproduct, particularly under conditions of uncertainty, this mutually constrained error minimization may produce increased polarization and extremism (Wheeler et al., 2020, p. 195).

updating that best accounts for the incompatibilities in question.[27] Yet, this progressive increase in the adjustment or 'resonance' of our conceptual informational systems with worldly processes is only possible in the long run and only on condition that the self-correcting tools of empirical inquiry are situated within an ambience of democratic social institutions based on an extended network of checks and balances (see also Chap. 10). This is needed for at least two reasons: (a) The absence of a direct method or criterion for comparing a conceptual framework with the world and its holistic and systematic nature makes it capable of—and even necessitates its—being shielded from empirical refutation as a way of retaining its socially coordinating role, which is its very *raison d'être* as an 'anticipatory world-story'. This shielding can be expressed in the form of interpreting empirical data in ways that conform with the basic principles of the system, by the ad hoc accommodation of incompatibilities between parts of the system, by allowing for persistent explanatory anomalies in peripheral domains while protecting the hard core of the system from significant disruption, by challenging the reliability of the source of the incoming empirical data, etc. (b) If this informational network is highly centralized and its patterns are systematically filtered by—and largely reflect the structure of—power societal relations, it can easily be socially entrenched (due to the social cohesion effects that it fosters in the short run) without being conceived as problematic by its users (since the informational network may well not conceive these problems as problems or it may conceive the features that make these phenomena problematic not as vices but as virtues both epistemically and morally). Thus, for ultimately evolutionary reasons, a conceptual informational network—and the model of the world

[27] Interestingly, this fact is built into the very logic of PP theories of cognition. We do not have a direct epistemic access to our senses and to the worldly causes that produce them. We only have access to these worldly causes indirectly, be devising provisional anticipatory 'world-stories' (this is the task of generative models) and by testing and adjusting them by appeal to hierarchies of lower-level incoming sensory signals which only propagate up sensory information that fails to be predicted by top-down expectations. Thus, the upper level in each pair of levels only 'knows' its own expectations, it is only told how these expectations are wrong and is never told directly what the level below informationally 'knows'. Similarly, the upper level never tells the lower level what message to pass on, it only ever anticipates what the lower level will tell it (Swanson, 2016). If the sensory signals provide enough reliable prediction errors, these error signals will propagate up the hierarchy to affect what we perceive. Importantly, this will only happen if alternative theories that explain the surprising sensory data can be formed, even if the perceiver did not originally expect these theories to be true.

it generates—can, to a large extent, remain stable and not directly refutable by maintaining a view of the world and ourselves replete with rationalizing ideological elements that while at times disrupting our allostatic balance, also forges collective order and provides orientation, a sense of belonging-together, and 'meaning' to our lives that compensates for this (epistemically opaque) loss of allostatic balance at least on the short run. So, it is no accident that generative models that systematically project to the world features that belong solely to the model have historically proliferated in our social world even if they ended up having catastrophic effects (ideologically justified religious wars, totalitarian regimes, genocides, witch hunt, etc.). And this is why the most reliable way for increasing our long-term collective resonance with the environment is by constructing and forging self-correcting epistemic, economic and political institutions based on an extensive network of checks and balances. For this cultural *niche* essentially works as an artificial tool or speeding up natural selection at the level of conceptual informational systems, and thus for filtering out at least the most problematic forms of systematically distortive and practically destructive conceptual generative models of ourselves-in-the-world.

Hyperstitional Reason

In this appendix I highlighted the a-rational element of hyperstitional processes. Yet, as I use the term, hyperstition is not exclusively connected to irrational or a-rational natural or social processes, but also, *pace* Land, to perfectly rational ones, namely, the collective practices that institute the space of reasons. The fact that hyperstitional processes, including reason-giving ones, do not directly represent the (natural or social) world does not necessarily render them irrational or ideological in a distortive sense. In fact, as we saw in Sect. 8.3.1 the conceptually articulated hyperstitional processes of giving and asking for reasons, while being resolutely non-representational, are a collectively instituted linguistic tool that makes us capable of achieving ever-increasing levels of resonance with natural and social reality by imposing, experimenting on and self-correcting epistemic norms that guide our perception, thought and action. Although the core norms of our reason-giving practices (objectivity, truth, warranted belief, autonomous action) are non-representational in that they do not resemble or correspond in form with natural and social reality, they are 'designed' to increase our collective attunement with the latter (e.g. by removing all kinds of unjustified cognitive and affective biases that hinder our collective welfare). Of course, this does not guarantee that the way these norms actually operate in our practices is purged off these biases (since the infusion of our reason-giving practices with ideological biases may well be advantageous in the short-term survival and reproduction of a community). But it is built in the very *meaning* of our reason-giving practices and norms that the entanglement of reason with ideology is problematic and something to be criticized and corrected since it is *itself* an exemplary case of a non-objective, biased viewpoint that effectively fosters the well-being of unjustifiably privileged societal groups.

In Chap. 9 we shall see in more detail how hyperstitional reason can be enacted in our political practices. Let me here just mention one such concrete case that shows how hyperstitional norms might work in the field of politics: consider participatory technology governance experiments, such as citizens' assemblies on artificial intelligence regulation (McKinney, 2024; Talpin, 2012). These initiatives bring together diverse actors (citizens, technologists, ethicists, policymakers) to collectively articulate normative frameworks for emerging technologies. What makes them hyperstitional is their performative temporality: they prefigure the democratic futures they seek to bring about (e.g. a future where technological development is governed democratically, inclusively and reflexively) by enacting them in the present. In this way, the process feeds back into itself by reshaping the political field in which it operates: by making deliberation more participatory and transparent, it increases public trust, expands legitimacy and influences how institutions view citizen input, thereby making the projected model more achievable in the broader political arena. In other words, these assemblies do not simply recommend policies, but instantiate in the present a vision or ideal of participatory, reflexive governance, thereby generating legitimacy for the very model they propose. Importantly, the legitimacy in question is *immanent to its performance*: the process justifies itself by showing in practice the possibility of the future it projects.[28] This recursive loop, where the process of deliberation itself embodies the political ideal it projects, allows these experiments to generate the conditions for their own future expansion and institutionalization, and thus to bootstrap their own normative authority.

[28] See, for example, Yang and Bachmann (2025) who discuss the case in which the assembly's outputs are co-produced by citizens and algorithms in an iterative, adaptive process. This recursive structure allows the assembly to bootstrap its own legitimacy: by showing real-time responsiveness to participants' inputs, it models the kind of dynamic, participatory governance it advocates, which in turn strengthens public and institutional acceptance.

In this example, the hyperstitional, *pace* Land, does not operate as an accelerative engine of late-capitalist deterritorialization, but as a way for generating reflexive universals that orient collective practice. As we shall see in Chap. 9, these universals open trajectories by aligning our conceptual, affective and institutional energies around shared emancipatory projects. In this way, hyperstitional politics emerges as a way of accepting the contingency of our futures while deliberately projecting and enacting the ones we aspire to create.

References

Althusser, L. (1971). Ideology and ideological state apparatuses. In *Lenin and philosophy and other essays*. Monthly Review Press.

Badiou, A. (1988). *Being and event*. Continuum.

Blumenberg, H. (2020). Theory of nonconceptuality. In C. Borck (Ed.), *History, metaphors, fables: A Hans Blumenberg reader* (pp. 259–299). Cornell University Press.

Brandom, R. (1994). *Making it explicit: Reasoning, representing and discursive commitment*. Harvard University Press.

Brandom, R. (2009). Reason, expression and the philosophical enterprise. In R. Brandom (Ed.), *Reason in philosophy* (pp. 111–129). Harvard University Press.

Brandom, R. (2019). *A spirit of trust: A reading of Hegel's phenomenology*. Harvard University Press.

Brassier, R. (2016). *Transcendental logic and true representings. Glass bead*. http://www.glass-bead.org/article/transcendental-logic-and-true-representings/?lang=enview

Castoriadis, C. (1975). *The imaginary institution of society*. Polity Press.

Christias, D. (2023). *Normativity, lifeworld and science in Sellars' synoptic vision*. Palgrave Macmillan.

Christias, D. (2025). The place of spirit in nominalistic metaphysics. In C. Sachs (Ed.), *Interpreting Sellars*. Cambridge University Press.

Chisholm, R., & Sellars, W. (1956). *Intentionality and the mental: The Chisholm-Sellars correspondence on intentionality*. https://www.ditext.com/sellars/sccor-f.html

Clark, A. (2016). *Surfing uncertainty: Prediction, action and the embodied mind*. Oxford University Press.

Clark, A. (2018). A nice surprise? Predictive processing and the active pursuit of novelty. *Phenomenology and the Cognitive Sciences, 17*(3), 521–534.

Clark, A. (2019). Replies to critics search of the embodied, extended, enactive predictive (EEE-P) mind. In M. Colombo, E. Irvine, & M. Stapleton (Eds.), *Andy Clark and his critics* (pp. 266–302). Oxford University Press.

Clark, A. (2023). *The experience machine: How our minds predict and shape reality.* Pantheon Books.
Dawkins, R. (1976). *The selfish gene.* Oxford University Press.
Deleuze, G. (1968). *Difference and repetition.* Columbia University Press.
Feenberg, A. (2017). *Technosystem: The social life of reason.* Harvard University Press.
Frege, G. (1964). On the scientific justification of a concept-script. *Mind, 73*(290), 155–160.
Friston, K. (2010). The free-energy principle: A unified brain theory? *Nature Reviews Neuroscience, 11*, 127–138.
Haslanger, S. (2017). *Resisting reality: Social construction and social critique.* Oxford University Press.
Hohwy, J. (2013). *The predictive mind.* Oxford University Press.
Kukla, R. (2000). Myth, memory and misrecognition in Sellars' 'empiricism and the philosophy of mind'. *Philosophical Studies, 101*, 161–211.
Land, N. (2011). *Fanged noumena: Collected writings 1987–2007.* Urbanomic.
McDowell, J. (1994). *Mind and world.* Harvard University Press.
Railton, P. (2000). Facts and values. *Philosophy and Public Affairs, 29*(2), 105–138.
Raynolds, J. (2024). Framing the predictive mind: Why we should think again about dreyfus. *Phenomenology and the Cognitive Sciences.* https://doi.org/10.1007/s11097-024-09979-6
Rorty, R. (1979). *Philosophy and the mirror of nature.* Princeton University Press.
Sachs, C. (2022). A cybernetic theory of persons: How Sellars naturalized Kant. *Philosophical Inquiries, 10*(1), 49–72.
Sacilotto, D. (2024). *Structure and thought: Toward a materialist theory of representational cognition.* Northwestern University Press.
Sellars, W. (1949). Language, rules and behavior. In S. Hook (Ed.), *John Dewey: Philosopher of science and freedom* (pp. 289–315). Dial Press.
Sellars, W. (1956). *Empiricism and the philosophy of mind.* Harvard University Press.
Sellars, W. (1963). Philosophy and the scientific image of man. In W. Sellars (Ed.), *Science, perception and reality* (pp. 1–40). Routledge.
Shelby, T. (2003). *We who are dark: The philosophical foundations of black solidarity.* Harvard University Press.
Simondon, G. (2020). *Individuation in light of notions of form and information.* University of Minnesota Press.
Simonelli, R. (2022). Sellars' ontological nominalism. *European Journal of Philosophy, 30*(3), 1041–1061.
Srnicek, N., & Williams, A. (2015). *Inventing the future: Postcapitalism and a world without work.* Verso.
Stanley, J. (2015). *How propaganda works.* Princeton University Press.
Wolfendale, P. (2019). The reformatting of homo sapiens. *Angelaki, 24*(1), 55–66.
Harari, J. (2024). *Nexus: A brief history of information networks from the stone age to AI.* Random House.

Kurzweil, R. (2024). *The singularity is nearer: When we merge with AI.* The Bodley Head.

McKinney, S. (2024). Integrating artificial intelligence into citizens' assemblies: Benefits, concerns and future pathways. *Journal of Deliberative Democracy, 20*(1), 1–12.

Pinker, S. (2021). Rationality: What it is, why it seems scarce, why it matters. .

Swanson, L. R. (2016). The predictive processing paradigm has roots in Kant. *Frontiers in Systems Neuroscience, 10*, 1–13.

Talpin, J. (2012). When democratic innovations let the people decide: An evaluation of co-governance experiments. In B. Geissel & K. Newton (Eds.), *Evaluating democratic innovations: Curing the democratic malaise?* (pp. 184–206). Routledge.

Voshughi, S., Roy, D., & Aral, S. (2018). The spread of true and false news online. *Science, 359*(6380), 1146–1151.

Wheeler, N., Allidina, S., Long, E., Schneider, S., Haas, I., & Cunningham, W. (2020). Ideology and predictive processing: Coordination, bias, and polarization in socially constrained error minimization. *Current Opinion in Behavioral Sciences, 34*, 192–198.

Yang, J., & Bachmann, F. (2025). Bridging voting and deliberation with algorithms: Field insights from Taiwan and Kultur Komitee. In *The 2025 ACM conference on fairness, accountability, and transparency*. ACM. https://doi.org/10.1145/3715275.3732205

PART IV

Freedom and Politics in an Immanently Fractured World

CHAPTER 9

Fractured Freedom and the Emancipatory Force of Universals

In this chapter we shall explore the consequences of our fractured immanent ontology on the issue of freedom. It will be argued that freedom is expressed in a fractured form: (1) as 'normative freedom', that is, freedom as self-determination (with Kant, Sellars and Brandom being their main representatives), and (2) as 'non-normative freedom', that is, as an increase of the power to act, as opposed to the 'power to dominate' (the main representatives here being Spinoza and Deleuze).

While the concept of normative freedom, at least as I use it, is closer to what is traditionally called 'positive freedom' (the freedom *to* do or become something, to have control over something), it does also encompass important aspects of 'negative freedom' (freedom *from* external constraints on my will or desire). This is because, as I shall suggest, freedom as self-determination does not only aim to cover individual autonomy (the overcoming of impulses through reason, the rejection of external authorities) or moral autonomy in our social relations with other people in the context of a community. Through the descriptive and explanatory tools of science—which express freedom as self-determination at the level of 'theoretical reason'—it also aims at the emancipation from *causal* constraints on the exercise of our individual and collective epistemic and practical capacities (where 'causal' here covers bodily limitations, limitations of our knowledge of the world due to ignorance, ideological or depth-psychological factors, historical facts, evolutionary built-in biases or

© The Author(s), under exclusive license to Springer Nature Switzerland AG 2025
D. Christias, *Towards a Hyperstitional Process Nominalism*, Palgrave Perspectives on Process Philosophy,
https://doi.org/10.1007/978-3-032-06350-2_9

constraints, etc.). This emancipation from the above kinds of broadly causal constraints, in turn, implies the *expansion* of our capacity for perception, thought and action, and the creation of *novel* desires, interests and purposes that go hand in hand with this expansion (which are aspects of *positive* freedom). Thus, in what follows, I shall not discuss the issue of positive vs negative freedom (e.g. as traditionally formulated in Berlin, 1969). For my purposes, I shall adopt MacCallum Jr. (1967) unitary conception of freedom, which views freedom as a triadic relation, which specifies *what* is free or unfree, *from* what it is free or unfree (constraints or limitations of freedom), and what it is free or unfree *to do* or *become* (what counts as a purpose that the agent can be described as either free or unfree to carry out). This unitary conception of freedom encompasses elements of both positive and negative freedom. Yet it is consistent with another, to my mind more philosophically interesting, distinction between normative and non-normative freedom as defined above. For the purposes of this book, I shall frame the discussion on freedom based on the distinction between normative and non-normative freedom, as opposed to the more traditional one between positive and negative freedom—since the latter is orthogonal to the former and to my concerns here. With that said, it will be argued that normative freedom and non-normative freedom are intimately related in that the former is epistemically and practically necessary for achieving the latter, while the latter is the ultimate *raison d'être* of the former, in the sense that the very motivational power of individual and collective autonomy essentially depends on the non-normative kinds of freedom it elicits.

In this connection, it will also be suggested that the well-known link, emphasized by German Idealism, between freedom and reason, in which freedom is conceived as 'constraint by norms' (Brandom, 1979), and which lies at the heart of our conception of freedom as self-determination, expresses a profound truth albeit only in connection with a non-metaphysical, normative and critical rendering of *universals* which shows their emancipatory role in our reason-giving practices. Specifically, building on Srnicek and Williams (2015) and Christias (2023), it will be proposed that the surplus 'ideal' dimension of universals, stripped of its metaphysical connotations and understood in normative and practical terms, can be thought of as an indispensable subversive tool of *immanent critique* of the existing semantic and epistemic *status quo*. From this point of view, universals are construed as projections of emancipatory ideals regarding our collective epistemic and general welfare that function as an axis for coordinating collective projects in the direction of fulfilling these ideals. This, among other things, shows how our non-traditional ontological nominalism not only accommodates but makes essential use of universals, by reconceiving them as semantic and epistemic tools of immanent critique. Thus, the at first glance non-nominalistic fact to the effect that the semantic content of universals goes beyond how universals are actually used to describe concrete particulars in the context of our everyday practices is not understood as a metaphysically existing unrealized potential or essence, but as a way of representing, calibrating and critically adjusting the structure of the inferential space of the universals in question.

9.1 Normative Freedom as Rational Constraint by Norms

An essential insight of Kant and German Idealism, rejuvenated by Sellars, McDowell and Brandom in twentieth-century analytic philosophy, is that freedom, despite appearances to the contrary, demands the presence or imposition of certain kinds of (rational) *constraints* between agents for it to be so much as possible. According to this conception, freedom's enabling constraints are constraints by *norms* as opposed to causal constraints (which can be external or internal—bodily, psychological ones). To be free is to believe, think and act on the basis of *reasons* that can potentially overtake our desires, dispositions, drives and impulses, and are independent of the causal force of tradition, authority or coercion. Importantly, from this point of view, the difference between being free or unfree becomes itself a *social normative* difference rather than an outright ontological one (recall that, as we saw in Chap. 7, the realm of causes and the realm of reasons are both conceptually articulated, and their difference is not one between incommensurable ontological realms). The distinction between freedom and unfreedom in this social-normative sense is, at bottom, a distinction between two ways of treating someone's or something's behavior. We treat someone or something as free insofar as we consider its behavior subject to the norms inherent in the social practices conformity to which is the criterion of membership in our community. The norms in question are, above all, those that make meaningful discourse possible. By contrast, insofar as we cope with someone or something in terms of the *causes* which objectively constrain it, rather than the norms which constrain it via our space-of-reasons practices (practices of meaningful and reasoned discourse), we treat him as an object, and thus as unfree (Sellars, 1963a, pp. 39–40, Brandom, 1979, pp. 192–193). Freedom in this resolutely normative sense is thus a *social status*, the possession of which depends on our normative *attitudes* toward patterns of behavior. As we shall see below, since according to our analysis (Chaps. 5 and 8) normative attitudes are genuine emergent processes that make a real difference to the world, there is a sense in which to be free cannot be exhausted by the possession of a social status and has also an ontological dimension that does not exclusively depend on the existence of rational constraints on our desires. This is the dimension of non-normative freedom that will be discussed in Sect. 9.4. For the time being, I shall focus on the fact that the (rational) constraints on our desires and impulses imposed by reciprocally recognized normative attitudes (holding and treating oneself and others as capable of undertaking commitments about one's beliefs, thought and actions, and thus as willing to risk or sacrifice one's other desires), while in a sense diminishing our negative freedom (e.g. the free expression of whatever desire or impulse happen to be more intense at a time), are what alone *make possible* novel forms *positive* freedom, namely, freedom *to* do something we could not do before or could not even *conceive* as something that we could intend to do. They thereby increase our expressive freedom even while simultaneously diminishing aspects of our negative freedom.

This increase in our expressive freedom can be readily explained if we realize that the kinds of positive freedom made possible by committing oneself to collectively binding norms are not limited to classic examples drawn from the political realm involving agents that are capable of controlling their own destiny in accord with their own interests (Berlin, 1969), but also include far more fundamental and important kinds of freedom. Consider, for example, the fact that as a result of normative *linguistic* constraints, namely, our collective commitment to bind ourselves to rules for the correct use of language, it becomes possible for us to create novel conceptual contents (meanings), purposes, interests and even novel needs. Indeed, our very *existence* as persons, capable of undertaking and attributing commitments and entitlements, authority and responsibility for our beliefs, thoughts and actions, over and above our behavioral habits, desires and impulses, and the causal force of tradition and societal conventions, is itself a case of a novel form of positive freedom made possible by mutually recognitive attitudes of constraint over the intensity, direction and flow of our multifarious short-term impulses, inclinations and desires. Absent such collective commitment to rules that function as standards or relevance and correctness of language use, our negative freedom could be said to typically 'increase' only at the expense of demolishing the network of social (linguistic) institutions that make meaning, personhood, social cooperation, social institutions and reasoned discourse possible. Think, for example, how many individual and collective purposes, interests, needs—in short, novel forms of expression and freedom—presuppose the existence of linguistic practices, and the existence of social institutions of money, education, democracy or law (which are themselves built on the foundation of linguistic practices). Consider also how many forms of social cooperation and collective action presuppose complex divisions of labor, mediated chains of engagement and abstract institutional structures (bureaucracy, political parties, government, markets), that is, processes that induce positive freedom by using alienation as a mode of enablement. Of course, all this is compatible with the view that normative constraints that make possible novel forms of positive freedom also make possible many novel forms of oppression, repression and alienation (masterfully analyzed by Deleuze, Foucault, Adorno and others) that would not otherwise exist as such and that are detrimental both to our normative and our non-normative freedom.[1] We shall revisit this issue in Chap. 10.

9.2 Normative Constraints as Homogenizing Universals and as Necessary for Novelty

It is true that the normative constraints of our actual practices—be they semantic, epistemic, moral or aesthetic—do diminish our non-normative freedom in various ways. Perhaps the most inconspicuous way in which normative constraints diminish non-normative freedom is by having an essentially 'identitarian' dimension or a homogenizing function, which in

[1] For example, in the *Dialectic of Enlightenment* (Horkheimer & Adorno, 1944). Horkheimer and Adorno argue that while the objectifying and rationalizing element of the cultural process of disenchantment was in a sense emancipatory it also yields non-emancipatory results. The completely formal character of this disenchanting process makes reason indifferent to the 'qualitatively' justified social injustices of pre-modern societies, and in this respect, the very formality and homogenizing dimension of reason turns out to be emancipatory. Yet this very same feature, as 'internalized domination', generates new and more sophisticated forms of self-oppression and impoverishment of experience: the repression of joyful, libidinal and creative dimension of our affective experience for the sake of compulsive productivity (see also Adorno, 2006; Marcuse, 1969).

effect sacrifices the unique, non-repeatable dimension of particulars for the sake of the general, repeatable characteristics of the universal to which they can be subsumed. In this way our creativity and experience are seriously impoverished, thereby diminishing essential aspects of our non-normative freedom (Adorno, 1969; Deleuze, 1968). However, I think that this is only part of the story and that when the whole story is told and the identitarian and homogenizing dimensions of normative constraints are understood in a non-metaphysical sense and as having a non-representational function, we can begin to see these pejorative terms in a different light. Thus, it can be argued that, at bottom, the identitarian and homogenizing dimension of normative constraints, has a *pragmatic* function. It is an *indispensable stabilizing infrastructure* of our semantic practices, an *axis of coordination* that alone makes possible the construction and evaluation not only of ordinary linguistic claims but, crucially, of *novel* claims or concepts when this is pragmatically demanded by the circumstances (e.g. in the case of 'recalcitrant experience', of the emergence of systemic anomalies in the system).

This point is intimately connected to Kuhn's well-known view to the effect that creativity and novelty presuppose the recognition of *anomalies* within a conceptual system (i.e. a violation of our expectations as the latter are projected by the semantic and epistemic rules of the system), which, in turn, presuppose *normalcy*—what Kuhn calls 'normal science', where scientists, by following the explicit and implicit rules of the scientific Paradigm, expand it to new regions and make it more determinate and specific in its claims. This kind of conformity to the rules of the Paradigm is necessary for the formulation of *determinate expectations* about what the behavior of the phenomena that the Paradigm examines. This is necessary for determinately recognizing a phenomenon as problematic, as an anomaly, and this recognition, in turn, is necessary for motivating the thinking process that creates novelty, that thinks 'out of the box'. As Kuhn himself puts this point: 'Novelty emerges only for the man who, knowing with precision what he should expect, is able to recognize that something has gone wrong' (Kuhn, 1962, p. 65). This shows why identitarian thinking is a necessary condition of enhancing our expressive freedom, especially in cases of theory-laden knowledge: the fact that observation or affection do not constitute given knowledge, and that they are always theory-laden and not transparent as to their meaning independently of the categorial framework of which it is part shows that a pre-categorial affective 'feel' of a situation or phenomenon as problematic, as a problem in need of a solution, is highly inchoate and becomes progressively determinate only if we possess the theoretical resources (sanctioned by the relevant community) to understand more precisely what *kind* of thing the encountered anomalous phenomenon is and for what *reasons* and in what *sense* are we frustrated with it (see also Dewey, 1938). And in the absence of conformity to these socially sanctioned categorial-normative implicit rules and practices that define a whole Paradigm for perceiving, thinking and acting, it is difficult to see how we could so much as *begin* to construct novel concepts with the features that are needed (specification, determinacy, explanatory scope, fertility, relevance) to overcome the encountered anomalies.

In fact, identitarian conformity and out-of-the-box creativity are even more tightly bound up than the above suggest. In our view of language and norms, that Brandom succinctly summarizes in discussing Rorty's conception of language, norms, freedom and creativity (Rorty, 1988), the distinctive feature of language as a tool is that it *maintains itself only by the generation of novelty that transforms it.* The inferential norms of language function as a synchronically stabilizing semantic infrastructure that enables the formulation and orients us in our assessment of novel claims: but, as Brandom puts it, 'since the inferential norms that govern the use of concepts are not handed down to us on tablets from above [and] they are not guaranteed in advance to be complete or coherent with each other, *they are at best constraints that aim us in a direction when assessing novel claims.* They neither determine the resultant vector of their interaction, nor are they themselves immune from alteration as a result of the collision of competing claims or inferential commitments that have never before been confronted with one another' (Brandom, 2000, p. 176, emphasis mine). This is a thoroughly dynamic and processual conception of language in which the elements that impart coherence to our linguistic practices and our concepts—namely, the inferential norms that govern their use in their diachronic dimension—act as *enabling constraints* of freedom and novelty and vice versa: the diachronic coherence of our concepts is preserved not despite but precisely *through* the continuous transformation of the inferential norms in question (i.e. the construction of novel conceptual models, analogies and categorial structures for accommodating collisions between competing inferential commitments). (Indeed, as we shall see in Sect. 9.6, universals, far from denoting traditional metaphysical essences exemplified in worldly particulars, can be thought of precisely as enabling normative constraints for the articulation, evaluation and revision both of first-order empirical beliefs about the world and of second-order empirical concepts themselves.)

9.3 Normative Freedom as Distinguishable Yet Inseparable from Non-normative Freedom

Normative freedom, or freedom as self-determination, also plays an important role in our descriptive and explanatory practices. This becomes obvious if we realize that the descriptive and explanatory tools of empirical inquiry and its sophisticated extension, science, are an essential part of our space-of-reasons practices. Our descriptive and explanatory practices attempt to identify and understand the nature of the causal constraints on our behavior (both external—environmental, evolutionary—and internal—bodily, historical or psychological—ones) with the aim of transforming them into elements that can be taken into consideration in our *reason-giving* practices. This is a self-correcting process that enhances expressive freedom by increasing our individual and collective self-determination, that is, our control over unknown causal constraints that impart our survival and well-being. This is particularly made clear in cases where our descriptive and explanatory practices attempt to unearth the (evolutionary, historical, depth-psychological, socially induced) causal factors that underpin our reason-giving practices *themselves*. As was mentioned in Chap. 7, in this process, our reasons for belief and action are sometimes revealed in the course of empirical inquiry and through our explanatory practices as mere *rationalizing causes* of our behavior, while the ensuing knowledge from this understanding of the causal constraints in question is reincorporated in the space of reasons as an element to be taken into consideration in judging if our reasons for believing, thinking or acting are sufficiently justified. In this way, the causal constraints in question, by becoming visible and understood as new knowledge about our behavior and by bringing to light the ideological elements in the function of space-of-reasons concepts, are transformed from unquestioned background assumptions integral of our self-identity into arbitrary impositions by nature, convention, authority or tradition that can no longer be accepted as part of our self-conception. This process thereby transforms our very self-conception and increases our self-determination, our control over ourselves and our environment.

This shows the complicated nature of freedom as self-determination. Its theoretical dimension, in the guise of the descriptive and explanatory practices of ongoing empirical inquiry, while itself a resolutely *rational* (and not merely causal) process, and thus an exemplary case of normative freedom, ultimately aims at something that *also* thereby increases our *non*-normative freedom: our emancipation from *causal* constraints on the exercise of our individual and collective epistemic and practical capacities. This emancipation from the above kinds of broadly causal constraints, in turn, makes possible the *expansion* of our capacities for perception, thought and action, and, ultimately, the creation of *novel* desires, interests and purposes that go hand in hand with this expansion—thus, again increasing non-normative freedom. We see here the complex dialectic between normative and non-normative freedom: normative freedom exercised through our descriptive and explanatory practices is necessary for achieving non-normative freedom, while the latter is the ultimate *raison d'être* and ontological underpinning of the former.

Indeed, I do not see any decisive reason for limiting the above point to the domain of our descriptive and explanatory practices. I take it that emancipation from all the above sorts of causal constraints is a regulative ideal operative *across* our practices in general. For example, arguably, the construction of novel forms of expressive freedom resulting from the overcoming of causal constraints in our perceptual and imaginative faculties is part of a regulative ideal built into our current aesthetic practices. And even in the case of our moral and political practices, where normative freedom (autonomy) is admittedly a constitutive value, it does not seem plausible to suggest that freedom as autonomy is a self-justifying value independently of its contribution to non-normative non-moral values and goods (see Sect. 9.4), such as social cohesion and improvements in standards of living that increase our collective welfare. For while it might be true that normative freedom ought to be pursued as a value even when the abovementioned kinds of non-normative freedom are compromised at the level of the *individual*, it does not seem equally plausible to say that the same holds at the collective level. A compromise of these non-normative values at a collective level would make ethico-political normative freedom lose its social significance, its motivational power and its material underpinnings. Thus, I think that the view to the effect that normative freedom finds its ultimate *raison d'être* in non-normative freedom can be legitimately extended beyond the realm of our descriptive and explanatory practices. The increase in freedom as individual and collective self-determination is not a self-standing ultimate purpose or value but derives its significance from the fact and to the extent to which it thereby contributes to an increase in our overall 'power to affect and to be affected'—in our collective *Conatus*.

Yet, even if the above analysis is the right direction, things get more complicated if we take into account the fact that the exercise of our conative powers sometimes is thought to include our power to dominate over others in the coercive sense of the term. In what follows, I will suggest that a proper account of non-normative freedom must sharply distinguish between freedom as an increase of our capacity to act (depending on our power to affect and be affected) and freedom as the power to dominate over others. It is thus time to turn to a brief analysis of non-normative freedom.

9.4 Non-normative Freedom as the Power to Act

As was mentioned earlier, the conception of freedom in normative terms, as collective self-determination through the imposition of normative constraints on patterns of behavior, while correct so far as it goes, is incomplete unless it is supplemented with a corresponding non-normative, naturalistic account, in which it can be understood in terms of increase or decrease of our non-conceptually characterized *affective powers*, and more specifically, as an increase in what Spinoza calls 'joyful passions' and a decrease in our 'sad passions' (Spinoza, 2018). This is a conception of freedom as an *occurrently existing power* (an expression of the occurrently existing virtual power discussed in Chap. 4) that materially underpins normative freedom conceived as a social status. Importantly, freedom in this sense does not operate on the basis of the imposition of normative constraints of any sort (commitments, entitlements, authority, responsibility) in our behavioral patterns—though, of course, the imposition of these normative constraints through our normative attitudes necessarily effects changes in our power to affect and to be affected at the non-normative level.

Moreover, as was briefly mentioned in the former section, I take it that non-normative freedom as an affective material power is intimately related to what we might call *non-moral goods*. Indeed, it can be argued that in our form of life—and indeed in sentient beings in general—non-normative freedom is expressed in the form of non-moral goods. A non-moral good is something that human beings pursue and find desirable even if no moral credit is accrued from pursuing or possessing it. Exemplary cases of non-moral goods are joy, happiness, creativity, self-actualization, security, trust, love, friendship, a sense of belonging, concern for others (sympathy) and physical health. Correspondingly, exemplary cases of non-moral 'evils' are suffering, cruelty, insecurity, misery, anxiety, resentment and illness. Non-moral goods contrast with moral goods such as justice, virtue, fulfillment of duty, fairness, autonomy, etc., whose value stems from the conscientious following of certain rules and principles which tell us what we *ought* to do irrespectively of our desires and impulses. Obviously, moral goods go hand in hand with normative freedom. Note also here that our distinction between non-moral goods and evils is deliberately meant to roughly correspond to Spinoza's distinction between joyful and sad passions or Nietzsche's similar distinction between active and reactive forces (Deleuze, 1983).[2]

I submit that the non-moral goods that are relevant to the exercise of non-normative freedom are those that increase our *power to act* and decrease our *power to dominate*. Let me explain these importantly different senses of power. Deleuze & Guattari (1980, p. 106) distinguish between power as *puissance* and power as *pouvoir*. Power in the first sense is the power *to* do or create something. It increases the individual and collective power to affect and to be affected (recall that this is our criterion for something to make an ontological difference in the world, and thus to count as existing in the 'really real' world of processual becoming) and amounts to the empowerment of individual or collective potentialities of action. It

[2] The correspondence is not fully exact although it is exact enough for my purposes. Consider, for example, the fact that the class of non-moral goods can be divided into what might be called 'altruistic' non-moral goods, such as love or concern for the well-being of others, and non-moral goods such as joy, happiness, physical health and creativity, which are not intrinsically altruistic. Yet, altruistic non-moral goods, especially the concern for the well-being for others, do not seem to fit neatly into either Nietzsche's category of active forces or to Spinoza's joyful passions (Christias, 2023, p. 274). But even here the divergence is less than might seem at first sight. It might be true that Nietzsche and Spinoza considered affective feelings such as sympathy for miserable, poor or bereaved people as an expression of sad passions, but not all cases of concern for others are of the above kind: if one is concerned for the well-being of one's family, friends, lovers, comrades, fellow human beings due to conceiving them as members of a we-community to which he identifies with, then one might well thereby express and trigger joyful passions individually and collectively. I will not take a stand over the difficult issue of whether, for example, sympathy for miserable people increases or decreases our non-normative freedom. Our analysis of non-normative freedom is not meant to offer a full theory of joyful or sad passions, as this is not demanded for the purposes of the argument developed in this chapter.

marks an increase in our individual and collective ability to carry out our desires as a result of an increase in our material capacity to act.[3] Power as *puissance* is also expressed by the formation of emergent desiring unities (within the individual, in symbiotic relationships or in collectives) that nonetheless respect the heterogeneity of their components. By contrast, power as *pouvoir* is power *over* something or someone that severs its potentialities to act. It is the power to dominate over something or someone, where external objects, animals or human beings are viewed as matter to be formed by the command of a superior. From this point of view, only the first form of empowerment is an expression of joyful passions. The second is ultimately tied to sad passions, to Nietzschean *ressentiment*, and does not empower one or the whole in the *puissance* sense.[4]

Importantly, as I see it, the concept of domination at work here must not be equated with the concept of *control* or *mastery* over things or humans, but rather with the severing of their potentialities of action and capacities to affect and be affected. Not all kinds of control have these latter consequences. There are kinds of control over things or humans that are *non-coercive* or at least have a non-coercive dimension. And the coercive dimension they might have need not necessarily be such as to sever the potentialities of action and capacities to affect and be affected of the coerced individuals. For example, while the process of learning all sorts of skills, from speaking a language to playing tennis or mastering various crafts or arts arguably 'coerces' the free expression of desires and impulses, it does not thereby sever our overall capacity to act. To take another

[3] Consider, for example, that our capacity to act is hindered under conditions of poverty, starvation, homelessness, unemployment, inequality, low-quality healthcare, education, lack of free time, etc. All these are materially coercive circumstances that stand in the way of our freedom in its non-normative dimension.

[4] But why is the power to dominate ultimately an expression of sad passions? Is it not the case that—sometimes at least—one joyfully exerts one's power over others? It can be argued that the power to dominate ultimately expresses sadness, however joyful and empowering for one's self-confidence it might be, in the sense that the compulsion to organize one's life around relations of domination over others reveals one's underlying weakness, lack of skill, anxiety and lack of self-confidence in confronting others under other-than-dominating interpersonal relations. One thereby—most often subconsciously—tries to weaken the power of others (*puissance*) so as to compensate for one's own lack of it. And this is a form of sadness however joyful, pleasurable or empowering it might be to one. Of course, this is compatible with the fact that this 'joyfully sad' passion might well be a successful evolutionary strategy for survival and reproduction and thus a deep-seated impulse of individuals that pervasively affects their social relations (see also Chap. 10).

example, think of a painter's succeeding in 'conquering' color (Deleuze & Parnet, 1996), which does not of course mean that the painter thereby enslaves it or limits its potentialities, but rather exactly the opposite. The same goes for a scientist who understands (and thus 'conquers') the causal mechanisms, underpinnings and constraints of a natural phenomenon or of human behavior (including his own) and uses this knowledge to unleash hitherto unknown potentialities in natural processes, human bodies, minds and societies. Consider also the case of a teacher whose teaching methods and paradigm activate the cognitive, affective and creative powers of his students in such a way that the latter can productively interact in non-dominating ways. Or consider an analogous situation in a non-hierarchical social relationship, for example, a relation between friends, where one, through one's paradigm and actions, attempts and succeeds in motivating the other to be cooperative and non-antagonistic.

Of course, it is true that certain forms of control over ourselves or others, by constraining or repressing the uninhibited expression of one's impulses or desires, in effect decrease aspects of our non-normative freedom and activate sad passions to a certain degree. Yet, the above-discussed forms of control that work in this way do not constitute control as coercive domination since their involvement in the activation of sad passions is internally connected—and indeed compensated by—a considerably greater increase of non-normative freedom both in its negative and positive aspects by making possible novel, joyful and creative forms of expression of our sensory and intellectual powers.

Note also that non-normative freedom in the above sense, since it is understood in terms of an increase of one's power to affect and to be affected, also represents an increase in one's self-determination, though in a very different sense from that in which normative freedom is self-determining: non-normative self-determination does not depend on the imposition of normative constraints and the emerging 'self' does not have the normative unity of the Kantian or Hegelian self. Rather, it is a 'center of gravity', a dynamic locus of field of forces that binds together neurophysiological, bodily and environmental processes in the form of 'monadic perspectives' and whose unity is a temporarily stable result of relations of dynamic allostatic balance holding between the co-occurring processes that comprise it (see also Chap. 4). An increase in one's power to affect and to be affected at once entrenches, diversifies and extends one's dynamic allostatic balance with the environment.

9.5 Fractured Freedom

As we saw in Sect. 9.3 the relations between normative and non-normative freedom are complex. On the one hand, the *ontological* cash value of normative freedom and its enabling constraints is provided by their ability to affect *non*-normative ('material') freedom. One of the consequences of this is that the normative statuses that express freedom as self-determination by the imposition of normative constraints are causally efficacious due to their enactment in *affective attitudes* that express non-normative freedom as the power to affect and be affected. On the other hand, normative freedom is a necessary *epistemic* condition of the recognition, and proper individual and collective regulation of non-normative freedom. The spontaneous flow of desires and impulses cannot self-regulate in a manner that would always or typically result in an increase of our power to act and a decrease of our power to dominate (whether individually or collectively). Desires, impulses and drives did not evolve for the optimal regulation of the freedom of the individual or the community (indeed, sometimes they are highly destructive at both levels). The imposition of normative constraints is precisely an evolutionarily developed strategy for regulating the expression of desiring flows in ways that enable the survival and social reproduction of increasingly larger cohesive social networks and communities. And while these normative constraints do not, of course, univocally increase our power to act over our power to dominate, they alone make it possible to thematize this distinction and thus to make it an object of evaluation and criticism. Moreover, this regulatory power of normative freedom (i.e. of normative constraints) over non-normative desires and impulses does not only recombine, connect, separate, occlude, inhibit, highlight or enhance already fixed non-normative contents, but also thereby creates *novel*—hitherto inconceivable—such contents. Normative constraints make possible and salient possibility spaces of exercising forms of positive and negative freedom that cannot even be conceived as such independently of these normative enabling constraints. Of course, this does not mean that an increase in normative freedom directly implies a respective increase in non-normative freedom. This is because the vehicles of positive freedom in its normative dimension, namely, normative enabling constraints and the normative attitudes in which they are enacted, increase non-normative freedom (power as *pouissance*) only by *simultaneously decreasing aspects of it*, both because they necessarily involve the suppression of short-term desires and impulses inherent in all our normative commitments, and because they make possible novel forms of oppression, suppression, coercion or alienation that would not exist independently of the imposition of normative constraints.

Our conception of normative and of non-normative freedom as distinguishable yet inseparable and complementary dimensions of freedom, in line with the main argument of this book, points toward an essentially *immanent yet fractured* conception of freedom. The power of normative freedom to 'intervene' and alter the order, sequence and qualitative character of sensory-affective patterns in the flow of the non-normative level does not derive from special emergent causal powers of normative freedom but by its being a subtractive manner of appearance of a *self-fracturing* of the non-normative affective level itself through processes of productive and subtractive imagination (described in Chaps. 5 and 8).

Our conception of fractured freedom here is similar—though not identical—to Adorno's conception of freedom. According to him, the origins of our sense of freedom as autonomy (normative freedom) can be traced to its separation from uncontrolled—potentially destructive—impulses. In a process that made civilization possible, human beings individually and collectively, under evolutionary pressures, developed new means of self-preservation by coming to partially tame, control and repress the spontaneous flow and intensity of desires and impulses within and between individuals. This was the birth—and subsequent glorification—of reason and freedom as autonomy. At the same time, normative freedom was not altogether cut off from its origins, and indeed preserves, in a subterranean manner, essential elements of these latter (now nearly forgotten, pre-egotic, archaic) somatic impulses and forces, namely, their potentially uninhibited, spontaneous and joyful character (non-normative freedom). This preservation—if even in quasi-suppressed form—is necessary since without this somatic material dimension of freedom, autonomous freedom would lose its ultimate motivational force[5] (Adorno, 1969).

Our model differs from—or better, structurally clarifies—Adorno's in that it does not view the level and distinctive mode of existence of normative freedom as something situated on top of more archaic levels (while at the same time retaining suppressed remnants of the latter) but as a novel world-making power which exists as a result of a subtractive self-fracturing

[5] As Adorno characteristically puts this point: 'Without an anamnesis of the untamed impulse that precedes the ego -an impulse later banished to the zone of unfree bondage to nature- it would be impossible to derive the idea of freedom' (Adorno, 1969, pp. 221–222).

of the non-normative level of somatic drives, impulses and affects itself. Thus, from our point of view, the level of normative freedom, just as the level of normativity *tout court*, does have world-making powers, yet not because it is a transcendent emergent level *over and above* the non-normative and non-conceptually articulated level of affective processual individuals, but by being an immanently emerging manner of appearance of the self-fracturing of the latter. In this way, the essentially subtractive procedures expressing this self-fracturing (spatiotemporal contraction, enhancement, inhibition, highlight, occlusion, separation, recombination of sensory-affective patterns) institute *novel salience points* in these patterns thereby making possible sensory and affective patterns with novel (conceptual, normative) articulation, modes of self-relation, contents and general orientation.[6] To repeat one of the main themes of this book: essentially subtractive processes can *as such* produce something genuinely new, without any need for postulating a metaphysically transcendent realm of *sui generis* normative or 'noumenal' causation to account for the irreducible reality of the novel features of the realm of the conceptual (the realm of reasons, the realm of freedom).

[6] Indeed, as we saw in Chap. 5, Adorno himself provides some resources for putting some flesh on the bare bones of this view when he argues (Horkheimer & Adorno, 1944) that conceptual rationality is *itself* a product of a change of orientation effected *within* non-conceptual 'mimetic' consciousness. To this end he introduces the concept of 'mimetic rationality' which expresses a change of orientation of our mimetic abilities initially geared toward an 'organic adaptation to otherness' (a direct immersion to the environment), into sublimated distanced forms of 'mimicking the invariant' for purposes of self-preservation (conceptual thinking based on substitution and on invariant features of worldly processes). Though I do not want to commit myself to the literal truth of this view, I find it interesting that this account of mimetic behavior and thinking chimes in with our account of non-conceptual thinking as based in our ability to 'see in what it is something other than it'. If we conceive of mimesis in non-representational terms, that is, not as the ability to passively mimic ready-made external forms, but as a kind of 'transducer' that non-conceptually responds to its affinity with nature by producing new relations of resemblance (Benjamin, 2005; Noppen, 2017), then this production of resemblances can be understood as an instance of the ability to 'see in what is something other than it', where in this case the 'other' that is seen 'in' what it is (the demon of the rain in water droplets, the meaning in marks and noises), is seen *in* what it is precisely because of the establishment of a direct relation of resemblance that both bodily unites and separates the perceived phenomena.

9.6 The Emancipatory Force of Universals

I shall end this chapter with a discussion of the role of universals and particulars in our space-of-reasons practices, exploring the contribution of this dialectic to our freedom as self-determination. As we shall see, properly reconceptualized, the traditional categories of universal and particular do retain a crucial and indispensable role in our practices and are not to be discarded as metaphysical relics of no use to philosophy. This is particularly important since this book opts for a non-traditional nominalism, whose successor concept of universals, namely, that of the virtual, does not seem to make room for any positive use of universals in the classic sense of the term. And without such use of universals, it is difficult to see how our 'nominalism of the virtual' can accommodate the—at first glance non-nominalistic—fact that the semantic content of universals has an 'ideal' dimension that goes beyond—and in effect orients—how universals are actually used to describe concrete particulars in the context of our everyday practices. In this section, it will be argued that our non-traditional nominalism can account for this fact by reconceiving universals as semantic and epistemic tools of immanent critique. In this way, the 'surplus' content of universals is readily accommodated, not as a metaphysically existing unrealized potential or essence, but as a way of representing, calibrating and critically adjusting the structure of the inferential space of the universals in question. This view does not attribute causal powers to abstract entities: universals, reconceived in the above manner, are not abstract entities representing metaphysically necessary facts. Thus, this view is fully compatible with nominalism and adds an immanently critical bite to it.

Before delving into the indispensable critical role of universals in our space-of-reasons practices, let me just mention what I take the non-metaphysical yet positive function of universals to be at the most general level of analysis. On this issue, I endorse Sellars' view according to which abstract universals such as properties, kinds, relations, contents, facts, etc. are, in the last analysis, linguistic or better, *conceptual* entities (Sellars, 1963b). This is obscured by the fact that semantic talk (in particular, truth-talk) seems to be world-involving: the truth of the sentence 'Peter is mortal' seems to be grounded by reference to the relevant abstract universal, the property of 'mortality' and its 'exemplification' in a particular existing being, Peter. Yet, reference to universals such as the property of mortality, and its exemplification, does no explanatory work. If we want to explain why Peter is mortal, which of course we can (at least in principle), we must appeal to empirical investigations, presumably to physics, evolutionary theory, genetics and biochemistry, not to the metaphysical property of mortality. Sellars takes it that those abstract universals are not external grounds of correctness of our practices, but rather ways in which these practices *represent to themselves* their inferential structure (Kraut, 2010, 2016). Abstract universals such as 'triangularity', 'redness', 'wisdom' or 'mortality' are *metalinguistic* devices which serve to functionally classify—and thus convey information about—the common perceptual-inferential-practical role that tokens of such expressions play in the context of our linguistic practices, abstracting from their otherwise radically different mode of material realization (in different systems of material signs and different empirical circumstances) (Sellars, 1963b). Categorially functioning universals such as powers, dispositions, laws of nature, natural kinds, properties, relations, facts, possible worlds receive the same treatment. They do not directly represent the *world* (they do not refer to an intelligible world of essences whose potential is 'inadequately' realized in actuality) but convey information about the perceptual-inferential-practical role of the fundamental *structural* features that make possible the descriptive and explanatory function of our first-order *empirical* concepts (cat, lion, copper, etc.) (Brandom, 2015). Thus, by this metalinguistic reification of the relevant patterns and possibilities in our perceptual-inferential-volitional behavior, both first-order universals (expressing our empirical concepts) and second-order universals (expressing our categorial concepts) *indirectly* exert influence on the real world of process. They do this by representing the *conceptual space* which semantically mediates our access to the world of process (O'Shea, 2007, see also Chap. 8).

We are now in a position to better understand this non-metaphysical yet indispensable role of universals in our space-of-reasons practices. We can begin to do so if we construe the universal-particular distinction not in metaphysical terms, as denoting a static metaphysical relation between essences and the realization of their potential in actuality, but in normative, critical and dynamic terms, as an essential structural element of the self-correcting process of our descriptive and explanatory practices. As we shall see, the meta-representational function of universals, namely, the fact that they represent the conceptual space which mediates our relation to the world of process, illuminates their role in our practices only if it is taken in conjunction with the essentially *normative* dimension of universals: the fact that they have a surplus 'ought' content that exceeds their de facto material realization in the empirical world of becoming. This normative surplus dimension of universals, especially in the form of the logico-epistemic demand that they exhibit a diachronic explanatorily coherent inferential profile, taken in conjunction with their meta-representational role, is what enables us to *recognize* incompatible commitments in our actual use of concepts and to initiate a process of self-correction of the very *concepts* that we apply in reality.[7] For the self-correcting dialectic of inquiry is arguably *itself* triggered and dynamically sustained by the double life of universals (a) as normatively individuated objects with a surplus *ought* dimension and (b) as actually used concepts and actually realized particulars with a descriptive '*is*' dimension (whose character and properties can be causally explained by empirical-scientific inquiry).

This surplus normative content of universals (specifying how things *ought* to be for the content of the universal to be fulfilled) that is not exhausted by the way in which the concept exemplifying the universal in question is actually used to describe concrete particulars in the context of our everyday practices is a projection of emancipatory ideals expressing historically evolving expectations regarding our collective epistemic and general welfare. Thus, far from pointing toward a *sui generis* metaphysical reality populated by universal essences, this surplus normative content functions as an *axis* for *coordinating* collective epistemic and practical projects in the direction of fulfilling the above emancipatory ideals.[8]

Let us see now in more detail how this self-correcting process of inquiry, expressed as a self-correcting dialectic between the universal and the particular, can function as an immanent critique of the existing semantic, epistemic and ethico-political *status quo*.

As we saw, the surplus content of universals is very general, homogenizing and schematic. This is no accident. If interpreted as having a

[7] The normative surplus dimension of universals comes first and explains our ability to recognize incompatible commitments in our practices in the order of *understanding*, not in the order of being. In this latter dimension, things are precisely the opposite: what comes first are (inchoately felt) problematic situations in our forms of life and the very existence and normative surplus dimension of universals is a (conceptual) *response* to these problematic situations that articulates progressively determinate conditions for conceiving the nature of the problems as well as for solving them.

[8] These epistemic and practical emancipatory ideals can be fused if they are understood in terms of aiming for constructing conceptual systems of increasing explanatory coherence the application of which increases our collective skillful coping in an ever-increasing range of physical and social environments (Christias, 2023, pp. 195–197).

normative, non-representational function, this surplus content has three important consequences:

1. By expressing and sustaining an open-ended, schematic and functional characterization of the concept-universal, it enables successive *changes* in the determinations of its sense (e.g. a change in the meaning of 'personhood' so that it includes higher animals or intelligent aliens, etc., the change in the meaning of 'mass' to include Einsteinian mass) as a result of the concept-universal's failure to measure up to the explanatory aspirations inherent to the practice of which is a part.[9] This failure, in turn, is itself registered as such only because these explanatory aspirations are themselves implicitly at work in the actual use of the universal in ongoing practice. In this way the very adequacy of a concept-universal in fulfilling the ideals inherent in its normative way of being is challenged from within. This is precisely a process of immanent critique of the existing semantic and epistemic *status quo*.[10]
2. This schematic ought-content enables the forming of *orienting narratives* with which to navigate forward in our attempt to cash it out (i.e. progressively determinate it) in actual evolving practice (Srnicek & Williams, 2015, p. 75). Thus, if we combine points 1 and 2, we see that this surplus ought-dimension of universals gives a *direction* to inquiry and *progressively generates criteria* to adjudicate which determinate applications of universals are adequate or inadequate.

[9] Sellars (1973) shows how successor concepts (e.g. of Einsteinian mass) in science can be *generically* similar though *specifically* different from their predecessors (Newtonian mass), due to the intra-theoretical functional role similarity between the two concepts, a similarity which is itself justified by their being involved in relevantly similar explanatory tasks in the context of the scientific practice of physics (O'Shea, 2007, pp. 159–160). Though the crucial notion of 'relevant similarity' here is not sharply defined and is context-sensitive (Dach, 2024), I contend that it can be used for justifying generic continuity between successor and predecessor determinate characterizations of concept-universals. The key move in this direction is to reject a Fregean conception of content determination as entailing sharp, fixed and complete boundaries, and instead adopt a Hegelian-Brandomian historical, dynamic and performative conception of content determination, and a corresponding historical and dynamic rendering of the sense-reference distinction, based on normative attitudes of diachronic reciprocal confession, recollective forgiveness and trust (Chap. 8).

[10] Arguably, immanent critique in this sense is a species of epistemological-cum-explanatory immanence (Chap. 2). For a comprehensive account of immanent critique see Jaeggi, 2018.

Importantly, these criteria do not have antecedently fixed content. What is fixed in the surplus dimension of universals and remains irreducible to its determinate specifications in actuality is its schematic character and functional characterization, not any particular supposedly invariable schematic or functional content. This means that the determinate specification of the 'ought' content of universals might well change in response to changes in the determinate specification of their 'is' (descriptive-explanatory) dimension. Thus, the criteria for adjudicating which applications of universals in actuality are adequate or inadequate are themselves subject to change.

3. This critical, regulative and orienting functions of universals are essentially normative functions, and since the way of being of the normative is hyperstitional (Chap. 8), the same goes for the above functions of universals: they are essentially *hyperstitional*. These functions are self-posited as expressions of something that does not yet exist that aims to transform itself into a truth by performative means. By exhibiting this hyperstitional structure, and by piggybacking on the schematic, homogenizing form and functional dimension of universals, which are themselves hyperstitionally sustained, the surplus content of universals enables a process of catalyzing dispersed affective states into novel, consolidating and expanding we-attitudes that can scale up and become a historical force of change, a world-making power with significant emancipatory potential (whose measure is the increase of our individual and collective autonomy and power to act).

More specifically, the critical function of universals reveals its emancipatory and subversive character when particular contents that were ignored as irrelevant or excluded from the range of application of a universal are shown to be essential 'determinate fillers' of its schematic surplus content, and thus essential part of the universal's unconditional demands.

To see how this works, consider the concept of universal human rights. Certain determinate versions of this concept-universal can be criticized as inadequate (i.e. as not properly representing its more skeletal and ideal content), due to their unjustified exclusion of particular persons and particular communities, such as women, blacks, homosexuals or immigrants, from the determinate universal in question. These people are recognized as persons, yet only to the extent that they are burdened with responsibility without having the corresponding authority, with duties and obligations without the corresponding rights. This is a situation of asymmetric recognition that fosters fundamentally unequal recognitive relations between persons and reveals a serious tension in the actual use of our concept-universal of personhood (Brandom, 2019). In situations like these, it seems that the determinate version of the universal 'person' excludes particulars and classes of particulars from its range of application not for principled reasons but due to non-rational causes (e.g. historically entrenched asymmetric power relations, superiority biases, fear, anxiety in the face of the 'other') or instrumentally guided reasons serving non-rational ends (e.g. the preservation of privileged positions in society). Now, the process by which a universal (human rights) is challenged as unjustifiably excluding particular individuals and communities from its range of application amounts to a doubly transformatory process between universal and (excluded) particular. The particular individuals and communities that object to their exclusion and demand their inclusion in the universal 'human rights' are thereby (hyperstitionally) transformed into *exemplary representatives* of the universal in question (changing its determinate content and 'orientation'). Conversely, the previously established universal (humanity) becomes revealed as a parochial particular ('western white heterosexual male'). In this way, by claiming to be embodied exemplary representatives of the universal 'humanity', the formerly excluded particulars 'woman', 'black', 'heterosexual', immigrant', as materially embodied in the hyperstitional processes they engender in society (socio-political movements), become a way to prosecute a critique against existing versions of the universal 'humanity', which in turn, becomes revealed as a particular lacking unconditional value (Srnicek & Williams, 2015, pp. 77–78, Jonsson, 2010). This is a process of immanent critique that *creatively extends* the determinate sense of a universal as a means of *elucidating* it, of staying *true* to it and its unconditional demands.[11]

[11] Of course, the same point about the subversive use of universals could be made with the use of different examples of excluded particulars ('blue collar workers', 'the precariat'), provided that their being treated as exemplary representatives of the universal under discussion results in the formation of practices (e.g. establishment of 'basic economic rights') that increase our normative and non-normative freedom. Note, moreover, that there are several historical examples of hyperstitionally functioning universals: the eighteenth- and nineteenth-century abolition of slavery, the Universal Declaration of Human Rights (1948) and the women's voting rights campaigns can be all understood as hyperstitional projections of universal norms that did not describe reality but gradually made themselves real by reshaping

Interestingly, this model is not limited to ethico-political universals but also finds application to natural-kind universals. For example, observable phenomena that were previously excluded from the natural-kind universal 'water' (e.g. its phase transitions into ice or gas) and are now explained on the basis of the behavior of chemical processes subsequently became exemplary representatives of the universal 'water', while its former defining observable properties (based on its liquid form) are now seen as surface manifestations of water in relation to the senses and the particular historical context and knowledge of particular human communities. Or, to take another example, consider the case of psychological phenomena (anxiety states, phobias, obsessions) and the defense mechanisms that explain them (repression, rationalization, sublimation, projection, introjection) that were excluded from the universal 'autonomous subject'. It can be argued that they have subsequently become relevant for criticizing prevailing versions of this universal by revealing it as a parochial particular. These defense mechanisms do this by claiming to be essential parts of the universal 'rational subject'. Freedom, reason and autonomy cannot be severed from desires and impulses but essentially presuppose—and even express—them in their own way of being (see Sect. 9.5). Thus, the conception of human rationality as in principle capable of completely severing its ties from desires and impulses is revealed as ideological, parochial and as not representative of the universal 'autonomous/rational subject'. Lastly, consider the contemporary debate over creativity and the status of generative AI, particularly large language models (LLMs). The growing integration of LLMs in cultural and intellectual production, where their outputs often exhibit novelty, stylistic innovation and contextual sensitivity, puts pressure on the universal 'creativity' from within, at least to the extent to which we think of creativity as an intrinsic property of individual intentional agents. It can be argued that the inclusion of LLMs to this universal would amount to reconceiving creativity as a relational achievement, as a distributed, emergent and processual function of what we might call 'sympoietic assemblages' involving humans, machines and environments.[12]

law, culture, and economic practice, purging them of their parochial and unjustifiably exclusive elements.

[12] This redefinition does not eliminate the normativity of creativity but redistributes it: LLMs are creative only within a socio-technical assemblage that essentially includes (reciprocally recognized) persons that take up and integrate LLMs outputs into their hyperstitionally constructed worlds.

Note that in all these cases, the critique of existing versions of universals and their exposure as parochial particulars is conducted from a standpoint that aspires to be *faithful* to the universal in question. In this respect, it is completely opposed to the currently fashionable post-modern anti-universalist attitude which sees the above critique of universals as entailing their elimination in favor of a series of diverse, incommensurable particulars (individuals, communities, etc.).[13] Of course, this eliminativist attitude toward universals, especially in the ethico-political sphere, is explainable and even understandable up to a point. For example, the fact that in modernity the European way of life presented itself as a universal is not unrelated to the fact that all other non-Europeans were seen as residual particulars that ought to be subsumed under the European way of life. And we are all familiar with the way in which this homogenizing and subsumptive attitude is related to European colonialism (Mingolo, 2011). From our point of view, this undeniably oppressive feature of universals is connected to their being conceived as having a (big-R) *representational* function, that is, as representing natural-kind essences (in this case 'human nature' or 'rationality') from which we can draw normative conclusions about how the particulars related to this essence *ought* to be or ought *not* to be. By contrast, our resolutely non-representational and non-subsumptive conception of universals does not have such inferential (and practical) consequences.

We can conclude this chapter by taking a step back and considering the argument of this section in relation to the overall philosophical orientation of this book. If successful, the argument shows how a process nominalism in ontology is not only consistent but complementary with an uncompromising universalism in semantic, epistemic and ethico-political contexts, provided that this universalism is construed as resolutely non-representational, critical and hyperstitional in nature. From this process

[13] This view thereby has interesting consequences regarding our contemporary ethico-political situation: the critical function of universals consists in highlighting the particular, the divergent, what is excluded from the dominant universalist discourse (women, immigrants, ethnic minorities, the LGBT community, the precariat) as a focal point for hyperstitionally constructing new 'hegemonic signifiers' that claim to be more faithful representatives of the unconditional demands and ideal dimension of the universal in question (e.g. human rights, human dignity, human freedom). In this respect, critical universalism is opposed to a particularistic form of 'identity politics' (the right to identify as a member of an oppressed group independently of universalist demands). Or better, it is true to its underlying spirit, while thoroughly criticizing its self-understanding and implementation in practice.

nominalist point of view, the schematic surplus content of universals and its progressive determination through the feedback-loopy self-correcting mechanisms of inquiry can be understood as novel (i.e. conceptualized) experiential ways for comprehending 'problems', for finding 'solutions' to these problems and for instituting a dialectic between them (see Chap. 4 for the problem-solution dialectic). This internal structure of universals also amounts to *a novel way of organizing temporality*: it transforms the 'really real' yet semantically and epistemically opaque modal dimension of virtuality through which the past and the future are directly effective in the present, into a progressively determined series of conceptual contents through which the present becomes directly effective both to the past (by retrospectively determining it) and to the future (by prospectively determining it). This novel reframing of temporality and of 'problem solving' mobilizes for this purpose semantic tools (e.g. negation, the logic of material incompatibility and consequence, modal categories of the object-property framework) that do not directly reflect the (non-conceptual) articulation of temporality and of the 'problem-solution' dialectic at the level of process nominalistic reality. However, conceptual problem-solving dialectics does make a real difference at the non-conceptual level, since the we-attitudes that sustain the practice of using and progressively determining the schematic content of universals are *themselves* denizens of the nominalistic process world. It's just that the exact way in which the conceptual level of progressive articulation of universals makes a difference to the non-conceptual level of nominalistically articulated problem-posing-cum-problem-solving processes is not semantically and epistemically transparent (Given) and can be brought into semantic self-consciousness only through the process of self-correcting inquiry.

References

Adorno, T. (1969). *Negative dialectics.* Bloomsbury Publishing.

Adorno, T. (2006). *History and freedom. Lectures 1964-1965.* Polity Press.

Benjamin, W. (2005). On the mimetic faculty. In M. W. Jennings et al. (Eds.), *Selected writings, part 2: 1931-1934* (pp. 720–722). Harvard University Press.

Berlin, I. (1969). Two concepts of liberty. In I. Berlin (Ed.), *Four essays on liberty* (pp. 118–172). Oxford University Press.

Brandom, R. (1979). Freedom and constraint by norms. *Pacific Philosophical Quarterly, 16*(3), 187–196.

Brandom, R. (2000). Vocabularies of pragmatism: Synthesizing naturalism and historicism. In R. Brandom (Ed.), *Rorty and his critics* (pp. 156–182). Blackwell.

Brandom, R. (2015). *From empiricism to expressivism: Brandom reads Sellars.* Harvard University Press.

Brandom, R. (2019). *A spirit of trust: A reading of Hegel's phenomenology.* Harvard University Press.

Christias, D. (2023). *Normativity, lifeworld and science in Sellars' synoptic vision.* Palgrave Macmillan.

Dach, S. (2024). *Wilfrid Sellars on truth: Between immanence and transcendence.* Routledge.

Deleuze, G. (1983). *Nietzsche and philosophy.* Continuum.

Deleuze, G. (1968). *Difference and repetition.* Columbia University Press.

Deleuze, G., & Parnet, C. (1996). *L'Abécédaire.* Wayne State University.

Deleuze, G., & Guattari, F. (1980). *A thousand plateaus: Capitalism and schizophrenia.* University of Minnesota Press.

Dewey, J. (1938). *Logic: The theory of inquiry.* Henry Holt and Company.

Horkheimer, M., & Adorno, T. (1944). *Dialectic of enlightenment.* Stanford University Press.

Jaeggi, R. (2018). *Critique of forms of life.* The Belknap Press of Harvard University Press.

Jonsson, S. (2010). The ideology of universalism. *New Left Review, 63*(2), 115–126.

Kraut, R. (2010). Universals, metaphysical explanations and pragmatism. *The Journal of Philosophy, 107*(11), 590–609.

Kraut, R. (2016). Norm and object: How Sellars saves metaphysics from the pragmatist onslaught. In J. O'Shea (Ed.), *Sellars and his legacy* (pp. 60–80). Oxford University Press.

Kuhn, T. (1962). *The structure of scientific revolutions.* University of Chicago Press.

MacCallum, G. C., Jr. (1967). Negative and positive freedom. *Philosophical Review, 76*, 312–334.

Marcuse, H. (1969). *An essay on liberation.* The Penguin Press.

Mingolo, W. (2011). *The darker side of western modernity: Global futures, decolonial options.* Duke University Press.

Noppen, P. F. (2017). Adorno on mimetic rationality: Three puzzles. *Adorno Studies, 1*(1), 79–100.

O'Shea, J. (2007). *Wilfrid Sellars: Naturalism with a normative turn.* Polity Press.

Rorty, R. (1988). *Contingency, irony and solidarity.* Cambridge University Press.

Sellars, W. (1963a). Philosophy and the scientific image of man. In W. Sellars (Ed.), *Science, perception and reality* (pp. 1–40). Routledge.

Sellars, W. (1963b). Abstract entities. *Review of Metaphysics, 17*(1), 67–90.

Sellars, W. (1973). Conceptual change. In G. Pearce & P. Maynard (Eds.), *Conceptual change* (pp. 77–93). D. Reidel.

Spinoza, B. (2018). *Ethics*. Cambridge University Press.

Srnicek, N., & Williams, A. (2015). *Inventing the future: Postcapitalism and a world without work*. Verso.

CHAPTER 10

The Politics of Fractured Immanence

In Chap. 9 we saw that gaining mastery over the world, ourselves and others through the imposition of all sorts of normative constraints on the latter is a vehicle of normative freedom but always comes with a reduction of non-normative freedom and the emergence of novel forms of oppression, suppression and alienation. We can see this clearly in the political sphere: different sorts of normative constraints (political institutions, bureaucracy, police, state-law) and related policies (taxation, decrease of economic inequalities) increase certain aspects of non-normative freedom while decreasing others. What is the proper content and blend of normative constraints for the optimal channeling of individual and collective non-normative desires, drives and impulses? This seems one way of formulating the fundamental question that animates political theory and practice.[1] Of course, my purpose here is not to offer a theory of justice in political philosophy or a substantial theory about the content and means of achieving and maximizing collective welfare. This would require all the dialectical abilities, interdisciplinary knowledge and philosophical genius of a Rawls or a Marx. Instead, my far more modest aim is to sketch what I take to be the consequences of a fractured immanent view of

[1] I take the traditional fundamental problem of political philosophy, namely, the justification of state power over individuals, to be a further specification of the above most general question and problem.

© The Author(s), under exclusive license to Springer Nature Switzerland AG 2025

D. Christias, *Towards a Hyperstitional Process Nominalism*, Palgrave Perspectives on Process Philosophy,
https://doi.org/10.1007/978-3-032-06350-2_10

persons-in-a-(processual)-world in the ethico-political sphere in light of the fact that normative and non-normative freedom presents us with two *equally* fundamental yet *not* mutually entailing or complementary regulative ideals.

Very roughly, it can be argued that the regulative ideal at the heart of our conception of normative freedom is expressed by the 'universalist' impulse for *collective (semantic, epistemic, moral) self-mastery* while the regulative ideal of our non-normative conception of freedom is expressed by the 'particularist' impulse for *unhinged monadic material empowerment* (think of Spinoza's *Conatus*). The difficulty in reconciling these regulative ideals can be immediately grasped if we realize that the impulse for collective self-mastery is a way of fostering our collective welfare by being oriented toward normatively *equalizing* or 'commensurating' the different monadic worldly perspectives of human beings by the construction of a *universalizable* shared semantic, epistemic and moral space representing a common world, while the impulse for monadic material empowerment is directed toward the empowerment of the *non-equalizable*, incommensurable and *non-universalizable* dimension of us as monadic perspectives.[2] Thus, the (non-egalitarian, non-universalizable) impulse for monadic empowerment and the (egalitarian, universalizable) impulse for collective self-mastery pull us in two significantly different directions and it is not easy to see how we can make them resonate together. Presumably, to say that these utopian freedom-impulses can resonate together is to envisage a regulative ideal in which our space-of-reasons practices, both in the theoretical realm (descriptive-explanatory practices of empirical-scientific inquiry) and in the practical realm (fundamental moral principles binding for all rational beings) are such as to enact we-attitudes of reciprocal recognition between persons that enable the joint empowerment of their monadic perspectives (i.e. their empowerment by means that do not unnecessarily sever anyone's capacity to act). In what follows, I shall suggest that these two fundamental freedom-impulses can be best made to resonate together in the context of a post-human community of persons organized around 'democratic prometheanist' ethico-political principles.

[2] Recall that monadic perspectives do not amount to fully formed individual selves (standing over against society and a common objective external world), although, obviously, they are intimately related to them. The formation of individual selves presupposes the existence of a conceptual space in which common semantic, epistemic and moral norms are already operative.

10.1 Toward a Post-Human Community of Persons

Let me briefly explain the rationale for this view, and how it fits with our processual and fractured immanent view of the world.

I shall begin with the denizens of space-of-reasons communities, namely, persons, and show why, if that to which we most deeply identify includes the above utopian freedom-impulses as they find expression in the respective regulative ideals of personhood, then we, as persons, must not be definitionally equated to human beings considered as a biological kind.

As we saw in Chap. 8, persons can exist in a world of nominalistic becoming as norm-governed occupants of functional roles in communities structured by reciprocally recognized normative attitudes. Ontologically, we are complex physical processes. There is no distinctive ontological category of personhood. Persons are not natural kinds (e.g. defined by our bio-genetic profile). From this point of view, to be a person is to be a being characterized by a special kind of normative-cum-functional unity of material structures and patterns, namely, those that instantiate the functional capacities of experience, thought and action. More specifically, to be a person is to be a recursively self-maintaining metabolic physical system whose pattern-governed behavior can be interpreted as a) changes in the internal environment of the complex physical system as a result of the latter's being affected by external and internal stimuli (sensing, feeling), b) retainings of past events (remembering), c) propensities to be in certain kinds of representational states that are followed by (or are followed by the absence of) certain other kinds of representational states (inferring) and d) changes in the external environment so as to satisfy the system's short- or long-term metabolic needs (willing, acting). And as we have seen, this kind of functional organization constitutes a person only if it forms a system of habits that are reciprocally recognized and normatively sanctioned (as commitments and entitlements, rights and duties, 'can's and 'ought's) from the standpoint of *other* such physical systems, and where these mutual recognitive relations comprise a *community* of such systems. These are the only 'essential' properties that a complex physical system must have to count as a person. Thus, to the extent to which us, human organisms, do have these normatively-cum-functionally individuated abilities and exhibit these properties, we thereby *are* persons. Note, of course, that from this it follows that any complex physical process that has these abilities and exhibits these properties thereby counts a person. Thus, the way is open for understanding persons as characterized by functional capacities that can be multiply realizable and thus as potentially including non-human or post-human complex physical systems.[3] Relatedly, this non-metaphysical, normative-functional rendering of persons not only expands our conception of what counts as a person but also shows that when we conceive *ourselves* as persons we thereby implicitly identify with an embracing community of sentient-cum-sapient beings *in general*, and not only with the community of specifically biological *human* beings. Indeed, it can be argued that our commitment to personhood carries with it (through the regulative ideals implicated in our semantic, epistemic and moral practices) the seeds of overcoming our current material embodiment—at least to the extent to which the latter falls short of the regulative ideals built in our conception of personhood (Christias, 2020). This is connected to the fact that an *essential* part of our commitment to personhood is to

[3] For a similar view see Wennemann (2013). Roden (2014) calls this view 'anthropologically bounded post-humanism' because it sets a priori conditions and limits to post-human persons based on normative-cum-functional constraints of human beings (e.g. the fact that rationality is necessarily expressed in the form of a language situated in a social space sustained by normative we-attitudes), and argues that it is empirically possible for there to be 'weird' forms of post-human persons that lack language or sociality altogether. I do not have the space to fully address this objection here (but see Negarestani, 2018, pp. 95–108). I shall limit myself to the following brief remarks: (a) The requirement that persons must be discursive beings is functional and broad enough to include any information processing and exchange among complex self-maintaining systems that can be functionally differentiated into language-entry transitions (perception), intra-linguistic transitions (inference) and language-exit transitions (action) (Sellars, 1954), and, since this abstracts from all material conditions that realize this discursive structure, except its dependence on a world that is partially out of our control but to which we can intervene, I do not see how this is anthropomorphic in any problematic way. (b) The requirement that persons must be social mutually recognitive beings directly follows from Wittgenstein's private language argument (1958), as well as from Hegel's and Sellars' attack to the myth of the given (Hegel, 1807; Sellars, 1956). Again, in light of these arguments, I do not see how a being could distinguish between objective and subjective standards of correctness for evaluating the reliability of its 'generative model' or the content of its own states solely on the basis of privately registered experiential regularities (see also Rosenberg, 2002). This seems to me to hold even in the case of Roden's (2014) hyperplastic beings (i.e. beings capable of predicting the effects of arbitrarily fine-grained self-interventions upon their physical and functional structure). If these beings picture the world through an analogue of PP generative models and if these models are not transparently Given to these beings by sensory or intellectual intuition, how can they distinguish between what *seems* to be the case and what *is* the case bereft of a social scorekeeping practice that keeps track and compares different perspectives from different generative models?

scrutinize it and elaborate its determinate content, and when necessary, *revise* the network of its incompatibility and consequence relations with other commitments or ours (Brandom, 2008; Negarestani, 2014, 2018). As we have already seen in the case of universals in general (Chap. 9), this is essentially a self-correcting process. It is the project of unpacking and updating our commitments about what we are in the face of incompatibilities—produced in the process of progressively determining these commitments—between our normative-functional conception of personhood and our evolving descriptive-explanatory renderings of its material embodiment. And in this self-correcting process nothing is sacred. Any view about the way in which our normative-functional 'essence' is realized in worldly structures—including the seeming platitude that the ideal medium for actualizing the essence of personhood is the human body in its present form or the fact that certain innate mental dispositions of a certain individual are part of its personality—is potentially revisable, the only criterion being explanatory coherence with our developing worldview.[4]

Indeed, these revisions would not only be possible but necessary if we take seriously the demand to satisfy the utopian impulse for freedom, either in its collective or in its monadic form: for, as we have seen in Chap. 9, the satisfaction of this impulse implies the quest for individual and collective emancipation from any broadly causal (evolutionary, historical, depth-psychological, socially induced) factor that unreflectively guides our reason-giving practices and hinders the free expression of our intellectual, sensible and practical capacities. As is obvious, the satisfaction of this regulative ideal would initiate a self-correcting process of successive reconceptualizations of the causal basis of what we take to be our *reasons* for believing, thinking and acting thusly-and-so, which would in turn revise our conception of the material basis of personhood, of its causal efficacy, of the ways it can affect and be affected by its physical and social

[4] Some of these beliefs about personhood—for example, the fact that our genetic heritage is a central component of our personal identity, and the foundation of human dignity and equal freedoms that exist among persons of equal birth (Fukuyama, 2002; Habermas, 2003)—may well be part of our current 'grammatical background certainties' around which other, less central and potentially revisable beliefs about personhood revolve (Wittgenstein, 1969). But these background beliefs are not diachronically sacrosanct. They could be replaced by successor post-human background certainties that better satisfy the regulative ideals of personhood and turn into empirical propositions, which, however pragmatically useful for everyday purposes within a certain physical and social environment, could turn out to be strictly speaking false from the point of view of the successor post-human framework.

environment, of the consequences of its actions and of the circumstances that hinder or enhance its powers to act. This deep-seated existential interest of ours in gaining theoretical and practical control over those broadly material conditions that hinder the optimal expression of our capacities and abilities at the collective or monadic level initiates a never-ending process of unearthing, thematizing, theoretically understanding and practically changing *whatever* material conditions stand in the way of our achieving this goal. And in this process, there is no fine line between things that function as enabling conditions of human freedom and things that block its further development. This is ultimately why no substantial conception about the essence of personhood can be considered as unrevisable, even if it is *currently* functioning as a background assumption (exempt from doubt) of our concept of personhood. Only two things would remain intact in such successive radical reconceptualizations of our substantive understanding of historical-psycho-social realization of personhood: the skeletal normative-functional unity of personhood as a locus of network of commitments and entitlements expressed in collective pattern-governed behavior in the form of language-entry, language-language and language-exit transitions, and what we might call our *structural finitude*, namely, the fact that the world in which persons have their being cannot be completely under their control under every aspect. One cannot be a person unless is situated within a partially independent external (and internal) world which s/he never made. Indeed, perceiving, thinking and acting depend for their very *existence* as structural features of personhood on the existence of a world that is partly out of our control and in which the success of our projects cannot be guaranteed. Yet, as we saw, the substantial content of those limits that are essential for making sense of personhood cannot be drawn in advance based on everyday lifeworld categories. The only existential limitation that might be thought as a necessary condition for a complex physical system to count as a person (as opposed, say, to an omnipotent God) is the always present possibility of failure of the life projects of a person, that is, the risk that they might turn otherwise than expected (Hägglund, 2020).[5]

[5] In my view—*contra* Hägglund (2020) and Heidegger (1927)—it is the *possibility*—not the inevitability—of loss (including death) that functions as an existential limitation necessary for making sense of persons capable of existentially committed to pursuing various worldly projects. This, among other things, means that we have no reason to dignify pain, suffering and death, however much we can learn from them (e.g. develop empathy for others, motiva-

Finally, it must be noted that the very normative-functional rendering of personhood that broadens our conception of persons and de-essentializes our humanity as a biological category also *embeds* us in the world of processual becoming. This is because, on this understanding of personhood, and precisely due to the self-correcting process of revising our substantive commitments about what persons 'really are' in the world of becoming, we begin to see that our distinctiveness, our non-identity and irreducibility to human beings, or to whatever materially embodies us in the actual world of becoming, is not a *metaphysical* one, grounded in some kind of ontological essence, but rather a result of our creation as complex self-maintaining systems, of a *distinctive tool* for representing our activity to ourselves for the purposes of regulating collaborative action (the score-keeping game of giving and asking for reasons), and whose ultimate *raison d'être* is to promote the general welfare of the community. Also, we begin to recognize the following existentially liberating-cum-frightening truth as a consequence of the fact that persons as normative entities are essentially *hyperstitionally* constituted: since persons ultimately exist as the bearers of reciprocally recognized normative attitudes (undertaking and attributing commitments and entitlements), their very existence and marching toward an ever-better theoretical and practical understanding of themselves-in-the-world is not automatic or a matter of the progressive unfolding of their 'metaphysical nature', but has to be sustained by their collective *binding* to, *enactment* of and *trust* in the multifarious normative attitudinal contexts that make them possible and the synchronic and diachronic entitlements and responsibilities that their binding in these contexts entail.

10.2 Toward Democratic Prometheanism

We saw that the community that realizes the regulative ideals of personhood is essentially characterized solely by normative-cum-functional constraints delineating its structural finitude and its nearly unlimited potential within the very broad confines of this finitude. Yet, we also saw that this is consistent, and indeed complementary, with viewing the denizens of this community as complex physical processes characterized by equally

tion for creativity), by taking them as necessary conditions for our life to have meaning or purpose.

deep-seated yet not always convergent 'freedom impulses'.[6] Thus, the question remains: How can we reconcile—even if only in the realm of thought—the essentially non-egalitarian, non-universalizable impulse for monadic empowerment with the essentially egalitarian, universalizable impulse for collective self-mastery? I submit that in view of the ontological priority of non-normative freedom over normative freedom and the epistemological priority of the latter over the former (Chap. 9), we must use the egalitarian and universalizable dimension of the conceptual tools of theoretical and practical reason alike as a *means of collective coordination* for achieving the *non-instrumental 'end'* or (non-moral) 'good' of *jointly* maximizing the power (*puissance*) of monadic perspectives. And since a necessary condition for achieving this end is to be sensitive to the distinctive non-universalizable and non-interchangeable particularity of monadic perspectives, it follows that the universalizable and interchangeable elements inherent in the very *form* of conceptual rationality both in its theoretical and practical dimension must be such as to somehow *bring forth* or *make manifest* the non-universalizable and non-interchangeable character of monadic individuals. It can be suggested that this can be done by combining a *process nominalist* language and ontology (since the process nominalist conceptual framework is constructed precisely with an eye to unearth the unique individuality of a process as opposed to considering it as an individual substance subsumed to kind-universals) with an essentially *cooperative* model of conceptual rationality. From this point of view, theoretical and practical reasonings are not understood on the model of means-ends rationality through which individual subjects aim at maximizing their utility, but on that of a collective and cooperative form of rationality in which we, as members of an embracing community of all rational beings and thus as bound by relations of *symmetric recognitive attitudes*, formulate

[6] Is this antagonism between these freedom impulses an idiosyncratic feature of our specifically human form of life or it should be considered as a structural feature of all rational beings, including, for example, echo-locating Martians? To the extent to which a necessary condition of being a being capable of rule-governed behavior is participation in *communal practices* (think here of Wittgenstein's rule-following considerations and his private language argument), I do not see how any non-human rational form of life could be composed of persons independently of their being placed in a recognitive community of other such persons (with differing monadic perspectives). And if this is correct, I do not see how the antagonism between the two freedom impulses could fail to occur or be somehow self-regulated without the use of space-of-reasons rationality in those alien forms or life, especially if we take seriously the fact that all forms of life are a result of broadly evolutionary processes that work by trial and error and preclude pre-established harmonies of the above sort.

and self-correct our individual and collective intentions, in our quest to discover the appropriate means and conditions to jointly maximize monadic welfare, while uncovering along the way what exactly that welfare consists of and what kinds of normative constraints must be implemented for the non-coercive joint realization of monadic welfare perspectives. In this way, we can envisage the universalizable and egalitarian dimension of conceptual rationality not as something that occludes, hides or suppresses the monadic, non-interchangeable dimension of monadic perspectives, but as universalizing and equalizing—to the extent to which it is practically possible—the *sensitivity* for and *visibility* of this non-interchangeable monadic dimension in our quest for collective self-mastery.[7]

Another way to put the above point is to view the potential tension between the impulse for collective self-mastery and the impulse for the empowerment of monadic perspectives as a tension between a collective-cum-normatively constrained and a monadic-cum-normatively unconstrained vision of what we might call 'promethean freedom'. From this point of view, the optimal resolution of this tension would require putting the satisfaction of this fundamental need for promethean freedom, whether collectively or monadically construed, to the service of our general welfare, where the latter is understood as distributed among monadic perspectives and as expressed by a community characterized by forms of cooperative (non-means-ends) rationality and relations of reciprocal recognition. And this, transposed to the political realm, implies that the impulse for promethean freedom should be put under democratic control and channeled through social institutions and avenues that foster transparency, accountability and criticism and are sensitive to the emergence and reproduction of systematically unequal distribution of individual, social and political power.[8]

[7] This would mean, for example, that each monadic perspective would be taken equally seriously, independently of contingently existing asymmetric power relations that might diminish its visibility, suffering, needs, aspirations and negotiating power as a perspective. To my mind, this egalitarian stance is not justified by the existence of unconditional ethical-cum-metaphysical facts such as human dignity (Kant, 1993) or the infinite otherness of the human face (Levinas, 1979), but by its contribution to collective self-mastery by enlarging 'the circle of sympathy' between monadic perspectives.

[8] Those socio-political functional constraints are abstract, general and flexible enough to find application even to post-human communities of persons in which, for example, the socially produced and distributed goods are our very bodily and mental abilities. These constraints function as regulative ideals and must not be confused with the *causal* (broadly social-scientific) project of finding the most efficient way in which existing institutions could

At this point it must be remembered (Chap. 9) that the semi-heteronomous status of social institutions (government, political parties, state, rule of law, bureaucracy, markets, science, etc.) is pragmatically indispensable for enabling novel forms of freedom at the level of the individual and in this sense is not incompatible with the existence of a democratically organized free and autonomous we-community. The view that there can be total fusion between our institut*ing* activity and the institut*ed* social structures in which we have our social being is a myth (one of the many versions of the myth of the Given). Indeed, as Hegel (1807) and Dewey (1939) have convincingly argued, the social institutions around which our political and economic life is organized can be considered as enabling empirical conditions of the origin, development and realization of freedom as autonomy. Social institutions can be seen as evolutionarily emerging and culturally transmitted large-scale information networks for controlling (processing, collecting, distributing, signaling) socially significant information. The abstract, hierarchical, homogenizing and subsuming dimension of the function of social institutions is precisely what enables them to solve social coordination problems (conflicts, antagonisms, exchange of goods, implementation of collective projects), while at the same time and by the same token producing new forms of alienation and coercion for individuals. Perhaps, by developing novel, scientifically updated, more powerful and context-sensitive AI-based tools for social information processing (e.g. real-time dynamic coordination of market prices or decision-making deliberative processes in large-scale citizens' assemblies), we could dispense at least with some of these—currently empirically necessary—alienating dimensions of social institutions. Yet, this would not amount to altogether dispensing with their heteronomous character. For, most probably, scientific and technological advances in information processing would *transfer* part of this heteronomy to the algorithmic network that would replace the function of a certain social institution (e.g. markets, law, etc.). This is related to the fact that the abstractive, formalizing, homogenizing, impersonal and mediating (i.e. not directly transparent) dimensions of 'social institutional machines' are

be modified or revised so as to contribute to the realization of these ideals without sacrificing vitally important existing material goods and social rights, such as the present level of material growth, the existing level of respect for human rights, individual and political freedom, etc. The latter causal project is a broadly empirical matter and cannot be settled in an a priori manner.

necessary features and tools for their being able to compress, integrate and stabilize socially significant information, which in turn is necessary for making this otherwise perceptually opaque social information *predictably comprehensible*, and thus *trustworthy* at a scalable level. Thus, this heteronomous dimension of these 'social information processing machines' essentially functions as a relatively trustworthy *extension* of the senses and of the evolutionarily ingrained information-processing limitations of human thinking and memory in increasingly large and complicated social contexts. The human mind has not been adapted by evolution to process complex social information at the scale of complex social organizations such as the modern metropolis. Social institutions are precisely the solutions developed and selected by cultural evolution that partly compensate for this biological deficiency. And the abstractive, formalizing, homogenizing and mediating bureaucracies associated with these social institutions that develop impersonal rule procedures for the dissemination of the relevant social information partly compensate for the fact that reliability, transparency, accountability and equality of access to socially significant information cannot be scaled up at a societal level in an immediate and spontaneous bottom-up manner due to deep-seated individual differences in expertise, goals, preferences, values and affective investments (Weber, 1921).

Of course, all the above do not detract from the fact that at a de facto level, our actual social institutions might not 'do their best' to foster the development and realization of freedom as autonomy, to the extent to which they function in tandem with power structures that systematically extract and harness the creative powers of individuals, ignore or discourage non-profit-related abilities and systematically reproduce conditions of unequal access for the majority of humanity to forms of socio-political power and individual well-being. From this point of view, the key to successful collective self-government lies in taking advantage of the possibilities of individual and collective freedom opened by social institutions while at the same time being vigilant to prevent the systematic sedimentation of unjustified inequalities that nearly inevitably will occur as a result of the fact that the deference of authority through the division of intellectual labor to social institutions can better solve vital collective coordination problems. Precisely because it is nearly inevitable to erase all power inequalities that result from the heteronomous nature of social institutions without at the same time eradicating those institutions themselves, thereby radically compromising vitally important sources of individual and political freedom, it is exceedingly important that democratic self-governance goes hand in hand with wide networks of checks and balances of the distribution of privileges and power and built-in self-correcting mechanisms both for democratic institutions themselves and for the relatively autonomous institutions (economy, science, law, education) that make the smooth operation of democracy possible.

Obviously, the fact that in existing societies the new knowledge about ourselves obtained through science and technology in the quest for promethean freedom could be utilized and harnessed as a means for capitalist profit and as an opportunity for an unprecedented increase of inequalities in political, social or individual power is a reason for being extremely vigilant in actually implementing such a vision in practice. Yet, it is not a reason for considering it incoherent or theoretically problematic. In fact, the promethean impulse, for example, for human enhancement and for full automation of production can be the basis for raising universalist emancipatory demands that are in direct conflict with current capitalist orthodoxy. For example, our ability to alter our physical and mental characteristics in the context of human enhancement projects could lead us to question the moral justification of the inequality of distribution of natural skills and talents of individuals. And, to take a more down-to-earth example, the intensification of automation could challenge the value of wage labor through the demand for a universal basic income, which could in turn foster attitudes that challenge current models of economic growth (Srnicek & Williams, 2015).

Palpably, there is no guarantee that promethean freedom elicited by theoretical reason (our descriptive-explanatory practices) and practical reason (our collectively committal moral attitudes) as they de facto operate in our current societal context will jointly empower individuals irrespectively of their privileges and their place in the network of societal power relations. The only 'certainty' in this regard is that the transformation of nature, individuals and society effected by disenchanting reason in its continuing effort to increase promethean freedom is such that the stakes of preserving and reproducing asymmetric relations of recognition between persons will increasingly become a matter of existential significance. Indeed, we can see even from our own viewpoint that, for example, increasing techno-scientific literacy, sophistication and knowledge sharing ('digital commons') would make the minimization of domination and the achieving of symmetric recognition a live option for everyone, while at the same time making it possible for those in positions of power (corporations, governments) to align humanity to their interests directly at the bio-psychical level (e.g. by valuing and reinforcing profit-related abilities, feelings, thoughts and behaviors) (Birkinbine, 2020; Zuboff, 2019). In such circumstances, the demand for equality in possession and distribution of socially significant goods would lose its ideal and semi-abstract status and would be motivationally salient and concrete for (almost) everyone. This heightened sense of the imminent utopic and dystopic possibilities that are simultaneously opened for us would force us to understand at the concrete, experiential level that the general (economic, political) structure of society is *directly connected* to our everyday concerns. It has been said that the utopian impulse can—and must—be released from its neoliberal-individualist ideological shackles in order to expand the space of the possible, mobilize a critical perspective on the present and cultivate new desires (Srnicek & Williams, 2015). Under such circumstances, this utopian impulse for equal access to promethean freedom would also lose its ideal status as a mere abstract regulative principle of political organization and would be concretely expressed as a motivationally salient existential commitment necessary for the very *survival* and minimal well-being of the community of persons. For these circumstances would make explicit the ever-increasing destructive potential of the deep-seated human tendency to seek asymmetric recognition (authority without responsibility) by obtaining a higher social status than other individuals and to systematically reproduce the conditions that maintain this status. Ironically, perhaps this intensification of the existential risks involved might be the most effective way for us to see through the invisible cage of ideology, to face the truth of our individual and social existence and to take sides in this existential struggle for survival and well-being. But again, nothing guarantees that the outcome of this struggle will benefit humanity (or our post-human descendants) as a whole.

10.3 Promethean Freedom as Controlled Interdependence

In the above I have explicitly attempted to disentangle my version of prometheanism and promethean freedom from what I consider to be the most implausible features of this general view both theoretically and practically. The quest for liberating ourselves from broadly causal (evolutionary, historical, psychological, socially induced) constraints is only meaningful in the context of our structural finitude, namely, the 'transcendental' fact that perceiving, thinking and acting depend for their very existence as structural features of personhood on the existence of a world that is partly out of our control and in which the success of our projects cannot be guaranteed. It was also pointed out that our promethean freedom is misunderstood if it is conceived in abstraction from our social institutions, their normative constraints and their heteronomous dimension. All these are *enabling* constraints on our freedom and not (only) disabling ones. Lastly, I argued that, as an ideal of practical reason, promethean freedom can only thrive in a democratically organized community of persons, where democracy is also conceived as an enabling constraint: by being based on wide networks of checks and balances and self-correcting mechanisms it functions as a filtering mechanism for selecting increasingly workable policies that foster the common good and individual autonomy while at the same time openly embracing antagonisms, conflicts and different viewpoints on what this common good is or what our 'real needs' really are, thus rejecting any privileged access (Givenness) with respect to their objective content. Thus, our view here might be described as a *self-constrained* and *self-correcting* prometheanism (*contra*, e.g. Bastani, 2019). It is now time to discuss one more very important feature of prometheanist freedom, which again distances it even from prometheanist accounts that are otherwise closely related to the view developed here (e.g. Brassier, 2014; Negarestani, 2018).

In my view, promethean freedom, and it core ideals of *liberation* from causal constraints that hinder the free expression of our sensory and intellectual capacities and of collective *mastery* over the world and ourselves as a necessary condition for achieving this liberation, must be understood as a form of *dependence*. More specifically, I suggest that promethean freedom, and the liberation through mastery over nature and ourselves that goes along with it, must be understood in terms of *controlled interdependence*. This is related to the fact that our promethean need to be free from any kind of causal constraint that hinders our ability to think and act is itself materially conditioned for its existence and efficaciousness on the establishment, maintenance and reproduction of dynamic allostatic relations between our bodies and our social and natural environment. Consider, for example, the fact that the overwhelming majority of our purposes, interests and needs presuppose for their free expression the existence and stability of social institutions that themselves depend on the cooperative activity (intellectual and manual labor) of vast numbers of people, as well as the existence and stability of a host of highly interconnected ecosystemic regularities (Khandker, 2021). This shows the plausibility of understanding freedom, even—and indeed especially—in its promethean form, in terms of controlled interdependence. From this point of view, freedom as self-mastery would serve as a means to *control*, regulate and/or coordinate the complex interdependence relations that form the background of our individual and collective well-being, namely, the socio-technical infrastructure and the complex network of environmental regularities that serve as the background material conditions for the free exercise of our sensory and intellectual powers. And freedom as *pouissance* would include these interdependence relations as an essential part of its *content* at the level of the monadic processual individual. Thus, to determine whether our total power to act increases or decreases within a certain socio-environmental context (and to what extent it does so) we would have to take into account the dynamic interdependence and the specific ways in which socio-technical and ecosystemic relations work as enabling constraints for the exercise of our sensory and intellectual powers as well as the effects of our actions on these complex interdependence relationships.

This understanding of promethean freedom in terms of controlled interdependence is not put forward here just because it happens to chime in with currently fashionable ethico-political issues, such as climate change. It is something that goes hand in hand with a *process* view of the world as the latter is to be applied in our understanding of the organism-environment relation. From this standpoint, organisms (including ourselves) are not to be conceived as substances, categorial unities, or as 'centers of reference' directed to the external world or to environmental affordances, where the latter are understood as categorially distinct and external to the organism in question. Instead, both organisms and their environments are to be conceived as *co-selected processes* standing in a mutually symbiotic relationship and ultimately as constitutive sub-patterns of a larger self-sustaining (life) process that incorporates, sustains, selects and reproduces both of them. Relatedly, our very sensory and motor surfaces are not to be understood as doors or windows through which we are 'directed' to external environmental affordances, but as functional points that, in their interaction, organizationally close us as sensorimotor systems through the environment. From this point of view, our environmental *niche* (and, *a fortiori*, our socio-technical *niche*) itself is to be understood as a functional component of our sensorimotor system, rather than something external to it, to which we may be 'intentionally directed' (Christias, 2023, 209–210, Villalobos & Ward, 2016). This radically different understanding of the organism-environment relationship has important ethico-political implications: the conception of organism and environment as co-selected process standing in mutually symbiotic relationship would presumably make more salient for us the number and complexity of the strong interdependence relations that exist between (a) our material and socio-technical bodies, and in particular between the formation and satisfaction of our individual needs and the socio-technical infrastructure that makes them possible, and (b) between our biological and socio-technical form of life, other living beings and the surrounding ecosystemic relations and processes in the context of which both us and animals can exist and thrive. This would make us more sensitive to the existential importance of resolving many humanly induced social problems, such as climate change or the proper use of artificial intelligence. It would also make us aware of the fact that ecosystemic and planetary constraints that limit our natural resources and the extent of our ability to harness nature are not static but are *themselves mutable* since, to an ever-increasing extent, these constraints co-evolve and are co-produced in tandem with our forces, means and technological methods of production (Moore, 2015).[9]

Here we must immediately pause to avoid certain misunderstandings of the above eco-prometheanist view. It is true that the combination of factual ignorance about ourselves, the world and the long-term consequences of our actions, coupled with an obsessive competence-based model of growth fixated on short-term profit drawn from relentless exploitation of

[9] I do not have the space to compare and contrast my democratic eco-prometheanist view with Srnicek and Williams' (2015) left accelerationist strand of prometheanism based on reason's ability to techno-scientifically reshape the world and ourselves in accord with rational ethico-political ideals, and Land's right accelerationist view (2011, 2022) that opts for the liberation of 'machinic' desiring production from its representational confinement in human-all-too-human values and its subjugation to the ends of reason. Although my point of departure is, like Land's, a Deleuzian one (reason is not imprinted in the fabric of reality), my overall view, by asserting reason's power to reshape the world and ourselves to their mutual benefit (where reason is construed as a non-representational world-making power), is closer to left accelerationism. It differs somewhat from left accelerationism in its construal of reason as fractured desire, its emphasis on structural finitude as an enabling constraint of prometheanist reason, its explicit metaphysical grounding on a resolutely process metaphysics with broader ecological implications and its emphasis on democratic values as means for internally connecting and coordinating epistemic and social transformation.

natural and human resources, makes climate change an imminent existential risk to which we must immediately and collectively respond. But (*pace* Anders, 2016, Jonas, 1984) this collective response must not be based on fear (of extinction), as fear, being a particularly intense sad passion, is often epistemically and politically paralyzing and is fundamentally a reactionary force which merely attempts to preserve what can be salvaged from our culturally inherited moral understanding instead of creating novel ways of understanding ourselves and the world in the face of new existential predicaments. It might be objected that the fear of death and the impulse for self-preservation, far from being conservative or reactionary attitudes, are actually beneficial for rethinking and reinvigorating enlightenment, humanism and universal rights for the twenty-first century, since they arguably inform the so-called ethics of vulnerability which universally unites human beings and ethically binds us with all sentient beings (Diamond, 2003; Jonas, 1984; Pelluchon, 2021). Yet, this view often confuses control and objectification (e.g. turning human beings, animals or nature as an object of investigation) with domination and ends up provisionally blocking the multiple paths in which the revolutionizing self-correcting power of theoretical and practical reason can expand the field of possibilities for creatively expressing ourselves-in-the-world, if there is the slightest possibility that this creative expression might have unwanted consequences. But freedom from fear is the precondition of emancipation (Brassier, 2014), and an emancipatory politics rooted in fear ends up being self-undermining: it tends to offer conservative conceptual responses that attempt to preserve at all costs the culturally inherited epistemic and moral *status quo* of our ordinary lifeworld in the face of destabilizing and not directly comprehensible techno-scientific developments, it makes it easier to eschew responsibility and collectively submit to external authorities and it operates under the illusion that our substantial practical and moral ends as a community are antecedently known and fixed independently of the ways in which techno-scientific developments are embedded in our ordinary lifeworld. Techno-scientific developments are conceived merely as means for fulfilling antecedently known ends and purposes, and their adequacy or rationality is judged by appeal to the spontaneous understanding of science and technology that we have learned in our 'mother's knee', thereby hiding from view the fact that novel techno-scientific realities and potentials go hand in hand with *novel criteria of rationality*, constitute *novel ways of intelligibility* and thus demand the creation of *novel concepts* for properly understanding the potential, problems and opportunities that are opened in their wake.[10]

Relatedly, nothing in this existential predicament of ours in the age of Anthropocene implies that nature is 'good' while human intervention is 'bad' (vaccines are good but not natural, hurricanes, earthquakes, viruses and malaria are natural yet deadly), that there is an ontologically important distinction between 'natural' and 'unnatural' or an ethico-politically important one between 'natural' and 'artificial', that nature works on the basis of laws that we must respect 'because nature knows better' and which we can break only on pain of hybris and nemesis or that nature operates on laws exhibiting certain deep homeostatic regular rhythms or patterns which should not be disturbed on pain of catastrophic consequences (in truth, our main sources of energy as well as many existing species are the outcome of disturbances, disequilibria and catastrophes of previous homeostatic earthly *milieux*). Climate change is deeply problematic not because it disturbs natural rhythms that are inherently good or because it expresses humanity's hubristic attitude of breaching nature's vital limits, but because it causes unnecessary *suffering* to sentient beings, including humans. Interestingly, this criterion (*pace* Jonas and Pelluchon) in a sense turns things upside down, since it implies that *anything* that causes *unnecessary* suffering to sentient and sapient beings, *including nature itself*, is something which we must take issue with, even if this, in the last analysis, demands intervention (to the extent to which this is techno-scientifically feasible) into the very evolutionary patterns that are related to this unnecessary suffering. Indeed, the *normative* evaluation and injunction to

[10] Nehamas succinctly puts this point in the context of a discussion of the supposed authenticity and reliability of speech over (the then newly formed) writing technologies in Plato's *Phaedrus*: 'What is often true in such discussions is that we make an unfair comparison: we judge the new medium according to its ability to communicate the type of ideas for which the older one had been designed, and it is no surprise that it fails in that regard. Moreover, we tend to identify the ideas suited to the old medium and the manner in which that medium communicates them with what is rational. Accordingly, even if the new medium is sometimes judged to be successful in communicating its own ideas by its own methods, we are tempted to consider these ideas at best as inferior to the former, at worst as irrational and harmful' (Nehamas, 1995, p. 346). This also explains why the invocation of 'practical wisdom' in our effort to rationally decide how we must act in the face of novel challenges posed by techno-scientific revolutions or how exactly we must 'draw the limits' in applying the latter (Jonas, 1984; Pelluchon, 2021) is ultimately uninformative: because techno-scientific developments, by instituting novel ways of intelligibility, thereby also *change* what counts as 'practical wisdom'.

intervene and do something about climate change—and, in the limit, about anything that causes unnecessary suffering—originates in the fact that it is internally connected to the *general welfare* of a we-community bound by a network of epistemic and ethical rights and duties, authority and responsibility, commitments and entitlements that define what it is to believe, think and act thusly-and-so within a world not of our own making. And since this general welfare itself is inextricably interconnected with the welfare of sentient beings and the sustainability of patterned regularities in non-sentient inorganic nature, the normative injunction to minimize unnecessary suffering initiates a process of rational intervention to nature and ourselves internally attuned to the cultivation, thriving and expansion of this dynamic interconnectedness.

Can we perhaps go even further and say, for example, that 'extinction is a trauma to an ecosystem' or that 'the dying out of species causes an experience of loss that reverberates on all scales of nature'? (Khandker, 2021, 187). This might be an insightful phenomenological description of our experience of environmental degradation. Yet to the extent to which it implies that nature comes with ready-made normative constraints and demands over our conduct ('thou shall not extinguish any of the beings I bring to existence'), it overly anthropomorphizes nature and mislocates the real source of otherwise plausible demands ('we should prevent climate change', 'we should treat animals as ends in themselves'). From the point of view of an evolutionary process nominalist metaphysics, as I understand it, to treat people, animals or nature as ends in themselves amounts to highlighting their monadic particularity and dynamic interconnectedness (by taking into account their individual causal-evolutionary history), as opposed to considering them as instances of static and independently individuated general essences. It does *not* imply that we must preserve monadic perspectives at all costs or that their existence comes with unconditional ethical demands.[11]

Does our above view reintroduce inadmissible ontological hierarchies at the heart of immanent being, since the community of rational beings is the sole normative criterion from the standpoint of which all existing

[11] Is not the imperative to minimize unnecessary suffering such an unconditional ethical demand? It might sound so, and for all practical purposes we might treat it as such, but, strictly speaking, it is not so, since it is not justified by the existence of metaphysically binding ethical facts such as Kantian dignity or 'the infinity of the Other's face', but by its contribution to monadic and collective empowerment (increase of joyful passions, decrease of sad passions).

beings are to be judged for their 'ontological' worth? Not at all. Our democratic prometheanist view does normatively distinguish persons only to reincorporate them ontologically to immanent being. For, as we saw in Sect. 10.1, our view implies that humans are essentially *normatively* defined as *persons*, and not as a distinct and hierarchically superior *metaphysical* category called 'humanity'. From this point of view, humans as complex natural processes are ultimately conceived as being in a mutually symbiotic relationship with the social and environmental *niches* and processes to which they are intertwined. And this change of focus comes with a higher sensitivity (systematically toned-down by an obsessive competence-focused model of social organization) to the strong interdependence relations that exist between our form of life, those of other living beings and the ecosystem as a whole. It turns out that it is precisely by asserting the value-laden *normative* distinction between treating something as a person and treating is as a mere object that we ultimately arrive at a metaphysically *non-hierarchical* picture of ourselves-in-the-world which stresses the interconnectedness of human and non-human forms of life and their dependence on the ecosystem as a whole. Thus, paradoxically, humanity loses its privileged place in the great chain of being precisely at the same time and to the extent to which it *universally asserts* its normatively distinctive character as a maximally inclusive community of persons-in-the-world (as opposed to less inclusive viewpoints according to which only humans are worthy of being treated as ends in themselves). It is precisely by epistemologically differentiating and maximally distancing ourselves from nature that we become concretely able to ontologically reunite with it. For the very *raison d'être* of our normative conception of a post-human community of persons whose sapient welfare is coordinated with their sentient welfare is its material cashing out in terms of an increasing ability for collective skillful coping ('resonance') with the complex networks of social and worldly interdependencies that affect their survival and flourishing.

The further question of how exactly we should act in order to most successfully prevent climate change cannot be adequately addressed in the context of this book, for, obviously, it is to a large extent a broadly empirical question. I shall limit myself to the following brief remarks of a more general nature. Often, climate change debates take the following form: Should we opt for a degrowth economic model with limited use of science and technology or rather mobilize large-scale techno-scientific means (geomechanics, geoengineering, nuclear fusion) trying to balance sustainability and growth? The degrowth model has the virtue of highlighting the capitalist relations of production as one of the main roots of the problem (Saito, 2023). Yet, by refusing to utilize advanced techno-scientific means (since it views them as the material expression of the capitalist predatory attitude toward nature) it needlessly avails itself of powerful and easy to scale-up tools in our attempt to prevent climate change, and by neglecting the issue of forging collective social networks of resistance that can rapidly and solidly increase their 'power to act' (so as to be able to confront capitalist reactionary forces), it has virtually zero probability of success. I think that what is needed in this regard is to disentangle the notion of control from that of domination (Chap. 9), while simultaneously attempting to put scaled-up techno-scientific knowledge and power under democratic control in a way that would instill fundamental democratic values (accountability, transparency, criticism, sensitivity to the emergence and reproduction of systematically unequal distribution of individual, social and political power) to the very *form* of our techno-scientific practices, organization of production and labor.[12] For, in our increasingly globalized and interconnected world, only a socio-political organization whose social and techno-scientific power can scale-up to the level of today's neoliberal-capitalist power can present itself as a serious threat—and thus alternative—to the present *status quo*.

From this point of view (*contra* Castoriadis, 1975), it is not enough for a social imaginary that seeks to replace the neoliberal-capitalist imaginary to merely denounce the latter's obsession with unlimited growth and domination over nature and others. It must also utilize science and technology and democratically repurpose them toward the development of a material power that can enable us to *collectively control* this very domination and unleash the productive (sensory and intellectual) powers that

[12] For a similar view, yet without our prometheanist emphasis on the importance of scaled-up socio-technical power, see Feenberg (2017). See also Simondon (1965) who warns for the danger in the other direction, namely, in the intention to limit technical innovation and possibilities according to cultural norms, which amounts to bringing a potential evolution to a halt and viewing the *status quo* as something that already enables us to define a kingdom of ends, a final code of values. Simondon (1958) also calls for the need of integrating technology with the other domains of culture by developing a 'technical culture', in which technical objects would turn from mere instruments in the service of human goals to meaningful elements of existence, and where technical education would be considered as essential as literacy to meaningful participation in society. This would be a key step in democratizing techno-scientific practices and the organization of production without undermining their prometheanist potential to expand the space of the possible and cultivate new desires, interests and goals.

foster collective growth. Thus, this new democratic prometheanist social imaginary must be seen not as aiming at mastery, but at *mastering (controlling and repurposing) the dominating and coercive powers* that produce an increasingly uninhabitable environment and hinder collective liberation[13] (Brassier, 2020). And it must do so by productively combining its acute awareness of the fact that there are no metaphysical guarantees of success, no invariantly meaningful ends or eternal values (beyond Deweyan 'growth'), with an attitude of joyfully affirming and harnessing the transformative power of hyperstitional reason 'here and now', thus viewing the world and ourselves at once as something to be affirmed and as something to be overcome.

10.4 Conclusion

The conceptual resources of fractured immanence, hyperstitional normativity and processual nominalism invite us to think differently about our place in the world: not as detached observers or sovereign agents, but as co-composing participants in a world of expressive becoming. From this point of view, our collective existential wager is not best described as the triumph of humanity over nature, but rather lies in coordinating our material, conceptual and affective powers into practices of world-building that are at once emancipatory and entangled with nature. It is the wager that freedom, understood as a process of co-creative empowerment that controls and marginalizes domination, can be immanently cultivated through the self-correcting projections of our collective norms. Indeed, this conception of practical freedom as collective material empowerment is inseparable from freedom as self-knowledge, namely, the capacity of thinking to emancipate itself by identifying and rectifying its unwarranted biases, reifications and projections—in short, its transcendental illusions. Thus, the big picture here is one of a self-differentiated and self-fracturing nominalistic process world in which complex physical systems coordinate themselves by collectively producing linguistic marks and noises, gradually attempting to seize the means of semantic production (make explicit,

[13] Of course, in this context 'mastery of mastery' does not mean freezing techno-scientific evolution into pre-given ends. Neither does it entail omniscient prediction or closure of future contingencies. Rather, it seeks to expand our collective capacities by instituting feedback-rich forms of governance for managing our interdependence with techno-scientific systems. Recall here that from our point of view collective self-mastery is itself a *hyperstitional* projection, which does not require exhaustive foresight.

regulate, control and critically change this patterned structure, purging it of its unwarranted biases) in their striving to obtain ever-increasing levels of collective resonance with the natural and social environment so as to freely exercise their sensory and intellectual powers as joyful ends in themselves. This vision, in turn, is not one of 'perfect adequation' with the world or of reaching an eternal universal equilibrium, but presupposes a metaphysics of openness: a mutabilist ontology of unfolding processes rather than fixed substances, a normativity that generates futures rather than representing essences and a semantics that expresses functional resonances rather than representational mirrorings. And an integral part of this vision is an understanding of ourselves not as exceptions to nature, but as nature's self-relating fold—its capacity to problematize, reorient and remake itself.

References

Anders, G. (2016). On promethean shame. In C. J. Müller (Ed.), *Prometheanism: Technology, digital culture and human obsolescence* (pp. 29–95). Rowman & Littlefield.

Bastani, A. (2019). *Fully automated luxury communism: A manifesto*. Verso Books.

Birkinbine, B. (2020). *Incorporating the digital commons: Corporate involvement in free and open source software*. University of Westminster Press.

Brandom, R. (2008). *Between saying and doing: Towards an analytic pragmatism*. Harvard University Press.

Brassier, R. (2014). Prometheanism and real abstraction. In R. Mackay, L. Pendrell, & J. Trafford (Eds.), *Speculative aesthetics* (pp. 156–173). Urbanomic.

Brassier, R. (2020). *The human*. https://www.scribd.com/document/552722242/The-Human-Ray-Brassier

Castoriadis, C. (1975). *The imaginary institution of society*. Polity Press.

Christias, D. (2020). Are persons human beings? *Res Philosophica, 97*(3), 363–385.

Christias, D. (2023). *Normativity, lifeworld and science in Sellars' synoptic vision*. Palgrave Macmillan.

Dewey, J. (1939). *Freedom and culture*. G. P. Putnam's Sons.

Diamond, C. (2003). The difficulty of reality and the difficulty of philosophy. *Journal of Literature and the History of Ideas, 1*(2), 1–26.

Feenberg, A. (2017). *Technosystem: The social life of reason*. Harvard University Press.

Fukuyama, F. (2002). *Our posthuman future*. Picador.

Habermas, J. (2003). *The future of human nature*. Polity Press.

Hägglund, M. (2020). *This life: Secular life and spiritual freedom*. Pantheon.

Hegel, G. (1807). *Phenomenology of spirit*. Oxford University Press.

Heidegger, M. (1927). *Being and time*. SUNY Press.

Jonas, H. (1984). *The imperative of responsibility: In search of an ethics for the technological age*. The University of Chicago Press.
Kant, I. (1993). *Grounding for the metaphysics of morals*. Hackett.
Khandker, W. (2021). *Process metaphysics and mutative life: Sketches of lived time*. Palgrave Macmillan.
Land, N. (2011). *Fanged noumena: Collective writings 1987–2007*. Urbanomic.
Land, N. (2022). *The dark enlightenment*. Imperium Press.
Levinas, E. (1979). *Totality and infinity*. Martinus Nijhoff Publishers.
Moore, J. (2015). *Capitalism in the web of life: Ecology and the accumulation of capital*. Verso.
Negarestani, R. (2014). The labor of the inhuman. In R. MacKay & A. A. (Eds.), *#Accelerate: The accelerationist reader* (pp. 427–466). Urbanomic.
Negarestani, R. (2018). *Intelligence and Spirit*. MIT Press.
Nehamas, A. (1995). The phaedrus. In A. Nehamas & P. Woodruff (Eds.), *Plato's 'Phaedrus'* (pp. xi–xxvi). Hackett Publishing Company.
Pelluchon, C. (2021). *Les Lumières à l'âge du vivant*. Editions de Seuil.
Roden, D. (2014). *Posthuman life: Philosophy at the edge of the HUMAN*. Routledge.
Rosenberg, J. (2002). *Thinking about knowing*. Oxford University Press.
Saito, K. (2023). *Marx in the anthropocene: Towards the idea of degrowth communism*. Cambridge University Press.
Sellars, W. (1954). Some reflections on language games. *Philosophy of Science, 21*(3), 204–228.
Sellars, W. (1956). *Empiricism and the philosophy of mind*. Harvard University Press.
Simondon, G. (1958). *On the mode of existence of technical objects*. University of Minnesota Press.
Simondon, G. (1965). Culture and technics. *Radical Philosophy, 189*(1), 17–23.
Srnicek, N., & Williams, A. (2015). *Inventing the future: Postcapitalism and a world without work*. Verso.
Villalobos, M., & Ward, D. (2016). Lived experience and cognitive science. *Constructivist Foundations, 11*(2), 204–212.
Weber, M. (1921). *Economy and society*. University of California Press.
Wennemann, D. (2013). *Posthuman personhood*. University Press of America.
Wittgenstein, L. (1958). *Philosophical investigations*. Blackwell.
Wittgenstein, L. (1969). *On certainty*. Blackwell.
Zuboff, S. (2019). *The age of surveillance capitalism: The fight for a human future at the new frontier of power*. PublicAffairs.

Index[1]

[1] Note: Page numbers followed by 'n' refer to notes.

© The Author(s), under exclusive license to Springer Nature Switzerland AG 2025

D. Christias, *Towards a Hyperstitional Process Nominalism*, Palgrave Perspectives on Process Philosophy,
https://doi.org/10.1007/978-3-032-06350-2

H

I

L

M

GPSR Compliance
The European Union's (EU) General Product Safety Regulation (GPSR) is a set of rules that requires consumer products to be safe and our obligations to ensure this.

If you have any concerns about our products, you can contact us on

ProductSafety@springernature.com

In case Publisher is established outside the EU, the EU authorized representative is:

Springer Nature Customer Service Center GmbH
Europaplatz 3
69115 Heidelberg, Germany

www.ingramcontent.com/pod-product-compliance
Lightning Source LLC
Chambersburg PA
CBHW070753220126
38433CB00024B/311
9783032063496